Springer Theses

Recognizing Outstanding Ph.D. Research

Aims and Scope

The series "Springer Theses" brings together a selection of the very best Ph.D. theses from around the world and across the physical sciences. Nominated and endorsed by two recognized specialists, each published volume has been selected for its scientific excellence and the high impact of its contents for the pertinent field of research. For greater accessibility to non-specialists, the published versions include an extended introduction, as well as a foreword by the student's supervisor explaining the special relevance of the work for the field. As a whole, the series will provide a valuable resource both for newcomers to the research fields described, and for other scientists seeking detailed background information on special questions. Finally, it provides an accredited documentation of the valuable contributions made by today's younger generation of scientists.

Theses may be nominated for publication in this series by heads of department at internationally leading universities or institutes and should fulfill all of the following criteria

- They must be written in good English.
- The topic should fall within the confines of Chemistry, Physics, Earth Sciences, Engineering and related interdisciplinary fields such as Materials, Nanoscience, Chemical Engineering, Complex Systems and Biophysics.
- The work reported in the thesis must represent a significant scientific advance.
- If the thesis includes previously published material, permission to reproduce this must be gained from the respective copyright holder (a maximum 30% of the thesis should be a verbatim reproduction from the author's previous publications).
- They must have been examined and passed during the 12 months prior to nomination.
- Each thesis should include a foreword by the supervisor outlining the significance of its content.
- The theses should have a clearly defined structure including an introduction accessible to new PhD students and scientists not expert in the relevant field.

Indexed by zbMATH.

Vincent Paul Flynn

Effective Non-Hermiticity and Topology in Markovian Quadratic Bosonic Dynamics

Doctoral Thesis Accepted by Dartmouth College, USA

Springer

Vincent Paul Flynn
Department of Physics and Astronomy
Dartmouth College
Hanover, NH, USA

ISSN 2190-5053 ISSN 2190-5061 (electronic)
Springer Theses
ISBN 978-3-031-52044-0 ISBN 978-3-031-52045-7 (eBook)
https://doi.org/10.1007/978-3-031-52045-7

This Springer imprint is published by the registered company Springer Nature Switzerland AG
The registered company address is: Gewerbestrasse 11, 6330 Cham, Switzerland

Supervisor's Foreword

Systems of "free" fermions or bosons have long played a paradigmatic role as tractable models for investigating both equilibrium and non-equilibrium many-body physics in settings where the underlying physical degrees of freedom are truly non-interacting or, more commonly, their interactions may be accounted for at an approximate (mean-field) level. While the phenomenon of Bose-Einstein condensation arguably represents the hallmark feature of bosonic physics at equilibrium, key differences between bosons and fermions also emerge in their dynamical behavior. Perhaps most remarkably, the Heisenberg equations of motion one may derive for the canonical creation and annihilation operators under a non-interacting (quadratic) bosonic Hamiltonian are expressed in terms of a non-Hermitian (NH) dynamical matrix, whenever particle-number conservation is broken. In other words, *effective NH dynamics* may arise for *closed* systems of non-interacting bosons as a sole consequence of quantum statistics, in the absence of any coupling to an external environment and with the physical Hamiltonian remaining perfectly Hermitian at the many-body level.

Effective non-Hermiticity endows free bosons with a variety of spectral properties and dynamical behavior that are inaccessible to free fermions. While a fermionic dynamical matrix is always diagonalizable by a unitary (Bogoliubov) transformation and possesses a purely real spectrum, a NH—more generally, *non-normal*—bosonic dynamical matrix can engender non-real eigenvalues and lose diagonalizability at *exceptional points* (EPs) in parameter space, where both eigenvalues *and* eigenvectors coalesce. EPs are at the root of exotic wave phenomena such as unidirectional invisibility and asymmetric mode-switching, and give rise to a stronger response to perturbations than any Hermitian degeneracy can, with implications for "EP-enhanced" sensing protocols. From a dynamical-system standpoint, quadratic fermionic Hamiltonians can only be dynamically stable. In contrast, *both* dynamically stable and unstable regimes can occur for bosons, resulting in non-trivial "dynamical stability phase diagrams" and ultimately enabling distinctive dynamical phenomena that range from signal amplification in cavity and circuit quantum electrodynamics to squeezing used for continuous-variable quantum information processing.

By the time Vincent Flynn started his PhD research, effective non-Hermiticity in bosons had been acknowledged, but no in-depth characterization was available, nor was a unified understanding of all its physical implications. Yet, a main factor conspired to make bosonic systems, and "NH physics" more generally, the focus of renewed, intense interest—namely, the influx of ideas from topology. Topology plays an extraordinarily important role in non-interacting fermions, by providing the essential ingredient for understanding their generic equilibrium phases as *symmetry protected topological (SPT) phases*, and eventually allowing a complete classification according to their internal symmetry classes—the celebrated "tenfold way." Of course, manifestations of topological physics in non-interacting bosons figure prominently in settings such as topological photonics or magnonics, whereby band structures with non-trivial topological invariants can be realized and, as a consequence of a suitably generalized "bulk-boundary correspondence," topologically mandated boundary-localized modes may emerge. These topological features, however, appear to be completely disconnected from the ground-state and "low-energy" properties that play such a central role in fermionic SPT phases from the point of view of their quantum *many-body* physics. To what extent (if any) can systems of free bosons exhibit topological physics that closely mirrors some of the rich SPT physics enjoyed by free fermions?

It is against the above motivational backdrop that the work Vincent describes in his doctoral thesis should be placed and understood. As it turned out, answering—in the affirmative—the above question required, as a stepping stone, the in-depth investigation of effective non-Hermiticity in quadratic bosonic Hamiltonians I alluded to before; ultimately, however, it also demanded a radical departure from the very Hamiltonian paradigm that is so deeply engrained in traditional condensed-matter physics. In essence, the strongest and, in my view, most innovative implication of Vincent's work is that fully embracing a *open many-body quantum system* paradigm—and the idea of "topology by dissipation"—could very well be the *only* pathway to bringing about SPT physics in bosonic systems without making them strongly interacting or, possibly, explicitly time-dependent. In the process of articulating this fundamental realization, Vincent establishes several remarkable results.

First, by skillfully combining tools from NH quantum mechanics and results from linear algebra in indefinite inner-product spaces with a particular type of stability theory of linear time-invariant dynamical systems—so-called *Krein stability theory*—he shows that every quadratic bosonic Hamiltonian possesses a *generalized parity-time (PT) symmetry*, and the onset of dynamical instabilities is *precisely* attributable to the spontaneous breaking of this symmetry. Generalized PT symmetry breaks if the relevant dynamical matrix approaches a point in parameter space where it loses diagonalizability (that is, an EP) or when, regardless of diagonalizability, its eigenvalue spectrum splits off the real axis into the complex plane, via a "Krein collision" (KC). While EPs are quintessential branch-point singularities that occur for generic NH matrices and engender polynomial dynamical instabilities, KCs are tied to the distinctive structural properties that valid bosonic dynamical matrices must obey. At a KC, the system is either dynamically unstable or at the cusp of instability, meaning that it can be destabilized by infinitesimally small perturbations. Dynamical phase boundaries that separate different stability regimes are then found to generally consist of loci of *both* EPs and KCs, and can be detected via the vanishing of a newly introduced indicator, called the *Krein phase rigidity*. Notably, exponentially localized zero modes are uncovered for a one-dimensional system of free bosons previously introduced as a bosonic analogue of the paradigmatic fermionic Kitaev-Majorana chain. While being tied to topological properties of the bulk, these "shadows" of bosonic Majorana zero modes can be shown to be necessarily intrinsically unstable, however—providing strong impetus for taking the search for SPT-like bosonic physics beyond the closed-system paradigm, as anticipated.

The second, impressive set of results does indeed pertain to open systems of free bosons, evolving under *explicitly* NH, dissipative quantum dynamics—governed by a Markovian Lindblad master equation, a quadratic bosonic *Lindbladian* for short. While the emergence of topological properties and zero modes had also been established in dissipative settings for non-interacting fermions, identifying proper signatures of SPT physics in bosons entails, again, profound differences: in particular, in stark contrast with standard finite-size scaling arguments, it becomes necessary to properly capture how the non-normality of the underlying dynamical generator may cause the spectral properties of the system in the thermodynamic (infinite-size) limit to dramatically differ from those of arbitrary finite-size truncations. Vincent deserves full credit for identifying *pseudospectral techniques* as the appropriate mathematical framework for doing that. Ultimately, this results in a novel, uniquely bosonic notion of *topological dynamical metastability*, whereby topologically non-trivial features are not to be found in the many-body ground or steady state of the system, but rather in a *transient* dynamical regime whose duration grows with system size. Most remarkably, tight analogues of Majorana fermions, so-called *Majorana bosons*, may then emerge, localized on opposite ends of the chain. In keeping with the breakdown of Noether's theorem for open quantum dynamics and the fact that no simple relationship exist in general between symmetries and conserved quantities, they consist of an approximate zero mode and a *generally distinct*, albeit canonically conjugate, generator of an approximate (Weyl) symmetry. Crucially, the existence of these topological bosonic edge modes is predicted to elicit divergent zero-frequency peaks in certain power spectra and increasingly stationary behavior of two-time quantum correlation functions as system size increases.

In closing, I cannot refrain from acknowledging how, as with the best graduate students, Vincent took—but brought much further than I ever imagined—an initial "hunch" I had on the necessity to decisively move into bosonic open-system dynamics. In doing so, however, he has greatly enriched my own understanding and contributed to shape our exploration by bringing in, as the above testifies, new mathematical tools and a wealth of fresh insight. To sum up, Vincent's thesis both provides an excellent entry point in an area—open many-body quantum systems—whose significance I can only foresee to continue expanding in the years ahead, and breaks important new ground for re-imagining topological many-body physics beyond equilibrium, likely with a host of fundamental and applied ramifications still awaiting discovery.

James Frank Family Professor of Physics
Dartmouth College
Hanover, New Hampshire, USA
January 2024

Lorenza Viola

Preface

This thesis covers various aspects of effective non-Hermiticity and the role played by topology in the physics of non-interacting bosonic systems. Here, I will take the opportunity to summarize both the scientific and personal aspects of the journey that led me to this point. During the latter part of my undergraduate studies at the University of Connecticut, I was privileged to aid in conducting experiments on exotic materials under the guidance of Prof. Jason Hancock. This experience sparked an initial fascination with condensed-matter physics, but more significantly, it helped me realize that my inclinations were (much) more aligned with theoretical pursuits. Specifically, I discovered that I derive greater satisfaction from utilizing new or overlooked mathematical techniques to gain valuable insights about concrete physical systems. More poignantly, I realized my physical presence in a laboratory was more of a liability than a benefit. As a result, Prof. Lorenza Viola was the clear choice for my PhD advisor when I arrived at Dartmouth in the Fall of 2017.

After a first year of courses, Lorenza was kind enough to let me sit in on several intense research meetings usually featuring some combination of Abhijeet Alase—Lorenza's senior graduate student at the time and current postdoctoral fellow at the University of Sydney, Emilio Cobanera—a former postdoc in the group and presently assistant professor of physics at SUNY Polytechnic Institute, Gerardo Ortiz—Professor of Physics at Indiana University, Bloomington, Qiao-Ru Xu—a PhD student of Gerardo's at the time, and, of course, Lorenza. The topics of discussion varied, but mostly consisted of further extensions of their work on a generalization of Bloch's theorem for clean, translationally invariant, non-interacting systems of fermions with arbitrary boundary conditions (BCs). Two immediate extensions they had in mind were (1) to non-interacting bosons and (2) to genuinely non-Hermitian Hamiltonians. In fact, they had several concrete models in mind that would offer valuable playground for testing the viability of the extensions. One such bosonic model was the bosonic Kitaev chain (BKC) that had just recently appeared in a preprint by McDonald *et al.* Lorenza posed a concrete problem for me to work on during my first summer: "Can you reproduce the analytical diagonalization of the BKC using our generalization of Bloch's theorem?" This was precisely the type of problem I wanted to sink my teeth into: applying novel mathematical tools to gain physical insight. Moreover, I could stick to the world of Hermitian Hamiltonians that I was most familiar with. I excitedly accepted the challenge and went to work dissecting and digesting all aspects of the theorem. Abhijeet, who had recently become my office mate, helped me learn both the practical aspects and deeper core idea of the techniques he had been developing over the years.

With a firm grasp on the tools, I was ready to begin. However, I soon encountered a peculiar obstacle that had been previously hinted at by the group, particularly by Emilio. In order to diagonalize a non-interacting bosonic Hamiltonian, it is often necessary to diagonalize a non-Hermitian matrix (or more generally, transform it into a Jordan normal form) often called the "dynamical matrix". Emilio had given me an early warning by highlighting a strange non-diagonalizable matrix that appears when the free particle Hamiltonian is cast into a bosonic operator basis. However, I had not fully grasped how the diagonalization of a Hermitian Hamiltonian could involve a non-Hermitian matrix until this moment. Nonetheless, a non-Hermitian matrix lay at the heart of the BKC. Of course, this had been understood—I was simply late to the party. There was no escaping non-Hermiticity, it seemed.

I was eventually able to reproduce the known features of the BKC and even extend the exact diagonalization to a family of non-trivial BCs. Notably, tuning these BCs resulted in a rich stability phase diagram. That is, certain configurations yielded stable evolution of observables, while others did not. The potential for dynamical instability in such quadratic bosonic Hamiltonians (QBHs) is intrinsically tied to non-Hermiticity: dynamically unstable Hamiltonians are precisely those whose dynamical matrices are either non-diagonalizable or possess non-real eigenvalues. Rather shockingly, the stable BCs corresponded precisely with the BCs that supported Majorana edge modes in Kitaev's original fermionic chain. In fact, this is what motivated Lorenza to suggest I investigate these BCs in the first place. I wanted to understand precisely why certain configurations provided stability while others did not and, given the strange correspondence with Majorana edge modes in the fermionic chain, if topology could be playing a role. Due to my new found interest in stability phase diagrams, Lorenza recommended I seriously engage with the (rapidly growing) literature on non-Hermitian quantum systems—more specifically, PT-symmetric quantum mechanics and a somewhat niche series of papers on a concept known as "phase rigidity". She had suspected that the conditions for the onset of dynamical instability should be equivalent to a symmetry-breaking condition, and that it may have something to do with these concepts. Simultaneously, we had jointly realized that these dynamical matrices were "pseudo-Hermitian" and, thanks to a series of papers by Hermann Schulz-Baldes *et al*, it was brought to my attention that there existed a mathematical theory of stability for these matrices. This stability theory, known as Krein stability theory, proved absolutely essential for characterizing the stability phase diagrams of these Hamiltonians. Bringing together all of these tools, with no shortage of help from Lorenza and Emilio, my first two papers emerged. The first presented an elaborate dissection (or "deconstruction") of the stability phase diagrams of QBHs, while the second focused on the role of pairing in dynamically stable QBHs.

These papers were very exciting, but there was still an itch I didn't quite scratch. Where was the topology hiding in all of this? During this time, I had contributed to a paper first-authored by Qiao-Ru that, in particular, seemed to show topology, or more precisely, symmetry-protected topological (SPT) physics, was forbidden in "thermodynamically stable" QBHs (i.e., those QBHs whose possess well-defined Gibbs thermal states). This was strange to me: there was absolutely *no* shortage of papers discussing the topological aspects of non-Hermitian systems, or even non-interacting bosonic systems! What was going on here? After many discussions, I was forced to finally appreciate the subtle distinction between "topological physics"—physical phenomena that can be convincingly associated to topological features of the system—and "SPT phases"—quantum phases of many-body systems defined (loosely) by global topological invariants. Emilio, in particular, had convinced me that much (in fact, all) of the topological phenomena attributed to non-interacting bosonic systems fell into the former category. In particular, strange "shadows" of Majorana fermions, that we found by generalizing the BKC in such a way that allowed us to change topological invariants, could not be associated to SPT physics in any meaningful way. It seemed there was no hope for SPT phases of non-interacting bosons.

Anyone who knows Lorenza will agree that she has a burning passion for *open* (or as she would say, *real*) quantum systems. She had been pushing this more condensed-matter-oriented side of the group toward the open world for years now (since Emilio was her postdoc), but no one had fully taken the bait. She and Emilio shared several papers regarding so-called "quadratic Lindbladians," which seemed to be the simplest extension of quadratic Hamiltonians into the dissipative realm. Furthermore, Lorenza had shared a number of papers covering "dissipative phase transitions" which, loosely, generalize the notion of a quantum phase transition to the open case. In short, the many-body energy gap is typically replaced with the dissipative, or Lindblad gap, and the many-body ground state would be replaced by the (usually, unique) steady-state. My interest in open systems grew after attending a few of Lorenza's lectures on the subject at the end of her advanced quantum course. I couldn't wait to get my hands dirty with the topic. After a long and winding diversion into this open-systems realm (with lots of growing pains!), we had all realized that there could be hope for *somehow* replicating SPT phenomena with non-interacting bosons undergoing Markovian dissipation described by quadratic bosonic Lindbladians (QBLs).

Using a dissipative variation of the BKC as my toy model, I discovered stable bosonic analogs of Majorana zero modes. When I presented my findings to Lorenza and Emilio, Emilio posed a seemingly innocent question: "Where are the canonically conjugate partners?" He had observed that these zero modes commuted under certain conditions and therefore were not mutually conjugate like the well-known Majorana edge modes of topological superconductors. After a bit of pondering, I managed to find these partners. Interestingly, they were not zero modes as we had anticipated, but rather symmetry generators. I was unaware at the time, but this actually reflected the breakdown of Noether's theorem in open quantum systems of which Lorenza and Emilio had already been well-aware. These discussions and calculations led me to two conclusions. Firstly, my derivation of these conjugate partners was quite general, and I was able to eventually establish a general correspondence theorem for zero modes and symmetry generators in these QBLs. Secondly, despite the existence of zero modes, the Lindblad gap of the finite system remained non-zero. Zero was not only absent from the spectrum, but far from it! After delving deeper into the non-Hermitian literature, I learned that this was a manifestation of "pseudospectrum." This concept extends the notion of the spectrum to encompass approximate spectra. Additionally, it became clear that the pseudospectrum differed significantly from the spectrum for highly non-normal matrices. I realized that this accounted for the simultaneous presence of these "Majorana bosons" (MBs) and a non-zero Lindblad gap. Ultimately, these results expanded into two more papers where MBs, and their number-symmetric partners, "Dirac bosons," provided the first convincing indications of SPT physics in a non-interacting bosonic framework. The key observation was that these signatures can only manifest in the *transient* dynamics of the evolution, rather than the steady-state behavior. Along the way, we introduced the concepts of dynamical metastability and topological metastability, which describe in a precise way anomalous transient dynamics in these systems.

My scientific journey could not have developed in this way without the guidance of Lorenza and Emilio. On multiple occasions, the (seemingly) innocent questions they posed to me have resulted in paradigm shifts in my research. Emilio is always quick to ground my occasional drift into strange mathematics with real physics and often manifests (out of thin air, it sometimes seemed!) concrete, non-trivial models for me to explore. His expertise in condensed-matter physics has saved me from making (sometimes laughably!) unphysical claims and, as a result, had impacted my path massively. Lorenza's impact is almost impossible to capture with words. First and foremost, she has built a research environment that rewards and promotes intellectual freedom. She has consistently remained open to engaging with, in particular, obscure mathematical techniques that only a graduate student would be naive enough to pursue. If not for this freedom, I would have never have found, and successfully applied, the various tools (PT symmetry, Krein stability theory, phase rigidity, pseudo-Hermiticity, pseudospectrum, etc.) that make up the foundation of my thesis. Beyond this, she has exposed me to, and perfectly embodied, the essence of Freeman Dyson's "frogs" and "birds." Like a frog, Lorenza is unafraid to jump into the mud and grapple with the minute details of a complicated calculation. Simultaneously, somehow, she is able to leverage her massively diverse body of experience to fly high above the details (like a bird) and see important connections between *a priori* disparate concepts and fields. More importantly, she knows where to reign in her expertise and allow for her students to uncover the rest.

Beyond Emilio and Lorenza, a number of people at Dartmouth have influenced me greatly during this time. Two of my committee members, Sekhar and James, taught two of my favorite courses: microscopic theory of solids and quantum information, respectively. Both assigned research projects that have had lasting impact on my work far beyond coursework. Similarly, Roberto Onofrio's course on phase transitions had a similar impact. Not to mention the many lively discussions (both scientific and otherwise!) over roasted potatoes and homemade pasta. Hours of conversations with the other members of Lorenza's group (Mariam, Francisco, and Joshuah, to name a few) have further influenced my work in many ways. I would also like to acknowledge and thank Prof. Steve Girvin, whose scientific work has influenced aspects of my future ambitions, for agreeing to be a part of my committee and for being exceedingly patient with the various scheduling hiccups.

Above all else, there is no person more deserving of my thanks and recognition than my wife, Andrea. Her unwavering support, companionship, and love over the last 12 years has kept me focused and propelled me into the privileged position I find myself in today. Without her, I would have never had the courage to pursue my love for physics. Additionally, my brother Will, his wife Kate, and Andrea's parents Jim and Doreen have supported and encouraged me nonstop through many of life's challenges. I would also like to thank my close friends Money, Drew, Hitchcock, and DJ for keeping me sane throughout grad school. I must also acknowledge my two cats, Schrödinger and Cleopatra, who have been a constant source of joy, entertainment, and unconditional love throughout the years. Finally, I want to thank my mom. Among everything else, she instilled in me a love for math and science at a young age that blossomed into the passion I have today. For that, and for the indiscriminate love and support that she provided me throughout her life, I am forever grateful.

Hanover, NH, USA Vincent Paul Flynn

Part of This Thesis

Parts of this thesis have been published in the following journal articles:

- **Vincent P. Flynn**, Emilio Cobanera, and Lorenza Viola, "*Topological zero modes and edge symmetries of metastable Markovian bosonic systems*", Physical Review B **108**, 214312 (2023).
- **Vincent P. Flynn**, Emilio Cobanera, and Lorenza Viola, "*Topology by Dissipation: Majorana Bosons in Metastable Quadratic Markovian Dynamics*", Physical Review Letters **127**, 245701 (2021).
- **Vincent P. Flynn**, Emilio Cobanera, and Lorenza Viola, "*Restoring particle conservation in quadratic bosonic Hamiltonians with dualities*", Europhysics Letters **131**, 40006 (2020).
- **Vincent P. Flynn**, Emilio Cobanera, and Lorenza Viola, "*Deconstructing effective non-Hermitian dynamics in quadratic bosonic Hamiltonians*", New Journal of Physics **22**, 083004 (2020).

Additional motivational results may be found in

- Qiao-Ru Xu, **Vincent P. Flynn**, Abhijeet Alase, Emilio Cobanera, Lorenza Viola, and Gerardo Ortiz, "*Squaring the fermion: The threefold way and the fate of zero modes*", Physical Review B **102**, 125127 (2020). Editor's Suggestion.

Contents

1 Introduction ... 1
1.1 Context and Motivation ... 1
1.2 Outline ... 8
References ... 12

Part I Effective Non-Hermiticity in Closed Bosonic Systems

2 Background: Quadratic Bosonic Hamiltonians ... 17
2.1 Multimode Bosonic Fock Space and Gaussian Transformations ... 19
2.2 Lattice Models ... 23
2.3 Closed Quadratic Bosonic Dynamics ... 26
2.3.1 Effective Non-Hermiticity ... 26
2.3.2 Notions of Stability and Their Criteria ... 30
2.3.3 The Indefinite Inner-Product Structure of Nambu Space and Reformulating Dynamics ... 32
2.4 Diagonalization of Quadratic Bosonic Hamiltonians ... 36
2.5 Bulk-Translationally Invariant QBHs ... 42
2.5.1 The Translationally Invariant Case ... 42
2.5.2 Boundary Conditions and the Generalization of Bloch's Theorem ... 43
References ... 46

3 Dynamical Stability Phase Transitions ... 47
3.1 Spontaneous Generalized PT Symmetry-Breaking ... 49
3.1.1 The Equivalence Between Pseudo-Hermiticity and Generalized PT Symmetry ... 49
3.1.2 Generalized PT Symmetry in Quadratic Hamiltonians ... 52
3.2 Classification and Detection of Stability Phase Transitions ... 53
3.2.1 Tools from Krein Stability Theory and Their Implications for QBHs ... 54
3.2.2 Krein Phase Rigidity ... 56
3.3 Examples ... 59
3.3.1 A Single-Mode Model ... 60

3.3.2 A Two-Mode Cavity QED Model 62
3.3.3 A Bosonic Kitaev Chain 65
References 81

4 The Role of Pairing in Dynamically Stable QBHs 83
4.1 A Number Conservation-Restoring Duality Transformation 85
4.1.1 Existence and Construction of the Duality 85
4.1.2 Physical Interpretation of the Duality 89
4.2 Examples 93
4.2.1 A Gapped Harmonic Chain and General Single-Band Pairing Chains 93
4.2.2 A Squeezing Duality for the Bosonic Kitaev Chain 96
References 98

5 Obstructions to SPT-Like Physics in QBHs 101
5.1 Topological Invariants in the Presence of Effective Non-Hermiticity 103
5.2 The Search for SPT Physics in QBHs 106
5.2.1 Primer: The Fermionic Kitaev Chain 106
5.2.2 Thermodynamically Stable Systems 109
5.2.3 Beyond Thermodynamic Stability: Bosonic Shadows of Majorana Fermions 110
References 113

Part II Signs of Genuine SPT Physics in Open Bosonic Systems

6 Background: Quadratic Bosonic Lindbladians 117
6.1 The Lindblad Formalism 119
6.1.1 Spectral and Convergence Properties 122
6.1.2 Symmetries and Conserved Quantities in Markovian Systems 123
6.2 Quadratic Bosonic Lindbladians 125
6.2.1 Equations of Motion 125
6.2.2 Stability Criteria and Asymptotic Relaxation 128
6.2.3 An Elementary Example 129
6.2.4 Translationally Invariant QBLs and Arbitrary BCs 130
References 132

7 Zero Modes, Weyl Symmetries, and QBL Design 133
7.1 A Partial Restoration of Noether's Theorem 135
7.1.1 The Exact Case 135
7.1.2 The Approximate Case 138
7.2 Two Protocols for QBL Design 141
7.2.1 Embedding Fermionic Hamiltonians in Bosonic Dissipators 141
7.2.2 Reservoir Engineering Pure Steady States via Dualities 142
References 145

8 Signatures of SPT Physics in 1D Bulk-Translationally Invariant QBLs 147
8.1 Beginning the Search 150
8.1.1 The Need for Bulk Instabilities and Non-normality 150
8.1.2 Primer: The Pseudospectrum 152
8.2 Anomalous Transient Dynamics in QBLs 154
8.2.1 Anomalous Relaxation and Dynamical Metastability 154
8.2.2 Divergence of the Transient Timescale in Dynamically Metastable Chains 156
8.3 Topological Metastability 158
8.3.1 Majorana Bosons and Quasi-Steady States 159
8.3.2 The Impact of Number Symmetry: Dirac Bosons 163
8.4 Relationships to Existing Work 166
References 167

9 The Realm of Possibilities 169
9.1 A Purely Dissipative Topologically Metastable Chain 171
9.2 A Dissipative BKC 174
9.2.1 The Parameter Regime $\mu = 0$ 175
9.2.2 The Parameter Regime $\Gamma = 0$ 176
9.3 A Dissipative BKC with a Pure Steady State 179
9.3.1 Constructing the Model 179
9.3.2 Relaxation Dynamics of the Quasi-Steady States 181
9.3.3 Transient Odd-Parity Behavior 183
9.4 Dirac Bosons in a Number-Symmetric Dissipative Chain 185
9.5 Observable Signatures 189
9.5.1 Two-Time Correlation Functions and Power Spectra 189
9.5.2 Signatures of Topological Dynamical Metastability 191
9.5.3 Distinguishing Split and Non-split Majorana Bosons 192
9.5.4 Signatures of Dirac Edge Bosons 196
References 197

10 Summary and Outlook 199
10.1 Summary of Key Results 199
10.1.1 Effective Non-Hermiticity in Quadratic Bosonic Hamiltonians 199
10.1.2 Manifestations of SPT Physics in Quadratic Bosonic Lindbladians 201
10.2 Outlook 204
References 207

A Spectra and Pseudospectra of Block-Toeplitz Matrices and Operators 209
A.1 The Non-block Case ($d = 1$) 211
A.2 Block Case ($d > 1$) 212

A.3 Jordan Canonical Form of Corner-Modified Banded Block-Toeplitz Matrices 213

B Miscellaneous Technical Calculations 219
B.1 Existence of a Bosonic Eigenbasis for Dynamically Stable QBHs 219
B.2 Eigendecomposition of the BKC Under Various BCs 220
B.2.1 Open BCs 220
B.2.2 Twisted BCs 222
B.3 An Isospectral Mapping Between the PDC and the DBKC 228
B.4 Persistence of MBs in a Disordered DBKC 229
B.5 Exact Time-Evolution of Bosonic Parity of a Dissipative Harmonic Oscillator Prepared in a Cat State 230
References 231

Curriculum Vitae 233

Table of Acronyms

Table of acronyms used throughout the thesis in no particular order

Term	Acronym
Symmetry-protected topological	SPT
Quadratic bosonic/fermionic Hamiltonian	QB/FH
Quadratic bosonic/fermionic Lindbladian	QB/FL
Bogoliubov-de Gennes	BdG
Gorini-Kossakowski-Sudarshan	GKS
(Open/Periodic/Semi-infinite/Bi-infinite) Boundary condition	(O/P/SI/BI)BC
Exceptional point	EP
Krein collision	KC
Zero mode	ZM
Symmetry generator	SG
(Generalized) Parity-time	(G)PT
(Krein) Phase rigidity	(K)PR
Quantum electrodynamics	QED
Majorana boson	MB
Dirac boson	DB
Purely dissipative chain	PDC
Fermionic/Bosonic Kitaev chain	F/BKC
Dissipative bosonic Kitaev chain	DBKC
Dissipative number-symmetric	DNS

Chapter 1
Introduction

Abstract We introduce and review the essential kinematical and dynamical differences between the two most commonly encountered examples of indistinguishable quantum particles. Namely, bosons and fermions. We discuss, in detail, paradigmatic examples of dynamical and thermodynamical instabilities arising bosonic systems that lack any fermionic counterpart. In the non-interacting setting, the emergence of these instabilities can be, in part, associated to an intrinsic non-Hermiticity present in the Heisenberg equations of motion. As such, we motivate the study of this effective non-Hermiticity from the perspective of realizing novel non-Hermitian physics known to exist in nonequilbrium and open quantum systems. A second key distinction between non-interacting fermionic and bosonic physics is the apparent lack of topology as a classifying agent for bosonic thermodynamic phases. We explain the reason for this and motivate the move into an explicitly open Markovian setting in order to obtain bosonic signatures of non-trivial topology reminiscent of their fermionic counterparts. We then summarize all of the main results presented in the thesis.

1.1 Context and Motivation

A large portion of the field of quantum statistical mechanics rests upon the simple-to-state indistinguishability principle: the microscopic constituents of a quantum many-body system should be treated as fundamentally *identical*. Any two physical states that differ only by a permutation of particles must be indistinguishable. Employing this principle consistently demands that arbitrary physical observables must be left invariant under arbitrary permutations of these identical constituents. It follows all observables and the dynamics must leave invariant certain subspaces of the physical Hilbert space on which the set (group) of all possible permutations acts irreducibly, with no interference between states of different permutation symmetry being possible. Moving one step further, the so-called *symmetrization postulate* asserts that precisely two of these subspaces are physical: The *symmetric* subspace, which consists of those state vectors left invariant under arbitrary perturbations, and the *antisymmetric* subspace, which consists of those state vectors which accumulate

V. P. Flynn, *Effective Non-Hermiticity and Topology in Markovian Quadratic Bosonic Dynamics*, Springer Theses, https://doi.org/10.1007/978-3-031-52045-7_1

a negative sign upon swapping two particles. Moreover, the former correspond to particles with integer spin—so-called *bosons*—while the latter correspond to particles with half-integer spin—so-called *fermions*. Examples of bosons are photons, phonons, magnons, and neutral atoms with an even odd number of neutrons, while examples of fermions are electrons, protons, muons, and neutral atoms with an odd number of neutrons. This difference in *exchange statistics* immediately elicits a major kinematical schism between the two types of particles. Namely, a fixed single-particle state can only be occupied by either 0, or 1, fermionic particles. This fact, known historically as the *Pauli exclusion principle*, is a direct consequence of the symmetrization postulate. For bosons, there is no such restriction on the occupation numbers. Thus, in a second quantization framework, whereby an arbitrary number of particles may be present, the landscape of potential many-body states differ dramatically. Beyond kinematics, the exclusion principle (or lack-thereof) leads to dramatic implications at the level of equilibrium physics. Consider, for example, a free gas whose constituents are either bosons, or fermions. In the degenerate regime (characterized by high-density, low-temperature, or some combination of the two), the free boson gas will necessarily manifest a macroscopic population of the lowest energy single-particle state. The resulting many-body state, called a *Bose-Einstein condensate* (BEC), was first predicted by Bose and Einstein in the 1920s [1, 2], and confirmed experimentally in 1995 [3, 4]. In sharp contrast, a free gas of N fermions must fill the N lowest-energy single-particle states up to the Fermi energy. This configuration, known as the *Fermi sea*, plays an essential role in describing the equilibrium behavior of electric insulators and conductors [5]. However, equilibrium differences between bosons and fermions are just the tip of the iceberg.

The unbounded nature of bosonic occupation numbers provide fertile ground for realizing *unstable dynamics* in bosonic systems. Consider, for example, parametric amplification of a quantum optical mode. By means of parametric down-conversion via a coherent pumping field and transforming to a rotating frame, one may achieve a simple effective Hamiltonian for the optical mode [6]. This Hamiltonian (loosely) mediates a balanced injection and removal of pairs of photons and, importantly, facilitates a *dynamical instability*: Regardless of the initial number of photons initially present, the average occupation number will diverge exponentially. Such behavior is *forbidden* in fermionic systems: a single-particle state (e.g., the optical mode) can only support, *at most*, one fermionic excitation. This capacity for instabilities is the main explanatory mechanism for, or in some cases, operating principle behind, a diverse set of distinct physical phenomena—including signal amplification in cavity and circuit quantum electrodynamics (QED) [7–15], squeezing used for continuous variable quantum information processing [16–19], decay mechanisms for atoms in optical lattices [20], and instabilities in BECs [21, 22]. Depending on the application, methods for tapping into this latent potential vary. Remarkably, accessing these high occupation numbers may be achieved in a coherent (i.e., unitary) and non-interacting fashion via *two-photon driving*, in the context of quantum optics (as in the example above), or *bosonic pairing*, in condensed-matter parlance. Methods for implementing these processes have been both proposed, and experimentally realized, in cavity- and circuit-QED platforms [11, 12, 23–32], microlasers and ring resonators [33, 34], optomechanical systems [35, 36], and vibronic lattices [37]. From a theoretical perspective, it has been understood that such behavior is attributable to *intrinsic*, or *effective, non-Hermiticity* built-in to the Heisenberg equations of motion for the elementary bosonic degrees of freedom [7, 38, 39]—a feature that is completely *absent* in the fermionic setting.

Thanks to a number of theoretical and experimental advancements, non-Hermitian physics, more generally, has received renewed attention in the last several years. On its surface, the phrase "non-Hermitian physics" encompasses an extraordinarily diverse series of classical and quantum phenomena, whereby a non-Hermitian operator plays a central role in the mathematical modeling [40]. Perhaps the most interesting among these phenomena arises when a non-Hermitian operator is employed as the generator of quantum evolution. For example, there have been considerable developments extending the standard Hermitian (or, more precisely, self-adjoint) Hamiltonian postulate for closed quantum dynamics to instead allow for PT-symmetric [41–43], or pseudo-Hermitian [44] Hamiltonians, which both, under certain circumstances, retain entirely real (and thus, physically meaningful) energy spectra. Importantly, in this case the resulting dynamics remains unitary, possibly with respect to a modified inner product. Beyond these more fundamental perspectives, non-Hermiticity can arise in a natural way when one considers non-isolated, or *open*, quantum systems. For example, modeling the dynamics of a system coupled to an external bath may result in a dynamical law for the system that features a non-Hermitian generator, one example of which being the *Lindbladian*, in the case where the dissipation is Markovian [45, 46]. Beyond this fully quantum treatment, one may further model dissipative quantum dynamics in a *semiclassical* way by adding appropriate non-Hermitian terms to the bare system Hamiltonian, in a way to phenomenologically encode loss and gain. Additionally, quantum measurement can itself, in some cases, be modeled in terms of non-Hermitian generators [9, 47].

Non-Hermitian—or more broadly, *non-normal*—operators have been long-known to exhibit a rich array of spectral properties that manifest in striking ways depending on the physical context in which they arise. One of the most notable of such features is the ability for non-Hermitian operators to sustain *exceptional points* (EPs), i.e., points in parameter space where the operator loses diagonalizability—a possibility that is completely *forbidden* by the spectral theorem in the Hermitian case. From a mathematical perspective, EPs engender a degree of sensitivity to perturbations that exceeds any similar response in Hermitian systems. This mathematical fact has gained physical relevance as the operating principle behind *EP-enhanced sensing protocols* [48, 49]. Another form of sensitivity arises when one considers many degrees of freedom coupled together in a non-Hermitian manner. One incarnation of this comes in the form of the so-called *non-Hermitian skin-effect* (NHSE) [50], whereby imposing boundary conditions (BCs) on the system causes the spectra to change dramatically and a macroscopic number of eigenstates to localize at the boundary. This *extreme sensitivity to BCs*, which has also been proposed for sensing applications [51], was later more specifically attributed to extreme *non-normality* of the Hamiltonian, or even Lindbladian [52, 53].

To better understand how such non-Hermitian phenomena may manifest in the physics of closed, non-interacting bosonic systems, and to make clearer those connections that we believe are lacking between the two fields, let us describe more precisely the origin of effective non-Hermiticity. The systems in question are modeled in terms of Hamiltonians that are quadratic in the bosonic creation and annihilation operators which define the physical degrees of freedom. These quadratic bosonic Hamiltonians (QBHs) *are* perfectly Hermitian. Nonetheless, one finds that the Heisenberg equations of motion for the creation and annihilation operators directly involve a non-Hermitian matrix, i.e., the so-called *dynamical matrix*. The emergence of non-Hermiticity turns out to be the consequence of a *violation of number conservation* by the Hamiltonian. In fact, the violating terms are precisely the ones responsible for amplification in the previously mentioned example. Notably, the fermionic analogue of this matrix remains Hermitian even when fermionic pairing, which violates number conservation, is present. This provides us with an opportunity to explore the dynamical consequences of the rich spectral behavior of non-Hermitian operators. For example, how are the many-body dynamics affected by the existence of a non-real eigenvalue, or the loss of diagonalizability? In short, it turns out that, if either of these two situations arise, the QBH will be *dynamically unstable*. That is, it will generate *unbounded evolution* of certain observables (recall the divergence of photon number in the parametric amplifier). But these possibilities are a small part of a much richer family of spectral behavior exhibited by non-Hermitian operators. For example, there exists entire bodies of mathematical literature on the *stability theory* of certain classes of non-Hermitian matrices [54, 55]. How do these uniquely non-Hermitian stability theories manifest in the bosonic setting?

Importantly, while number-non-conservation is essential for non-Hermiticity of the bosonic dynamical matrix, it does not immediately imply dynamical instability. In particular, the non-Hermitian dynamical matrix may still be diagonalizable and possess an entirely real spectrum. This fact about non-Hermitian operators has been appreciated for decades in the context of PT-symmetric and pseudo-Hermitian quantum mechanics. For example, it is known that such operators can be effectively "made" Hermitian by suitably modifying the Hilbert space metric. The implications of these results must then necessarily, if applicable, correspond to a *physical* statement about number-non-conservation in bosonic systems. There has been no *systematic* investigation of the implications of this in bosonic systems. Stated simply, to what extent are number non-conserving terms even necessary for describing dynamically stable QBHs? More generally, what, if any, connections can be made between the fields of PT-symmetric or pseudo-Hermitian quantum mechanics and closed, non-interacting bosonic systems? Let us synthesize the points we raised so far into the first central question that we shall attempt to answer in this thesis:

(1) What are the most salient consequences of the effective non-Hermiticity intrinsic to the equations of motion for closed, non-interacting bosonic systems?

The emergence of effective non-Hermiticity in closed, non-interacting bosonic systems is a striking dynamical consequence of exchange statistics—especially when compared to the fermionic counterpart. However, a more subtle, but equally potent consequence can be uncovered by closely examining the role played by *topology* in both cases. In the fermionic case, topology is a necessary ingredient for describing and classifying the possible *quantum phases* of non-interacting fermionic matter. Following the discovery of a peculiar class of phase transitions by Berezinskii et al. [56, 57], it became clear that the Ginzburg-Landau paradigm is insufficient for classifying all phases of matter. The systems they studied exhibited phase transitions *without* spontaneous symmetry breaking and featured only global, not local, order parameters of topological origin. These early works eventually lead to an explosion of discoveries of so-called “topological insulators” [58–61].

Topological insulators, which represented a brand new state of fermionic matter, are unified by two main features: (1) From the bulk properties of the system, one can compute *topological invariants* that cannot be altered without either breaking certain symmetries of the system, or crossing a quantum phase transition signaled by closing of the many-body gap; and (2) upon truncation of the system, edge, or surface, states emerge and provide dissipation-free conduction channels, thus circumventing the insulating behavior of the bulk. Moreover, the bulk invariants have been directly linked with the edge states and boundary invariants—a result now known as the *bulk-boundary correspondence* (BBC) [62]. Beyond insulators, it was latter realized that topology and, importantly, topologically-mandated edge states could also appear in superconducting systems [63, 64]. In these systems, known as *topological superconductors*, the insulating gap is replaced by the many-body superconducting gap [65]. Notably, however, the edge states in topological superconductors obeyed a different set of exchange statistics than the ones obeyed by standard *Dirac* fermions. Namely, the edge states took on the form of *Majorana fermions* [66].

Topological insulators and superconductors were both eventually understood as *symmetry-protected topological* (SPT) *phases*—which emphasizes, in particular, the protecting role played by the symmetries of the system. This concept, in tandem with topology, is *essential* for understanding the quantum phases of free-fermionic matter, in particular. The reason is simple: generically, these phases are gapped, lack a local order parameter, and maintain the same set of global symmetries on each side of the transition. This sparked major efforts towards classifying all possible SPT phases of free fermions [67], ultimately resulting in a complete classification. Within this scheme, known as the *tenfold way*, insulators and superconductors are classified by the presence of, or lack-thereof, a small set of protecting symmetries [68]. In the wake of these foundational works, there were significant pushes to develop both practical applications of SPT physics (most notably, Majorana-based quantum computation [69]) and extensions into the realms of non-equilibrium and engineered quantum matter. For example, the NHSE has been recently understood to have a *topological origin* [70]. It is not, a priori, clear if this manifestation of topology has any relationship with the topology of SPT phases of free fermions, however. To this end, a classification of non-Hermitian topological phases that utilizes a suitably generalized form of the free-fermionic tenfold way has been successful in describing non-Hermitian SPT phases of both explicitly non-Hermitian fermionic Hamiltonians [71] and open quantum systems of free fermions subject to Markovian dissipation [72–76]. Further developments have even begun to push this classification beyond the Markovian regime [77].

The discussion of topology has, thus far, been entirely grounded in fermionic systems. Manifestations of topological physics in non-interacting bosonic settings are numerous, but qualitatively different in significant ways. Topological photonics [11, 78], magnonics [79, 80], phononics [37], and amplification [11, 13–15, 81, 82] are, at this stage, well-established fields that feature a multitude of topological features in systems comprised of bosonic degrees of freedom. These systems often possess band structures with non-trivial topological invariants and, as a consequence of a suitable generalized BBC, they manifest edge or surface states upon truncation. However, claims that these phenomena result from an underlying bosonic SPT phase are not compelling. After all, topological band structure, and even the BBC, is an extremely general feature of wave equations [83] and, as such, need not correspond to any underlying quantum phase. In sharp contrast to fermions, there exists *no* clear connection between quantum phases of non-interacting bosons and topology.

It turns out that the search for any such connection is immediately met with complications. In addition to dynamical instabilities, QBHs may also exhibit *thermodynamical, or Landau instabilities*. Consider a fixed QBH H. The quantum phase of H is, by definition, linked to the properties of the ground state. However, can we be sure H even *has* a ground state? Suppose $H = \hbar\omega(a^\dagger a - b^\dagger b)$, with a and b bosonic annihilation operators and $\omega > 0$. The energy eigenvalues of H are unbounded in both directions, and thus, unless external restrictions are present (e.g., fixing the total particle number), H lacks a ground state. Even more severely, the thermal Gibbs state is ill-defined for any temperature. Once again, bosonic statistics allows for instability—in this case, however, it is thermodynamical in nature. If we cannot even guarantee the existence of a ground state, how can we consistently define the notion of quantum, let alone SPT, phases for *arbitrary* QBHs?

One work-around may seem obvious. Suppose that we restrict to only those QBHs that are thermodynamically stable, i.e., those that have well-defined ground states. Topological physics *can* still manifest in the presence of this constraint. Specifically, it is well-understood that thermodynamically stable systems can exhibit topologically non-trivial *high-energy* bands (i.e., bands beyond those surrounding zero energy), and thus, feature *high-energy* edge states [80]. However, the lower energy bands must always remain topologically trivial. So, can thermodynamically stable QBHs exhibit SPT phases? In monumental contrast to their fermionic counterparts (which are *always* thermodynamically stable, thanks to fermionic statistics), the answer, unambiguously, is "*No*". This fact is captured by three no-go theorems that, in addition to forbidding SPT phases, also eliminate the possibility for topologically mandated zero modes and parity switches (which are characteristic of topological phase transitions in fermionic systems) under the constraint of thermodynamical stability [84].

From here, one may simply forgo thermodynamical stability and thus, any conventional notion of a quantum phase. Certain aspects of non-interacting fermionic SPTs may survive. For example, many photonic incarnations of topological insulators exhibit thermodynamical instabilities. Such systems *can* support topologically non-trivial low-energy bands, and thus, topologically-mandated *zero modes* [80, 85, 86]. As it turns out, these modes are plagued by dynamical instabilities: *Arbitrarily small* perturbations can cause these modes to become unstable. Thus, any dynamically-oriented notion of robustness is absent. Nonetheless, the question "To what extent do these edge modes resemble their fermionic counterparts?" remains open.

Dropping thermodynamic stability completely deflates any convincing connections between topological bosonic physics and non-interacting fermionic SPTs. However, there may be hope if we drop a different constraint. Namely, *unitarity*. If we consider *open bosonic systems*, whose non-interacting nature must be maintained in a consistent way, can we uncover *tight* bosonic analogues to non-interacting fermionic SPT phases? First, we must specify the type of open, or dissipative, dynamics we should consider. The simplest extension beyond unitary dynamics is to allow for *Markovian* dissipation, in which case the dynamics are described in terms of a Lindblad master equation. Importantly, this allows us to maintain the non-interacting nature of the system by demanding that the Lindblad generator is quadratic in creation and annihilation operators [87–92]. That is, we take our generator to be a *quadratic bosonic Lindbladian* (QBL). From here, we will take inspiration from the works that have uncovered SPT physics in *open* fermionic systems [72–75] or, even more broadly, works that have solidified the concept of *dissipative* quantum phases [93, 94]. A common theme in these works begins by making the conceptual substitution of the many-body ground state with a (usually unique) many-body *steady state*. The notion of the many-body gap is then replaced by the *dissipative*, or *Lindblad gap*, which physically defines the asymptotic relaxation rate of the dynamics. Altogether, a dissipative phase transition can be, for example, loosely described by the closing of the dissipative, or Lindblad, gap and the emergence of critical behavior in the steady-state. With the conceptual framework laid out, we may ask: What aspects, if any, of non-interacting fermionic SPT physics can arise in a non-interacting system of bosons subject to Markovian dissipation? Moreover, what are their *dynamical* consequences?

In the last few years, an additional point of contact between non-Hermitian topological physics and open ("driven-dissipative") bosonic physics has emerged. This burgeoning research area, known as *topological amplification* [11, 13–15, 81, 82], leverages certain aspects of topological non-Hermitian physics to develop design protocols for robust quantum amplifiers. Owing to its topological origin, the amplification mechanism enjoys a degree of robustness to disorder and can be, at least mathematically, thought of as "protected" by certain transformations derived from *fermionic* symmetries. While these points of contact are intriguing, they fail to address a more fundamental question. Altogether, by synthesizing the various open questions we have since called attention to, we arrive at the second core question of this thesis:

(2) To what extent can closed, or open, non-interacting bosons manifest physics associated to SPT phases of non-interacting fermions?

The motivational structure for questions (1) and (2) is the same. In both cases, there exists two, largely disconnected, bodies of research. For (1), we seek to establish explicit links between the, mutually vast, fields of many-body non-interacting bosonic systems, and non-Hermitian physics. For (2), we wish to connect the former instead with the world of non-interacting SPT physics. Of course, these two questions are not independent. Remarkably, in fact, we will see that our attempts to address (2) will take us (by force!) into the world of topological non-Hermitian physics. In the next section, we will summarize the key contents and results of each chapter. A more detailed summary of each chapter's contents, including their primary work of origin, will be provided in their respective preambles.

1.2 Outline

The presentation of our contributions is subdivided into two main parts. The results in Part I largely address question (1) above, while Part II addresses question (2).

Part I presents a systematic deconstruction of the consequences of effective non-Hermitian dynamics in closed systems of non-interacting bosons. In Chap. 2, we lay down the foundational aspects of time-independent, quadratic bosonic systems. Following a brief description of the multimode bosonic landscape we center ourselves in, we will present the theoretical frameworks necessary for describing dynamics generated by a QBH. Of particular relevance is the intrinsic non-Hermiticity built into the Heisenberg dynamics of those systems described by QBHs without total number symmetry. Two distinct notions of stability emerge for QBHs: thermodynamical and dynamical stability. We will explicitly define these notions and present equivalent characterizations in terms of the relevant dynamical matrix. Following this, we will develop a Bogoliubov-de Gennes (BdG) framework that, in particular, helps reveal the underlying indefinite inner-product structure intrinsic to multimode bosonic Hilbert spaces. This will be specialized to closed (and later in Chap. 6, open) dynamics. We conclude the chapter with a discussion of translation invariance, the key implications of Bloch's theorem in this setting, and the basics of its generalization for diagonalization in the presence of arbitrary BCs.

In Chap. 3, we present an in-depth analysis of the dynamical stability phase diagrams of QBHs. The first key observation is that boundaries separating dynamically stable and dynamically unstable phases are defined via two distinct flavors of spectral degeneracies: *EPs* and *Krein collisions* (KCs). These two flavors of degeneracy are distinguished by whether or not the relevant dynamical matrix is diagonalizable. This difference of origin will entail tangible differences in the dynamical and algebraic features of the normal modes at the transition point. Despite their differences, we are able to unify them through the lens of spontaneous *generalized PT-symmetry (GPT) breaking*. We connect this unifying perspective to the long-established, and rapidly growing, field of PT-symmetric quantum mechanics. Moreover, we prove equivalence between GPT-symmetry and pseudo-Hermiticity, a property intrinsic to the dynamical matrices of QBHs. Our dissection of dynamical stability phase boundaries utilizes the mathematical techniques of Krein stability theory, which we describe along the way as necessary. Putting these tools to use, we introduce a numerical indicator for dynamical stability phase transition known as *Krein phase rigidity* (KPR). Our development of this indicator, along with the understanding of its behavior, is inspired by the previously established algebraic features of bosonic normal modes in the vicinity of stability phase transition. Despite this boson-centric perspective, we find that it is, in fact, an extension of *phase rigidity* (a quantity relevant to the study of EPs in certain non-Hermitian systems) to the pseudo-Hermitian realm. Three example Hamiltonians are studied in detail: a single-mode toy model, a two-mode cavity QED model, and a bosonic Kitaev chain (BKC) under a family of BCs. In all cases, the stability phase diagrams are computed and boundaries analyzed. The GPT-symmetry breaking and behavior of the KPR are studied in detail. Further general features regarding phase-dependent transport in QBHs are elucidated by means of the BKC.

Chapter 4 is dedicated to a set of results specialized to dynamically stable QBHs. In particular, we import mathematical techniques developed for pseudo-Hermitian quantum systems in order to construct a duality transformation that restores number symmetry in dynamically stable QBHs with pairing. As it turns out, the duality transformation is intrinsically related to the covariance matrix of the quasi-particle vacuum of the QBH. Implications for analogue simulation of non-Hermitian quantum dynamics, as well as topological invariants and edge states are discussed. We specialize our results to a general dynamically stable single-band models, exemplified by a gapped harmonic chain, as well as the BKC considered previously.

To conclude Part I, in Chap. 5, we apply the results of the previous two chapters to study certain topological aspects of QBHs. First, we leverage the duality of Chap. 4 to establish a connection between bosonic Berry phases and the more standard Berry phases encountered, for instance, in topological non-interacting fermionic systems. Following this, we recap the essential features of SPT phases in non-interacting fermions by means of a simple analysis of the fermionic Kitaev chain. We proceed to describe the practical and theoretical challenges towards realizing these features in QBHs. First, we describe the no-go theorems forbidding SPT phases in thermodynamically stable QBHs. Then, by forgoing the thermodynamic stability assumption, we explore the intrinsic instability of any zero energy edge modes that may arise. We illustrate these points by uncovering explicit bosonic "shadows" of Majorana fermions in a dynamically, and thermodynamically, unstable generalization of the BKC.

In Part II, we extend our analysis to non-interacting bosonic systems subject to Markovian dissipation and, ultimately, uncover convincing signatures of SPT physics in a bosonic setting. To open, Chap. 6 provides an extension of Chap. 2 to the Markovian setting. We first introduce the basic properties of quantum Markovian dynamical semigroups and, in particular, cover the basic notions of conservation laws, symmetries, and the breakdown of Noether's theorem. Then, we introduce the specific class of Markovian systems we are interested in, i.e., those systems whose Lindblad generator is a QBL. Such generators are defined by a QBH, paired with Lindblad dissipators that are linear in creation and annihilation operators. Our focus will be to highlight the key differences with the Hamiltonian formalism used up until this point. To this end, we emphasize the changes to the relevant equations of motion for linear and quadratic observables, and discuss the appropriate extensions of the stability notions in QBHs and their criteria. Of particular interest are the steady states, which we will characterize in detail. The chapter concludes with a simple single-mode example that exemplifies the key parts of the formalism, in addition to an account of the implications of bulk-translation invariance for these systems.

Chapter 7 contains the a set of new results related to general QBLs that will later prove essential to our search for non-interacting bosonic SPT physics. The first half of the chapter describes the breakdown of Noether's theorem in QBLs within the space of linear forms. For this, we define *ZMs* of QBLs. Additionally, we define *Weyl symmetry generators* (SGs), which are generators of displacements in phase space that leave the overall dynamics invariant. Despite no obvious connection between these two operators in an open context, we prove a *one-to-one canonical correspondence* between them: To each zero mode, there must exists a canonically conjugate Weyl SG, and vice versa. We further generalize this result to accommodate *approximate* zero modes and generators of *approximate* symmetries. The second half of the chapter presents two design protocols for QBLs with certain desirable properties. The first takes, as an input, a quadratic fermionic Hamiltonian (QFH) with edge-localized, Hermitian, (possibly approximate) zero modes and provides, as an output, *purely dissipative QBL* (i.e., a QBL whose system Hamiltonian vanishes) possessing bosonic analogues of these (possibly approximate) zero modes. The second takes, as an input, a dynamically stable QBH and utilizes the duality of Chap. 4 to engineer a QBL whose unique steady state is given by the corresponding quasiparticle vacuum.

Chapter 8 contains the central results of Part II and ultimately culminates in the discovery of *SPT-like bosonic edge zero modes and symmetries* that manifest, and persist, in a *transient dynamical regime whose duration increases with system size*. To arrive at this result, we begin by spelling out the necessary features any system supporting such entities must possess: *Non-trivial spectral topology*, *bulk-instabilities*, and *highly non-normal dynamical matrices*. These conclusions lead us to move beyond spectral considerations, and instead, invoke the mathematical theory of pseudospectra. Following a brief mathematical account of this theory, we proceed to uncover conditions for *anomalous transient dynamics in QBLs*. Two relevant dynamical phases emerge: an *anomalously relaxing* phase, and a *dynamically metastable* phase. The anomalously relaxing phase is characterized by exponential relaxation of generic observables in a two-step manner. The first step is characterized by a decay rate set by the bulk (infinite-size) Lindblad gap, while the second rate is set by the finite-size Lindblad gap which, remarkably, differs from its infinite-size counterpart by a non-zero amount, independent of system-size. A dynamically metastable system is one for which all finite open boundary truncations are dynamically stable, despite possessing an unstable infinite-size limit. Such systems possess bulk instabilities that are suppressed by imposing hard-wall boundaries. The evolution of dynamically metastable systems is characterized by a transient regime whereby generic observables are amplified, followed by an asymptotic exponential relaxation at a rate set by the finite-size Lindblad gap. Both the length of the transient, and the degree of amplification increase with system size. We proceed to zoom in on those dynamically stable QBLs that possess non-trivial bulk topological invariants, and we deem *topologically metastable*. In the simplest one-dimensional cases (which we always restrict to), these invariants are the winding numbers of the bulk bands. The combined assumptions of dynamical metastability and non-trivial topology facilitate the emergence of pairs of operators consisting of one approximate zero mode and one approximate SG. These modes are *Hermitian*, *macroscopically separated*, and *canonically conjugate*. We call the members of these pairs *Majorana bosons* (MBs) due to their tight similarities with Majorana fermions. One particular implication of their existence is the emergence of a manifold of long-lived *quasi-steady states*, that have a number of distinct features that explicitly require topological metastability to exists in a QBL. We then explore the implications of a global number symmetry on topologically metastable systems, resulting in the discovery of edge modes, which we call *Dirac bosons* (DBs), reminiscent of the Dirac fermion edge modes in topological insulators.

Chapter 9 is devoted to dissecting a series of four models that exemplify the general results of Chap. 8. The first model is a purely dissipative QBL derived by applying the first recipe of Chap. 7 to the fermionic Kitaev chain (FKC) Hamiltonian. The resulting QBL is topologically metastable in certain parameter regimes that overlap with the topological phase diagram of the FKC and, most importantly, sustains an MB pair that derive from the Majorana edge modes of the FKC. Interestingly, these provide an example of "non-split" MBs, i.e., MBs that are both ZMs *and* SGs. In this sense, these are the *tightest* bosonic analogues of Majorana fermions. The second model is a dissipative version of the BKC first explored in Chap. 3. By completely characterizing the five-parameter topological phase diagram, we uncover three relevant dynamical phases: an anomalously relaxing one, a non-topological dynamically metastable one, and a topologically metastable one. We explore the dynamical features of each phase in detail and, in particular, compute MBs that arise in the topologically metastable phases. The third model once again adds dissipation to the BKC Hamiltonian. However, in this model, the dissipator is constructed following the second recipe of Chap. 7, and thus, the resultant QBL possesses a unique, pure steady-state. Additionally, we find that this QBL has a metastable regime. This allows us to explore the interplay between MBs and pure steady states. In particular, we analytically compute the quasi-steady states and uncover surprisingly non-trivial parity dynamics of certain cat-state superpositions. The final model is a number-symmetric QBL that possesses a topologically metastable phase. We explicitly construct the DBs predicted in Chap. 8. We conclude the chapter with an analysis of certain multitime correlation functions in QBLs. We find that, generically, topological metastability may be characterized by the existence of long-live two-time quantum correlation functions between the macroscopically separated MB partners, in addition to the emergence of divergent zero-frequency power-spectral peaks. We further use two-time correlation functions to distinguish the unique features of the models of interest.

Chapter 10 concludes the main body of the thesis with an in-depth recounting of the key implications of each result presented so far, along with remaining open questions and future directions.

In addition to the main text, we include two Appendices. Appendix A covers the mathematical tools used to compute the spectra and pseudospectra of block-Toeplitz matrices and operators. In particular, we summarize the key aspects of a generalization of Bloch's theorem used throughout the thesis to analytically diagonalize (or more specifically, cast into Jordan normal form) various models featuring non-trivial BCs. Appendix B contains a series of secondary technical results and proofs that are separated from the main text in order to simplify the reading experience. Several of these technical results involve the analytical diagonalization of explicit models.

Part I is based largely upon publications [95] and [96]. Part II is based upon publications [97] and [98]. Several motivating results, specifically those found in Sect. 5.2, originate in Ref. [84].

References

1. S.N. Bose, Zeit. Phys. A **26**, 178 (1924)
2. A. Einstein, Quantentheorie des einatomigen idealen Gases. Sitzungsber. Preuß. Akad. Wiss. Phys. Math. Kl. **1924**, 261 (1924); *ibid.* 3 (1925)
3. M.H. Anderson, J.R. Ensher, M.R. Matthews, C.E. Wieman, E.A. Cornell, Science **269**, 198 (1995)
4. K.B. Davis, M.-O. Mewes, M.R. Andrews, N.J. van Druten, D.S. Durfee, D.M. Kurn, W. Ketterle, Phys. Rev. Lett. **75**, 3969 (1995)
5. N.W. Ashcroft, N.D. Mermin, *Solid State Physics* (Cengage Learning, 2022)
6. P.D. Drummond, M. Hillery, *The Quantum Theory of Nonlinear Optics* (Cambridge University Press, 2014)
7. D.F. Walls, G.J. Milburn, *Quantum Optics* (Springer Science & Business Media, 2007)
8. M.O. Scully, M.S. Zubairy, *Quantum Optics* (American Association of Physics Teachers, 1999)
9. H. Carmichael, *An Open Systems Approach to Quantum Optics: Lectures Presented at the Université Libre de Bruxelles, October 28 to November 4, 1991*, vol. 18 (Springer Science & Business Media, 2009)
10. A.A. Clerk, M.H. Devoret, S. Girvin, F. Marquardt, R.J. Schoelkopf, Rev. Mod. Phys. **82**, 1155 (2010)
11. V. Peano, M. Houde, C. Brendel, F. Marquardt, A.A. Clerk, Nat. Commun. **7**, 1 (2016)
12. A. McDonald, T. Pereg-Barnea, A.A. Clerk, Phys. Rev. X **8**, 041031 (2018)
13. D. Porras, S. Fernández-Lorenzo, Phys. Rev. Lett. **122**, 143901 (2019)
14. T. Ramos, J.J. García-Ripoll, D. Porras, Phys. Rev. A **103**, 033513 (2021)
15. C.C. Wanjura, M. Brunelli, A. Nunnenkamp, Nat. Commun. **11**, 3149 (2020)
16. D. Gottesman, A. Kitaev, J. Preskill, Phys. Rev. A **64**, 012310 (2001)
17. U.L. Andersen, G. Leuchs, C. Silberhorn, Laser Photon. Rev. **4**, 337 (2010)
18. V.V. Albert, K. Noh, K. Duivenvoorden, D.J. Young, R.T. Brierley, P. Reinhold, C. Vuillot, L. Li, C. Shen, S.M. Girvin, B.M. Terhal, L. Jiang, Phys. Rev. A **97**, 032346 (2018)

19. O. Krüger, *Quantum Information Theory with Gaussian Systems*, Ph.D. thesis, Technical University of Braunschweigs, 2006
20. J.P. Martikainen, Phys. Rev. A **83**, 013610 (2011)
21. R. Rossignoli, A.M. Kowalski, Phys. Rev. A **72**, 032101 (2005)
22. Y. Kawaguchi, M. Ueda, Phys. Rep. **520**, 253 (2012)
23. J. Koch, A.A. Houck, K. Le Hur, S.M. Girvin, Phys. Rev. A **82**, 043811 (2010)
24. A. Kamal, J. Clarke, M.H. Devoret, Nat. Phys. **7**, 311 (2011)
25. B. Abdo, K. Sliwa, L. Frunzio, M. Devoret, Phys. Rev. X **3**, 031001 (2013)
26. K.M. Sliwa, M. Hatridge, A. Narla, S. Shankar, L. Frunzio, R.J. Schoelkopf, M.H. Devoret, Phys. Rev. X **5**, 041020 (2015)
27. B.M. Anderson, R. Ma, C. Owens, D.I. Schuster, J. Simon, Phys. Rev. X **6**, 041043 (2016)
28. P. Roushan, C. Neill, A. Megrant, Y. Chen, R. Babbush, R. Barends, B. Campbell, Z. Chen, B. Chiaro, A. Dunsworth, A. Fowler, E. Jeffrey, J. Kelly, E. Lucero, J. Mutus, P.J.J. O'Malley, M. Neeley, C. Quintana, D. Sank, A. Vainsencher, J. Wenner, T. White, E. Kapit, H. Never, J. Martinis, Nat. Phys. **13**, 146 (2017)
29. C. Owens, A. LaChapelle, B. Saxberg, B.M. Anderson, R. Ma, J. Simon, D.I. Schuster, Phys. Rev. A **97**, 013818 (2018)
30. Y. Zhang, B.J. Lester, Y.Y. Gao, L. Jiang, R.J. Schoelkopf, S.M. Girvin, Phys. Rev. A **99**, 012314 (2019)
31. E. Kim, X. Zhang, V.S. Ferreira, J. Banker, J.K. Iverson, A. Sipahigil, M. Bello, A. González-Tudela, M. Mirhosseini, O. Painter, Phys. Rev. X **11**, 011015 (2021)
32. Y. Gong, L. Guo, S. Wong, A.J. Bennett, S. Soon Oh, Sci. Rep. **11**, 1055 (2021)
33. H. Zhao, P. Miao, M.H. Teimourpour, S. Malzard, R. El-Ganainy, H. Schomerus, L. Feng, Nat. Commun. **9**, 981 (2018)
34. M. Parto, S. Wittek, H. Hodaei, G. Harari, M.A. Bandres, J. Ren, M.C. Rechtsman, M. Segev, D.N. Christodoulides, M. Khajavikhan, Phys. Rev. Lett. **120**, 113901 (2018)
35. D. Malz, L.D. Tóth, N.R. Bernier, A.K. Feofanov, T.J. Kippenberg, A. Nunnenkamp, Phys. Rev. Lett. **120**, 023601 (2018)
36. A. Youssefi, S. Kono, A. Bancora, M. Chegnizadeh, J. Pan, T. Vovk, T.J. Kippenberg, Nature **612**, 666 (2022)
37. V. Peano, C. Brendel, M. Schmidt, F. Marquardt, Phys. Rev. X **5**, 031011 (2015)
38. J.H.P. Colpa, Diagonalization of the quadratic bosonic hamiltonian, Physica A **93**, 327 (1978); "with zero modes: I. Mathematical" *ibid.* **134** 377 (1986); "with zero modes: II. Physical" *ibid.* 417 (1986)
39. J.-P. Blaizot, G. Ripka, *Quantum Theory of Finite Systems* (MIT Press Cambridge, Cambridge, 1986), Chap. 2-3
40. Y. Ashida, Z. Gong, M. Ueda, Adv. Phys. **69**, 249 (2020)
41. C.M. Bender, S. Boettcher, Phys. Rev. Lett. **80**, 5243 (1998)
42. C.M. Bender, Rep. Prog. Phys. **70**, 947 (2007)
43. C.M. Bender, P.D. Mannheim, Phys. Lett. A **374**, 1616 (2010)
44. A. Mostafazadeh, J. Math. Phys. **43**, 205; *ibid. 2814*; *ibid.* 3944 (2002)
45. R. Alicki, K. Lendi, *Quantum Dynamical Semigroups and Applications* (Springer, Berlin, 1987)
46. H.-P. Breuer, F. Petruccione, *The Theory of Open Quantum Systems* (Oxford University Press, 2002)
47. C. Presilla, R. Onofrio, U. Tambini, Ann. Phys. **248**, 95 (1996)
48. W. Chen, Ş.K. Özdemir, G. Zhao, J. Wiersig, L. Yang, Nature **548**, 192 (2017)
49. J. Wiersig, Nat. Commun. **11**, 2454 (2020)
50. S. Yao, Z. Wang, Phys. Rev. Lett. **121**, 086803 (2018)
51. J.C. Budich, E.J. Bergholtz, Phys. Rev. Lett. **125**, 180403 (2020)
52. T. Haga, M. Nakagawa, R. Hamazaki, M. Ueda, Phys. Rev. Lett. **127**, 070402 (2021)
53. T. Mori, T. Shirai, Phys. Rev. Lett. **125**, 230604 (2020)
54. V.A. Yakubovich, V.M. Starzhinskiĭ, *Linear Differential Equations with Periodic Coefficients* (Wiley, New York, 1975)

55. I. Gohberg, P. Lancaster, L. Rodman, *Indefinite Linear Algebra and Applications* (Springer, 2005)
56. V.L. Berezinskii, Sov. Phys. JETP **34**, 610 (1972)
57. J.M. Kosterlitz, D.J. Thouless, J. Phys. C **6**, 1181 (1973)
58. K.v. Klitzing, G. Dorda, M. Pepper, Phys. Rev. Lett. **45**, 494 (1980)
59. D.J. Thouless, M. Kohmoto, M.P. Nightingale, M. den Nijs, Phys. Rev. Lett. **49**, 405 (1982)
60. Y. Hatsugai, Phys. Rev. Lett. **71**, 3697 (1993)
61. M.Z. Hasan, C.L. Kane, Rev. Mod. Phys. **82**, 3045 (2010)
62. B.A. Bernevig, T.L. Hughes, *Topological Insulators and Topological Superconductors* (Princeton University Press, 2013)
63. A. Kitaev, Phys. Usp. **44**, 131 (2001)
64. S. Ryu, Y. Hatsugai, Phys. Rev. Lett. **89**, 077002 (2002)
65. X.-L. Qi, S.-C. Zhang, Rev. Mod. Phys. **83**, 1057 (2011)
66. E. Majorana, Il Nuovo Cimento **14**, 171 (1937)
67. A. Altland, M.R. Zirnbauer, Phys. Rev. B **55**, 1142 (1997)
68. C.K. Chiu, J.C.Y. Teo, A.P. Schnyder, S. Ryu, Rev. Mod. Phys. **88**, 035005 (2016)
69. S.D. Sarma, M. Freedman, C. Nayak, NPJ Quant. Inf. **1**, 1 (2015)
70. N. Okuma, K. Kawabata, K. Shiozaki, M. Sato, Phys. Rev. Lett. **124**, 086801 (2020)
71. K. Kawabata, K. Shiozaki, M. Ueda, M. Sato, Phys. Rev. X **9**, 041015 (2019)
72. C.-E. Bardyn, M.A. Baranov, C.V. Kraus, E. Rico, A.İmamoglu, P. Zoller, S. Diehl, New J. Phys. **15**, 085001 (2013)
73. S. Diehl, E. Rico, M.A. Baranov, P. Zoller, Nat. Phys. **7**, 971 (2011)
74. E. Vernier, SciPost Phys. **9**, 049 (2020)
75. M. van Caspel, S.E.T. Arze, I.P. Castillo, SciPost Phys. **6**, 26 (2019)
76. S. Lieu, M. McGinley, N.R. Cooper, Phys. Rev. Lett. **124**, 040401 (2020)
77. A. Altland, M. Fleischhauer, S. Diehl, Phys. Rev. X **11**, 021037 (2021)
78. T. Ozawa, H.M. Price, A. Amo, N. Goldman, M. Hafezi, L. Lu, M.C. Rechtsman, D. Schuster, J. Simon, O. Zilberberg, I. Carusotto, Rev. Mod. Phys. **91**, 015006 (2019)
79. R. Shindou, R. Matsumoto, S. Murakami, J. Ohe, Phys. Rev. B **87**, 174427 (2013)
80. V. Peano, H. Schulz-Baldes, J. Math. Phys. **59**, 031901 (2018)
81. C.C. Wanjura, M. Brunelli, A. Nunnenkamp, Phys. Rev. Lett. **127**, 213601 (2021)
82. M. Brunelli, C.C. Wanjura, A. Nunnenkamp, arXiv:2207.12427v2 [quant-ph] (2022)
83. S. Raghu, F.D.M. Haldane, Phys. Rev. A **78**, 033834 (2008)
84. Q.-R. Xu, V.P. Flynn, A. Alase, E. Cobanera, L. Viola, G. Ortiz, Phys. Rev. B **102**, 125127 (2020)
85. R. Barnett, Phys. Rev. A **88**, 063631 (2013)
86. B. Galilo, D.K.K. Lee, R. Barnett, Phys. Rev. Lett. **115**, 245302 (2015)
87. T. Prosen, New J. Phys. **10**, 043026 (2008)
88. T. Prosen, J. Stat. Mech. **2010**, P07020 (2010)
89. T. Prosen, T.H. Seligman, J. Phys. A **43**, 392004 (2010)
90. A.E. Teretenkov, Infin. Dimens. Anal. Quantum Probab. Relat. Top. **22**, 1930001 (2019)
91. T. Barthel, Y. Zhang, J. Stat. Mech. **2022**, 113101 (2022)
92. D. Poletti, in *Infinite Dimensional Analysis, Quantum Probability and Applications: QP41 Conference, Al Ain, UAE, March 28–April 1, 2021* (Springer, 2022), p. 197
93. E.M. Kessler, G. Giedke, A. Imamoglu, S.F. Yelin, M.D. Lukin, J.I. Cirac, Phys. Rev. A **86**, 012116 (2012)
94. M. Fitzpatrick, N.M. Sundaresan, A.C.Y. Li, J. Koch, A.A. Houck, Phys. Rev. X **7**, 011016 (2017)
95. V.P. Flynn, E. Cobanera, L. Viola, New J. Phys. **22**, 083004 (2020)
96. V.P. Flynn, E. Cobanera, L. Viola, Europhys. Lett. **131**, 40006 (2020)
97. V.P. Flynn, E. Cobanera, L. Viola, Phys. Rev. Lett. **127**, 245701 (2021)
98. V.P. Flynn, E. Cobanera, L. Viola, Phys. Rev. B **108**, 214312 (2023)

Part I
Effective Non-Hermiticity in Closed Bosonic Systems

Chapter 2
Background: Quadratic Bosonic Hamiltonians

Abstract We present the foundational aspects of time-independent, quadratic bosonic systems. Following a brief description of the multimode bosonic landscape we center ourselves in, we discuss the theoretical frameworks necessary for describing dynamics generated by a QBH. Of particular relevance is the intrinsic non-Hermiticity built into the Heisenberg dynamics of those systems described by QBHs without total number symmetry. Two distinct notions of stability emerge for QBHs: thermodynamical and dynamical stability. We explicitly define these notions and present equivalent characterizations in terms of the relevant dynamical matrix. Following this, we develop a Bogoliubov-de Gennes (BdG) framework that, in particular, helps reveal the underlying indefinite inner-product structure intrinsic to multimode bosonic Hilbert spaces. This will be specialized to closed (and later in Chap. 6, open) dynamics. We conclude the chapter with a discussion of translation invariance, the key implications of Bloch's theorem in this setting, and the basics of its generalization for diagonalization in the presence of arbitrary BCs.

In this chapter, we lay down the foundational framework used throughout this thesis for studying QBHs.[1] As we have detailed in the introduction, the simplest examples of unstable phenomena (e.g., an amplifying optical mode) arise naturally when considering bosonic degrees of freedom. What is not clear, however, is whether or not unstable phenomena may arise in a *coherent* manner. That is, can a bosonic Hamiltonian be "unstable", and if so, what exactly are the properties of these

[1] *The primary contents of this chapter, excluding Sect. 2.3.3, are background materials synthesized from a number of key references on quadratic bosonic systems [1–6]. Additionally, the discussion of bulk-translation invariance in lattice models is adapted partially from the mathematical methods developed in Ref. [7] used to prove a generalization of Bloch's theorem for a class of fermionic systems with arbitrary BCs [8, 9]. This extension into a bosonic context is discussed in Appendix A.3 as well as Ref. [10]. Section 2.3.3 develops the mathematical formalism of Nambu space that we have first introduced in Ref. [10] and further developed in Refs. [11, 12] (all three of which were jointly co-authored with Emilio Cobanera & Lorenza Viola). Importantly, this formalism rests, in part, on the explicit identification of an indefinite inner-product (or Krein) space structure observed concretely, for instance, in Refs. [13–15].*

V. P. Flynn, *Effective Non-Hermiticity and Topology in Markovian Quadratic Bosonic Dynamics*, Springer Theses, https://doi.org/10.1007/978-3-031-52045-7_2

instabilities? It turns out that, not only do instabilities arise in bosonic Hamiltonians, they can arise in the simplest, *quadratic*, or "non-interacting", case.

QBHs can be grossly classified according to two stability criteria: *thermodynamic stability* and *dynamical stability*. Thermodynamically stable Hamiltonians are those whose Hamiltonians are bounded from below (or above, in which case an overall minus sign can be used to obtain a bounded-from-below system). Dynamically stable Hamiltonians are those that generate bounded evolution for arbitrary observables when prepared in arbitrary initial states. The possibility that a bosonic Hamiltonian can lack either of these two features is entirely thanks to Bose-Einstein statistics and the associated lack of an exclusion principle. Simply stated, states may support an indefinite number of bosonic excitations. As it turns out, dynamical instabilities, in particular, exist as a consequence of an intrinsic *effective non-Hermiticity* present in the Heisenberg equations of motions of the fundamental creation and annihilation operators. Perhaps unsurprisingly, this effective non-Hermiticity is explicitly tied to a loss of number conservation, and thus, the total number of bosonic excitations need not be bounded. What is surprising, however, is that number-non-conserving Hamiltonians can remain stable, in both senses of the word. The stability criteria come in the form of a spectral characterization of the so-called "dynamical matrix", which is a non-Hermitian matrix obeying a number of properties inherited from fundamental physical constraints. One property, in particular, is known as *pseudo-Hermiticity*, and arises as a consequence bosonic commutation relations. Ultimately, pseudo-Hermiticity allows us to recast the dynamical characterization of QBHs in terms of an linear time-invariant dynamical system defined on an *indefinite inner-product* (or *Krein*) *space*. As one last surprise, it turns out that the normal mode analysis of this non-Hermitian dynamical system is intrinsically tied to the problem of diagonalizing the Hamiltonian by means of a Bogoliubov transformation.

Our ultimate goal is to study, and uncover exotic features of, the dynamical properties of many-body, or more precisely, many-mode bosonic systems. The simplest instance of such a system is one that consists of infinitely-many modes coupled in a translationally invariant manner. In this case, we may leverage Bloch's theorem, which still applies in the presence of effective non-Hermiticity, to completely solve the dynamics of the system. One may then immediately wonder: can BCs be imposed in such a way that changes the stability properties of a given Hamiltonian? For example, can hardwall boundaries be used to suppress an amplifying system? In order to grapple with questions like these, we will discuss the applicability of the generalization of Bloch's theorem, originally developed for fermionic systems under arbitrary BCs, in this bosonic context. The final result is a clear methodology for exactly solving the relevant dynamics under a wide class of BCs.

The outline for this chapter is as follows. In Sect. 2.1, we identify the multimode bosonic Fock space upon which the dynamical systems we later study are defined, Gaussian states, and Gaussian transformations. In Sect. 2.2, we define the physical lattice spaces that our multimode systems exist within and present an in-depth account of translation operators in these spaces. In Sect. 2.3, we identify the closed-system dynamics we are interested in, i.e., those whose unitary propagators are generated by purely QBHs. We identify the effective non-Hermiticity underlying the equations of motion for any such Hamiltonian that breaks total number conservation, define and characterize dynamical and thermodynamical stability in these systems, and formally characterize the indefinite inner-product structure of the bosonic Nambu space via the introduction of the so-called "hat map" for linear and quadratic forms. In Sect. 2.4, we establish the mathematical link between the problems of diagonalizing a given QBH via Bogoliubov transformation and conducting a normal mode analysis on the Heisenberg equations of motion for the creation and annihilation operators. Finally, we conclude with a characterization of those QBHs that possess discrete translation symmetry, perhaps up to an arbitrary BC in Sect. 2.5.

2.1 Multimode Bosonic Fock Space and Gaussian Transformations

The background Hilbert space on which the majority of our work rests upon is a multimode bosonic Fock space. That is, we consider N (possibly infinitely-many) bosonic modes that are distinguished by a mode index, $j = 1, \ldots, N$. In the language of second quantization, we work within a bosonic Fock space $\mathcal{F}_N$ built upon N distinct single-particle states labeled by j, i.e., $\mathcal{F}_N \equiv \bigoplus_{n=0}^{\infty} \mathcal{H}_n^{(S)}$, where $\mathcal{H}_n^{(S)}$ is the symmetrized Hilbert space consisting of n particles distributed among the $N = \dim \mathcal{H}_1$ single-particle states. We may equivalently characterize $\mathcal{F}_N$ has a tensor product of N infinite-dimensional single-mode bosonic Fock spaces, i.e., $\mathcal{F}_N = \otimes_{j=1}^{N} \mathcal{F}_1^{(j)}$. To each mode, we associate a canonical creation and annihilation pair $(a_j^\dagger, a_j)$ satisfying the *canonical commutation relations* (CCRs), $[a_j, a_j^\dagger] = 1_{\mathcal{F}}$, with $1_{\mathcal{F}}$ being the Fock space identity. The inter-mode algebraic relationships follow as $[a_j, a_k] = 0$ and $[a_j, a_k^\dagger] = \delta_{jk} 1_{\mathcal{F}}$. These operators can be conveniently packaged into the bosonic Nambu array[2] $\Phi \equiv [a_1, a_1^\dagger, \ldots, a_N, a_N^\dagger]^T$. The bosonic algebra is then recast as

$$[\Phi_j, \Phi_k^\dagger] = (\boldsymbol{\tau}_3)_{jk} 1_{\mathcal{F}}, \quad \Phi^\dagger = (\boldsymbol{\tau}_1 \Phi)^T, \tag{2.1}$$

where we have introduced the matrices $\boldsymbol{\tau}_j = \mathbb{1}_N \otimes \boldsymbol{\sigma}_j$, with $\boldsymbol{\sigma}_j$ the usual Pauli matrices and $\mathbb{1}_N$ the $N \times N$ identity matrix. Associated to each mode are a set of dimensionless quadrature operators, $x_j = (a_j + a_j^\dagger)/\sqrt{2}$ and $p_j = -i(a_j - a_j^\dagger)/\sqrt{2}$,

[2] Within the literature, there are multiple conventions for the ordering of this array. For example, $\Phi' = [a_1, \ldots, a_N, a_1^\dagger, \ldots, a_N^\dagger]^T$ is commonly encountered. The convention we have chosen in this thesis, motivated by the work in [9], will later provide utility in analyzing bulk translation-invariant systems.

which satisfy the *Heisenberg-Weyl relations* (HWRs), $[x_j, x_k] = [p_j, p_k] = 0$ and $[x_j, p_k] = i\delta_{jk}1_{\mathcal{F}}$, with $\hbar = 1$ henceforth. Defining the quadrature array $R = [x_1, p_1, \ldots, x_N, p_N]^T$ allows us to identify

$$R = \mathbf{\Sigma}\Phi, \quad \mathbf{\Sigma} = \mathbb{1}_N \otimes \frac{1}{\sqrt{2}}\begin{bmatrix} 1 & 1 \\ -i & i \end{bmatrix}, \tag{2.2}$$

from which a symplectic structure emerges

$$[R_j, R_k] = i\mathbf{J}_{jk}1_{\mathcal{F}}, \quad \mathbf{J} = \mathbb{1}_N \otimes \begin{bmatrix} 0 & 1 \\ -1 & 0 \end{bmatrix} = i\boldsymbol{\tau}_2. \tag{2.3}$$

Unitarity of $\mathbf{\Sigma}$ allows us to think of moving to quadratures as a change of basis within the Nambu space, i.e., the complex linear span of the creation and annihilation operators. The symplectic form $\mathbf{J}$ then serves as the quadrature analogue of $\boldsymbol{\tau}_3$.

Each mode contributes a vacuum state $|0_j\rangle \in \mathcal{F}_1^{(j)}$, which consists of zero j-mode bosons satisfying $a_j|0_j\rangle = 0$. We may then characterize $\mathcal{F}_1^{(j)}$ as the complex span of Fock states

$$|n_j\rangle = \frac{(a_j^\dagger)^{n_j}}{\sqrt{n_j!}}|0_j\rangle, \quad n_j \in \mathbb{Z}_{\geq 0}, \tag{2.4}$$

This leads naturally to multimode Fock states

$$|n_1, \ldots, n_N\rangle = \bigotimes_{j=1}^{N} |n_j\rangle = \frac{(a_1^\dagger)^{n_1}\cdots(a_N^\dagger)^{n_N}}{\sqrt{n_1!\cdots n_N!}}|0\rangle, \tag{2.5}$$

with $|0\rangle \equiv \otimes_{j=1}^{N}|0_j\rangle$. From this definition and the CCRs, we have the familiar properties

$$a_j|n_1, \ldots, n_j, \ldots, n_N\rangle = \sqrt{n_j}|n_1, \ldots, n_j - 1, \ldots, n_N\rangle, \tag{2.6}$$

$$a_j^\dagger|n_1, \ldots, n_j, \ldots, n_N\rangle = \sqrt{n_j + 1}|n_1, \ldots, n_j + 1, \ldots, n_N\rangle, \tag{2.7}$$

$$a_j^\dagger a_j|n_1, \ldots, n_j, \ldots, n_N\rangle = n_j|n_1, \ldots, n_j, \ldots, n_N\rangle, \tag{2.8}$$

$$\langle n_1, \ldots, n_N|m_1, \ldots, m_N\rangle = \delta_{n_1,m_1}\cdots\delta_{n_N,m_M}. \tag{2.9}$$

Another ubiquitous family of states are the multimode coherent states. These states are defined as simultaneous eigenstates of the annihilation operators a_j, i.e.,

$$a_j|\vec{\alpha}\rangle = \alpha_j|\vec{\alpha}\rangle, \quad \vec{\alpha} \in \mathbb{C}^N. \tag{2.10}$$

which satisfy the overcompleteness and closure relationships:

$$\langle\vec{\alpha}|\vec{\beta}\rangle = e^{-\left(\vec{\alpha}^\dagger\vec{\alpha}+\vec{\beta}^\dagger\vec{\beta}-2\vec{\alpha}^\dagger\vec{\beta}\right)/2}, \quad \int |\vec{\alpha}|\,\langle\vec{\alpha}| \prod_{j=1}^{N} \frac{d\alpha_j\, d\alpha_j^*}{\pi} = 1_{\mathcal{F}}. \tag{2.11}$$

These states can be constructed from the vacuum via the multimode displacement operator (or Weyl displacement) $D(\vec{\alpha})$ defined as

$$D(\vec{\alpha}) \equiv \prod_{j=1}^{N} e^{\alpha_j a_j^\dagger - \alpha_j^* a_j}, \quad |\vec{\alpha}\rangle = D(\vec{\alpha})|0\rangle. \tag{2.12}$$

The displacement operator is also closely connected to the *bosonic parity operator*, $P \equiv e^{i\pi\sum_{j=1}^{N} a_j^\dagger a_j}$ [16]. In particular,

$$P = \frac{1}{2^N}\int D(\vec{\alpha}) \prod_{j=1}^{N} \frac{d\alpha_j\, d\alpha_j^*}{\pi}, \tag{2.13}$$

which may be checked, for instance, by comparing coherent state matrix elements.

Unlike Fock states (excluding the vacuum), coherent states are examples of Gaussian states. While many equivalent definitions exist, multimode Gaussian states are most compactly defined as those states ρ whose (Wigner) *characteristic functions*, $\chi_\rho(\vec{\beta}) \equiv \mathrm{tr}[\rho D(\vec{\beta})]$, are Gaussian. For example,

$$\chi_{|\vec{\alpha}\rangle}(\vec{\beta}) = e^{-\vec{\beta}^\dagger\vec{\beta}/2} e^{\vec{\alpha}^\dagger\vec{\beta}-\vec{\beta}^\dagger\vec{\alpha}} \tag{2.14}$$

is Gaussian in the $2N$ variables β_j and β_j^*. Among many other properties, this implies a Gaussian state ρ is uniquely defined by its *mean vector* $\vec{m}_\rho$ and *covariance matrix*[3] $\mathbf{C}_\rho$, defined via

$$\vec{m}_\rho \equiv \mathrm{tr}[\rho\Phi], \quad (\mathbf{C}_\rho)_{jk} \equiv \frac{1}{2}\mathrm{tr}[\rho\{\Phi_j, \Phi_k^\dagger\}] - \mathrm{tr}[\rho\Phi_j]\mathrm{tr}[\rho\Phi_k^\dagger]. \tag{2.15}$$

Importantly, the covariance matrix $\mathbf{C}_\rho$ is positive-definite.[4] In the case of a coherent state, $m_{|\vec{\alpha}\rangle} = [\alpha_1, \alpha_1^*, \ldots, \alpha_N, \alpha_N^*]^T$ and $\mathbf{C}_{|\vec{\alpha}\rangle} = \mathbb{1}_{2N}/2$. Since Gaussian states are uniquely defined by these two quantities, all other properties can be derived from

---

[3] In other contexts such as continuous variable quantum information, the mean vector and covariance matrix may instead be defined by replacing Φ with R, i.e., in the quadrature basis. These are related to the convention used in this thesis via the unitary transformation $\boldsymbol{\Sigma}$.

[4] Consider the matrix $\mathbf{C}_\rho + \boldsymbol{\tau}_3$. Given a vector $\vec{\alpha}$, we have $\vec{\alpha}^\dagger(\mathbf{C}_\rho + \boldsymbol{\tau}_3)\vec{\alpha} = \mathrm{tr}[\rho A A^\dagger] \geq 0$, with $A = \sum_j \alpha_j^*(\Phi_j - \langle\Phi\rangle_j)$. Thus, $\mathbf{C}_\rho + \boldsymbol{\tau}_3 \geq 0$. Using $\boldsymbol{\tau}_1\mathbf{C}_\rho^T\boldsymbol{\tau}_1 = \mathbf{C}_\rho$ additionally yields $\mathbf{C}_\rho - \boldsymbol{\tau}_3 \geq 0$. Combining these implies $\mathbf{C}_\rho \geq 0$.

them. For example, the *purity* of a Gaussian state ρ is given by

$$\mathrm{tr}[\rho^2] = \frac{1}{2^N \sqrt{\det[\mathbf{C}_\rho]}}. \tag{2.16}$$

Together with Gaussian states, *Gaussian transformations* (sometimes called "Gaussian operations" or "Gaussian quantum channels") are the canonical starting point of continuous variable quantum information [17], and arise naturally in the study of non-interacting (or mean-field), condensed-matter, quantum-optical, cavity- and circuit-QED systems. Such transformations are defined as those that map Gaussian states to Gaussian states. This state-centric description can be restated in terms of mappings on the bosonic operator algebra. Namely, a Gaussian transformation can be thought of as a linear transformation of the creation and annihilation operators:

$$\Phi \mapsto \Psi = \mathbf{T}\Phi + \vec{z}\, 1_{\mathcal{F}}, \tag{2.17}$$

with $\mathbf{T}$ a $2N \times 2N$ matrix and $\vec{z} \in \mathbb{C}^{2N}$. One may, in addition, require that this mapping is canonical, i.e., it preserves the CCRs. Consequently, $\mathbf{T}$ and $\vec{z}$ must then satisfy

$$\boldsymbol{\tau}_3\mathbf{T}^\dagger\boldsymbol{\tau}_3 = \mathbf{T}^{-1}, \quad \boldsymbol{\tau}_1\mathbf{T}\boldsymbol{\tau}_1 = \mathbf{T}^*, \quad \vec{z} = \boldsymbol{\tau}_1\vec{z}^*. \tag{2.18}$$

The implicit quadrature transformation takes the form

$$R \mapsto V = \mathbf{S}R + \vec{\eta}\, 1_{\mathcal{F}} = \boldsymbol{\Sigma}\Psi, \tag{2.19}$$

with $\mathbf{S} = \boldsymbol{\Sigma}\mathbf{T}\boldsymbol{\Sigma}^\dagger$ a real symplectic matrix (i.e., $\mathbf{S}^* = \mathbf{S}$ and $\mathbf{S}^T\mathbf{J}\mathbf{S} = \mathbf{J}$) and $\vec{\eta} = \boldsymbol{\Sigma}\vec{z} \in \mathbb{R}^{2N}$.

Thus far, we have not yet directly characterized the actual operators on Fock space that implements Eq. (2.17). Of particular relevance are those that are specified by unitary or antiunitary transformations. That is, consider those transformations in which $\Psi_j = U\Phi_j U^{-1}$, with $U : \mathcal{F}_N \to \mathcal{F}_N$ either unitary or antiunitary. By construction, they are canonical and, in many cases, represent physical symmetries of a given multimode system, in the sense of Wigner's theorem. In the unitary case, one finds that Gaussianity guarantees $U = e^{iQ}$, with Q a Hermitian operator that is at most quadratic in Φ. When Q is strictly quadratic, U implements a homogeneous transformation ($\vec{z} = 0$), that is well-known throughout physics as a *Bogoliubov transformation*. One commonly encountered example is the isotropic phase rotation, or number-symmetry operator,

$$U(\theta) = e^{i\theta \sum_{j=1}^N a_j^\dagger a_j}, \quad \theta \in \mathbb{R}, \tag{2.20}$$

which implements a mode-independent U(1) phase rotation $a_j \mapsto e^{i\theta}a_j$, or equivalently, $\Phi \mapsto \exp(i\theta\boldsymbol{\tau}_3)\,\Phi$. Another common unitary Gaussian transformation is that of single-mode squeezing:

$$S(z) = e^{\frac{1}{2}(z^* a^2 - z a^{\dagger 2})}, \quad z \in \mathbb{C}, \tag{2.21}$$

which implements $a \mapsto \cosh(|z|)a - (z^*/|z|)\sinh(|z|)a^\dagger$. Phase rotation and squeezing provide representative examples of two distinct flavors of Gaussian unitary transformations. Specifically, number symmetry is an example of an operation that is represented by a unitary matrix in Nambu space, while squeezing is not. The representation of single-mode squeezing is explicitly given by $\Phi \mapsto \mathbf{T}(z)\Phi$, with $\mathbf{T}(z) = \exp[-(\mathrm{Re}(z)\boldsymbol{\sigma}_1 + \mathrm{Im}(z)\boldsymbol{\sigma}_2)]$ manifestly non-unitary. This suggests that a unitary representation of the group of all Gaussian unitary transformations is impossible. In fact, this is the case, and can be heuristically explained by observing that the space on which these transformations act is infinite-dimensional. This is in sharp contrast to the case of fermions, whereby the finite dimensionality of the local (single-mode) Hilbert space always ensures the existence of a unitary representation of the Gaussian unitary group acting on any finite collection of fermionic modes. This is the first hint of so-called "effective non-Hermiticity," or in this instance, effective non-unitarity, that is embedded deeply within bosonic physics. We further note that every transformation of the form Eq. (2.17) satisfying Eqs. (2.18) can be lifted to a unitary transformation U on the Fock space by virtue of the Stone-von Neumann theorem [18].

A key focus of this thesis will be the study of continuous-time, autonomous, Markovian dynamical systems that implement Gaussian transformations on multi-mode Fock spaces. In the unitary (closed-system) case, such dynamics are defined via a unitary propagator $U(t, t_0) = e^{-i(t-t_0)H}$, $t \geq t_0$, with $H = H^\dagger$ a Hamiltonian that is at most quadratic in creation and annihilation operators. As it turns out, continuous families of Gaussian maps need not result only from unitary transformations on Fock space. For example, continuous Gaussian maps may arise in the more general *Markovian* case whereby states evolve according to a Markovian master equation $\dot{\rho}(t) = \mathcal{L}(\rho(t))$. Such dynamics implement Gaussian maps as long as the Lindblad generator $\mathcal{L}$ ("Lindbladian") is quadratic, in a sense to be defined in Sect. 6.2.1. In both cases, Gaussianity of the generators (enforced by the quadratic condition), ensures that Gaussianity of the states is preserved in time.

2.2 Lattice Models

We have so far not specified exactly what distinguishes individual bosonic modes. Since we will explore the consequences of discrete translation invariance, we will embed our bosonic modes onto a (possibly finite) lattice Λ in D dimensions. Generally, we allow for d_{int} internal degrees of freedom on each lattice site $\mathbf{r} \in \Lambda$. Thus, our mode index will be a tuple $(m, \mathbf{r})$, with $m = 1, \ldots, d_{\text{int}}$ the internal index and $\mathbf{r} \in \Lambda$ the lattice index. It is then convenient to introduce a local Nambu array,

$\phi_{\mathbf{r}} = [a_{1,\mathbf{r}}, a^\dagger_{1,\mathbf{r}}, \ldots, a_{d_{\text{int}},\mathbf{r}}, a^\dagger_{d_{\text{int}},\mathbf{r}}]^T$, with the full CCRs given explicitly by

$$[a_{m,\mathbf{r}}, a^\dagger_{m',\mathbf{r}'}] = \delta_{mm'}\delta_{\mathbf{r}\mathbf{r}'}, \quad [a_{m,\mathbf{r}}, a_{m',\mathbf{r}'}] = 0. \tag{2.22}$$

The local Nambu array can be used to identify the full Nambu array

$$\Phi = \sum_{\mathbf{r}\in\Lambda} \vec{e}_{\mathbf{r}} \otimes \phi_{\mathbf{r}}, \tag{2.23}$$

with $\vec{e}_{\mathbf{r}}$ the canonical basis of a $|\Lambda|$-dimensional vector space.

The relevant setting in this thesis will be one-dimensional ($D = 1$) lattices with one internal degree of freedom ($d_{\text{int}} = 1$) per site. The lattice, in this case, will take on one of three main configurations:

$$\text{Finite 1D chain:} \quad \Lambda = \Lambda_{(1,N)} = \{1, 2, \ldots, N\}, \tag{2.24}$$

$$\text{Semi-infinite 1D chain:} \quad \Lambda = \Lambda_{(1,\infty)} = \{1, 2, \ldots\}, \tag{2.25}$$

$$\text{Bi-infinite 1D chain:} \quad \Lambda = \Lambda_{(-\infty,\infty)} = \{\ldots, -1, 0, 1, \ldots\}, \tag{2.26}$$

where we note that we have normalized the lattice spacing to 1. In the finite case, the Nambu space is isomorphic, as a vector space, to $\mathbb{C}^N \otimes \mathbb{C}^2 \simeq \mathbb{C}^{2N}$. For the second two cases, we define the Hilbert spaces of square-summable infinite and bi-infinite sequences as $\ell(\mathbb{N})$ and $\ell(\mathbb{Z})$, respectively. Then, the Nambu spaces corresponding to the two infinite lattices are isomorphic, as vector spaces, to $\ell(\mathbb{N})\otimes\mathbb{C}^2$, and $\ell(\mathbb{Z})\otimes\mathbb{C}^2$, respectively. In all cases, the first factor indicates that there are two operators a_j and $a_j^\dagger$ per site, while the second encodes the lattice degree of freedom.

On each of the three lattices, we have a well-defined notions of left and right shift operators. For finite lattices, there are two natural definitions. First, we may consider the left and right shift operators that terminate at the boundaries, i.e.,

$$\mathbf{T}_N \vec{e}_j = \begin{cases} \vec{e}_{j-1}, & 1 < j \le N, \\ 0 & j = 1, \end{cases} \quad \mathbf{T}_N^\dagger \vec{e}_j = \begin{cases} \vec{e}_{j+1}, & 1 \le j < N, \\ 0 & j = N, \end{cases} \tag{2.27}$$

where we have made the implicit observation that the adjoint of the left-shift operator $\mathbf{T}_N$ is the right shift operator. These operators are quintessential examples of *Toeplitz matrices*, i.e., matrices that are constant along diagonals. In this case, $\mathbf{T}_N$ has 1's along the first upper diagonal and 0's everywhere else. Both matrices are nondiagonalizable and, for finite N, possess only one eigenvector with eigenvalue 0.

We may also consider the circulant shift operators

$$\mathbf{V}_N \vec{e}_j = \begin{cases} \vec{e}_{j-1}, & 1 < j \le N, \\ \vec{e}_N, & j = 1, \end{cases} \quad \mathbf{V}_N^\dagger \vec{e}_j = \begin{cases} \vec{e}_{j+1}, & 1 \le j < N, \\ \vec{e}_1 & j = N, \end{cases} \tag{2.28}$$

which may be interpreted as shift operators on a finite ring. Unlike the previous case, the circulant shift operators are unitary, i.e., $\mathbf{V}_N\mathbf{V}_N^\dagger = \mathbb{1}_N$ and they are diagonalized via Fourier transform, namely,

$$\mathbf{V}_N \vec{f}_k = e^{ik}\vec{f}_k, \quad \vec{f}_k = \sum_{j=1}^{N} e^{ijk}\vec{e}_j, \quad k \in \mathcal{K}_N, \tag{2.29}$$

where $\mathcal{K}_N$ is the finite Brillouin zone defined by[5]

$$\mathcal{K}_N = \begin{cases} \{0, \pm 2\pi/N, \pm 4\pi/N, \ldots, \pm\pi(1-1/N)\}, & N \text{ odd}, \\ \{0, \pm 2\pi/N, \pm 4\pi/N, \ldots, \pm\pi(1-2/N), -\pi\}, & N \text{ even}. \end{cases} \tag{2.30}$$

As suggested by the name, these operators are quintessential examples of circulant matrices, i.e, matrices whose (i, j)'th element depends only on $(i - j) \mod N$.

The most useful translation operator in the semi-infinite case are the unilateral left and right shift operators that terminate on the only edge, i.e.,

$$\mathbf{T}\vec{e}_j = \begin{cases} \vec{e}_{j-1}, & 1 < j \le N, \\ 0 & j = 1, \end{cases} \quad \mathbf{T}^\dagger \vec{e}_j = \vec{e}_{j+1}. \tag{2.31}$$

These shift operators are Toeplitz operators, i.e., the natural infinite-dimensional generalization of Toeplitz matrices. Interestingly, $\mathbf{T}_N^\dagger$ has no eigenvectors or eigenvalues. Instead, the spectrum, i.e., the numbers λ for which $\mathbf{T}^\dagger - \lambda\mathbb{1}$ is not invertible, is the entire unit disk $|\lambda| \le 1$. The spectrum of $\mathbf{T}$ is the same, however each element of the spectrum λ, with $|\lambda| < 1$ is actually an eigenvalue with eigenvector $\sum_{j=1}^{\infty} \lambda^j \vec{e}_j$.

Finally, the most useful shift operators for the bi-infinite case are the bilateral shifts

$$\mathbf{V}\vec{e}_j = \vec{e}_{j-1}, \quad \mathbf{V}^\dagger \vec{e}_j = \vec{e}_{j+1}. \tag{2.32}$$

These bilateral shift operators are *Laurent operators*, i.e., the infinite-dimensional generalization of circulant matrices. Just as in the finite case, these operators are unitary on $\ell^2(\mathbb{Z})$ and have the entire unit circle as a spectrum. However, they have

[5] While a much simpler choice of k values would be $2m\pi/N$, with $m = 0, \ldots, N-1$, we choose this slightly more complicated convention to better conform with the usual $k \in [-\pi, \pi]$ used throughout physics.

no normalizable eigenvectors; instead, we have the formal identity

$$\mathbf{V}\vec{f}_k = e^{ik}\vec{f}_k, \qquad \vec{f}_k = \sum_{j=-\infty}^{\infty} e^{ijk}\vec{e}_j, \quad k \in [-\pi, \pi], \tag{2.33}$$

with the understanding that the plane-waves $\vec{f}_k$ are not in $\ell^2(\mathbb{Z})$.

These various shift operators are used to construct physical translation operators. For example, consider the unitary, discrete left-translation operator L_N defined on a finite ring

$$L_N \phi_j L_N^\dagger = \begin{cases} \phi_{j-1}, & 1 < j \leq N, \\ \phi_N & j = 1, \end{cases} \tag{2.34}$$

with $\phi_j = [a_{1,j}, a_{1,j}^\dagger, \ldots, a_{d_{\text{int}},j}, a_{d_{\text{int}}}^\dagger]^T$ the local Nambu array on the ring. It follows that

$$L_N \Phi L_N^\dagger = \left(\mathbf{V}_N \otimes \mathbb{1}_{2d_{\text{int}}}\right) \Phi. \tag{2.35}$$

The matrix $\mathbf{V}_N \otimes \mathbb{1}_{2d_{\text{int}}}$ serves as the representation of U_N on Nambu space and may be readily checked to satisfy Eqs. (2.18) with the $\boldsymbol{\tau}$-matrices, in this context, taking the form $\boldsymbol{\tau}_j = \mathbb{1}_N \otimes \mathbb{1}_{d_{\text{int}}} \otimes \boldsymbol{\sigma}_j$. Replacing $\mathbf{V}_N$, with $\mathbf{V}$ generalizes to the unitary bilateral translation operators on a bi-infinite chain. However, if one replaces $\mathbf{V}_N$, with $\mathbf{T}$ or $\mathbf{T}_N$, unitarity of the many-body operation is lost. Ultimately, we will be interested in either finite or infinite systems that either (1) possess discrete translational symmetry or (2) have discrete translational symmetry broken only by BCs.

2.3 Closed Quadratic Bosonic Dynamics

2.3.1 Effective Non-Hermiticity

As mentioned, the closed-system dynamics we concern ourselves with are continuous-time and autonomous. Such dynamical systems are defined by a time-independent Hamiltonian $H = H^\dagger$ that is at most quadratic in bosonic creation and annihilation operators. The general expression for such a Hamiltonian is

$$H = \frac{1}{2}\sum_{i,j=1}^{N}\left(a_i^\dagger \mathbf{K}_{ij} a_j + a_i \mathbf{K}_{ij}^* a_j^\dagger + a_i^\dagger \boldsymbol{\Delta}_{ij} a_j^\dagger + a_i \boldsymbol{\Delta}_{ij}^* a_j\right) + \sum_{j=1}^{N}(z_j a_j + z_j^* a_j^\dagger), \tag{2.36}$$

with $\mathbf{K}$ an arbitrary complex Hermitian $N \times N$ matrix, $\boldsymbol{\Delta}$ an arbitrary complex symmetric $N \times N$ matrix, and z_j arbitrary complex numbers. This can be compactly summarized using the Nambu array:

$$H = \frac{1}{2}\Phi^\dagger \mathbf{H}\Phi + \vec{\zeta}^\dagger \boldsymbol{\tau}_3 \Phi, \tag{2.37}$$

with

$$\mathbf{H} = \begin{bmatrix} \mathbf{h}_{11} & \cdots & \mathbf{h}_{1N} \\ \vdots & \ddots & \vdots \\ \mathbf{h}_{N1} & \cdots & \mathbf{h}_{NN} \end{bmatrix}, \quad \mathbf{h}_{ij} = \begin{bmatrix} \mathbf{K}_{ij} & \boldsymbol{\Delta}_{ij} \\ \boldsymbol{\Delta}^*_{ij} & \mathbf{K}^*_{ij} \end{bmatrix}, \quad \vec{\zeta} = [z_1^*, -z_1, \ldots, z_N^*, -z_N]^T. \tag{2.38}$$

The matrix $\mathbf{H}$ is Hermitian and satisfies $\boldsymbol{\tau}_1 \mathbf{H}^* \boldsymbol{\tau}_1 = \mathbf{H}$. Further, the vector $\vec{\zeta}$ satisfies $\boldsymbol{\tau}_1 \vec{\zeta}^* = -\vec{\zeta}$. Since it will play a major role throughout the thesis, we also introduce the so-called dynamical matrix $\mathbf{G} = \boldsymbol{\tau}_3 \mathbf{H}$ which has two notable properties:

$$\boldsymbol{\tau}_3 \mathbf{G}^\dagger \boldsymbol{\tau}_3 = \mathbf{G}, \quad \boldsymbol{\tau}_1 \mathbf{G}^* \boldsymbol{\tau}_1 = -\mathbf{G}. \tag{2.39}$$

While Eq. (2.36) specifies the most general dynamical generator fitting our stipulations, we will further zoom in to the *purely quadratic* case, $\vec{\zeta} = 0$. The justification of this is twofold. Firstly, in almost all cases, one may effectively remove the linear term by making an appropriate shift to the Nambu array. Explicitly, if there exists a vector $\vec{\xi}$ satisfying $\mathbf{G}\vec{\xi} = \vec{\zeta}$, then one can verify that

$$H = \frac{1}{2}\Phi^\dagger \mathbf{H}\Phi + \vec{\zeta}^\dagger \boldsymbol{\tau}_3 \Phi = \frac{1}{2}\Psi^\dagger \mathbf{H}\Psi - \frac{1}{2}\vec{\xi}^\dagger \boldsymbol{\tau}_3 \vec{\zeta}, \quad \Psi = \Phi + \vec{\xi}\, 1_{\mathcal{F}}. \tag{2.40}$$

If this vector exists, then it satisfies $\boldsymbol{\tau}_1 \vec{\xi}^* = \vec{\xi}$, implying that H is unitarily equivalent to a purely quadratic Hamiltonian modulo a (real) constant shift. Such a vector is guaranteed to exist in the generic case where $\mathbf{G}$ is invertible, or even more generically, if $\vec{\zeta}$ is not in the kernel of $\mathbf{G}$. The second justification is to prioritize parity symmetry $[H, P] = 0$, thus setting the stage for direct comparison with the expansive fermionic literature—a recurring theme in this thesis.

In fermionic physics, parity symmetry plays a fundamental role: the parity superselection rule stipulates that all observables (including the Hamiltonian) must commute with the fermionic parity operator and that any physical state must be of strictly even or odd parity. Without these constraints, phase factors accumulated by particle exchanges would be detectable, ultimately violating causality [19]. Immediately, this forbids the existence of terms linear in fermionic creation and annihilation operators in quadratic fermionic Hamiltonians. No such superselection rule applies for bosons. At best, if the bosons are massive or charged, conservation laws may impose parity, or other related superselection rules. Nonetheless, bosonic parity retains physical relevance as a proxy for non-Gaussianity of states. Explicitly, if P is the bosonic parity operator, then Eq. (2.13) yields

$$\text{tr}[\rho P] = \frac{1}{2^N} \int \chi_\rho(\vec{\alpha}) \prod_{j=1}^{N} \frac{d\alpha_j \, d\alpha_j^*}{\pi} = W_\rho(0), \tag{2.41}$$

with W_ρ the Wigner function of the state ρ. The Wigner function, which may be understood as a Fourier transform of the characteristic function, is a useful indicator of non-Gaussianity and classicality. In particular, negativity of the Wigner function indicates that a quantum state is both non-classical and non-Gaussian.[6] Thus, if a state has a negative parity expectation value, we may automatically conclude that it is non-Gaussian. Despite these important differences, any faithful comparison between non-interacting bosonic and fermionic systems must require that they are placed on equal ground. Hereafter, our focus will thus be on strictly quadratic bosonic Hamiltonians, which we will, with a slight abuse of terminology, almost always abbreviate to simply QBHs.

Given a QBH, we construct the propagator $U(t) = \exp(-iHt)$ which implements a continuous family of homogeneous Gaussian unitary transformations

$$\Phi \mapsto \Phi(t) = U^\dagger(t)\Phi U(t) = e^{-i\mathbf{G}t}\Phi. \tag{2.42}$$

This may be understood as the solution to the Schrödinger-like equation of motion for the Nambu array derived in the Heisenberg picture as

$$\frac{d}{dt}\Phi(t) = i[H, \Phi] = -i\mathbf{G}\Phi(t). \tag{2.43}$$

We see that the dynamical matrix serves as the generator, or state matrix, of a linear time-invariant (LTI) dynamical system in Nambu space. Remarkably, this generator is generally non-Hermitian.[7] One may verify that $\mathbf{G} = \mathbf{G}^\dagger$ if and only if $\boldsymbol{\Delta} = 0$. Equivalently, the dynamical matrix is Hermitian if and only if H has total number symmetry, i.e., H commutes with the transformation Eq. (2.20). So, if we wish to study pairing Hamiltonians (those Hamiltonians with $\boldsymbol{\Delta} \neq 0$), as they are called in fermionic literature, we must confront a non-Hermitian equation of motion. However, once this equation is solved, we have effectively computed the Heisenberg

[6] Viewing χ_ρ as a characteristic function in the probabilistic sense, W_ρ can be formally understood as the associated probability density function. However, W_ρ can violate nonnegativity and thus the state cannot be associated to any meaningful classical probability density. For this reason, W_ρ is called a *quasi-probability distribution*. Notably, however, negativity of W_ρ (which is forbidden for Gaussian states, in particular) is not required for a state to lack a classical interpretation. Non-classicality may be identified by considering the nature of singularities in the related Sudarshan-Glauber P distribution [20]. It follows that Gaussianity need not exclude non-classicality. Squeezed states are an example of a non-classical Gaussian state.

[7] In fact, $\mathbf{G}$ is generically non-normal. Consequences of this fact for open bosonic systems will be explored in Sect. 8.1.1.

dynamics of any observable built up from products and sums of bosonic creation and annihilation operator. Such is a feature of unitary dynamics, whereby we have the multiplicative property $(AB)(t) = U^\dagger(t)ABU(t) = U^\dagger(t)AU(t)U^\dagger(t)BU(t) = A(t)B(t)$, as well as linearity. For example, the dynamics of a given number operator are computed according to

$$(a_j^\dagger a_j)(t) = a_j^\dagger(t)a_j(t) = \left(e^{-i\mathbf{G}t}\Phi\right)_{j+N}\left(e^{-i\mathbf{G}t}\Phi\right)_j. \tag{2.44}$$

Our primary goal will thus be to diagonalize (or more precisely, cast into Jordan normal form) $\mathbf{G}$, and hence completely determine the dynamics of the system.

With the importance of the dynamical matrix clearly spelled out, we will examine the implications of the intrinsic properties all such matrices possess, i.e., Eqs.(2.39). The first property, $\boldsymbol{\tau}_3\mathbf{G}^\dagger\boldsymbol{\tau}_3 = \mathbf{G}$, constitutes a generalization of Hermiticity called pseudo-Hermiticity[8] [21]. A matrix $\mathbf{M}$ is called pseudo-Hermitian if and only if there exists a Hermitian, invertible matrix η such that $\eta^{-1}\mathbf{M}^\dagger\eta = \mathbf{M}$. The terminology originates from generalizations of inner-product spaces. Consider the space $\mathbb{C}^n$ paired with the product $(\vec{v}, \vec{w})_\eta \equiv \vec{v}^\dagger\eta\vec{w}$. One may verify that, like an inner-product, this product is linear in the second argument and conjugate symmetric. However, this product need not be positive-definite nor non-degenerate. Specifically, if η, which we call the *indefinite metric*, is not positive-definite, then $(\vec{v}, \vec{v})_\eta$ can be positive, negative, or zero. Such a space is called an *indefinite inner-product, or Krein, space* [22], and the pseudo-Hermitian matrices are precisely those that exhibit the property $(\vec{v}, \mathbf{M}\vec{w})_\eta = (\mathbf{M}\vec{v}, \vec{w})_\eta$ for all $\vec{v}, \vec{w} \in \mathbb{C}^n$. In the special case where $\eta = \mathbb{1}_n$, the inner-product is the usual one and pseudo-Hermiticity coincides with Hermiticity. Like Hermitian matrices, these objects can be given a Lie algebra structure via the usual commutator bracket. The associated Lie group consists of the pseudo-unitary matrices $\mathbf{T}$, which satisfy $\eta^{-1}\mathbf{T}^\dagger\eta = \mathbf{T}^{-1}$, or equivalently, $(\mathbf{T}\vec{v}, \mathbf{T}\vec{w})_\eta = (\vec{v}, \vec{w})_\eta$, for all $\vec{v}, \vec{w} \in \mathbb{C}^n$.

Refocusing on bosonic systems, we observe that dynamical matrices $\mathbf{G}$ are always pseudo-Hermitian with metric $\boldsymbol{\tau}_3$. Furthermore, Gaussian canonical transformations are built-up from $\boldsymbol{\tau}_3$ pseudo-unitary matrices (see Eq. (2.18)). While identifying this indefinite inner-product structure will prove valuable in later sections, we can already conclude from this that, for each eigenvalue ω of $\mathbf{G}$, we have that ω^* is also an eigenvalue. This follows because pseudo-Hermiticity is a statement of similarity between a matrix and its Hermitian conjugate. In addition to symmetry of the spectrum about the real axis, there is also a symmetry about the imaginary axis. This follows from the second equation of (2.39), which allows us to conclude that if ω is an eigenvalue of $\mathbf{G}$, then so is $-\omega^*$. In fact, if $\vec{\psi}$ is an eigenvector, corresponding to ω, then $\boldsymbol{\tau}_1\vec{\psi}^*$ is also an eigenvector with eigenvalue $-\omega^*$. Altogether, the spectrum of an arbitrary dynamical matrix derived from a

[8] Other terminologies such as "para-Hermiticity", "quasi-Hermiticity", and "η-Hermiticity" exist within the literature. While these often coincide with pseudo-Hermiticity, there may be instances where a slightly different definition is used.

QBH enjoys a built-in fourfold symmetry, i.e., eigenvalues come in quadruplets $\{\omega, \omega^*, -\omega^*, -\omega\}$. Introducing the notation $\sigma(X)$ for the spectrum of a linear operator X, we can restate this symmetry property as

$$\sigma(\mathbf{G}) = \sigma(\mathbf{G})^* = -\sigma(\mathbf{G})^* = -\sigma(\mathbf{G}). \tag{2.45}$$

This begs the question: what is the physical meaning of a non-real eigenvalue? Moreover, what if $\mathbf{G}$ is not even diagonalizable in the first place? Such spectral properties are tied to various notions of stability that will be explored in the next section.

2.3.2 Notions of Stability and Their Criteria

In the study of any dynamical system, various notions of stability naturally arise. For QBHs, two distinct notions are relevant; thermodynamic stability, and dynamical stability. A QBH is said to be *thermodynamically stable* if the Hamiltonian is either bounded from above, or bounded from below.[9] Thermodynamically unstable Hamiltonians are precisely those that lack well defined thermal (Gibbs) states $\rho_{\text{th}} \sim e^{-\beta H}$, even if one allows for negative temperatures. This is closely related to the concept of a Landau instability [23], whereby a system with no lower bound can never minimize its free energy, and thus, never reach thermal equilibrium. The simplest example of a thermodynamically stable QBH is that of the harmonic oscillator $H = \omega a^\dagger a$, with $\omega \in \mathbb{R}$. In contrast, the two-mode Hamiltonian $H = \omega_1 a_1^\dagger a_1 - \omega_2 a_2^\dagger a_2$, with $\omega_1, \omega_2 > 0$, is thermodynamically unstable since the energies $n_1\omega_1 - n_1\omega_1$ with $n_1, m_1 \in \mathbb{Z}_{\geq 0}$ are unbounded in both directions. In particular, thermodynamically unstable Hamiltonians lack *absolute* ground states.[10] Despite this physically unattractive property, these Hamiltonians can arise as mean-field approximations of thermodynamically stable systems (see e.g., Sect. 3.3.2).

In contrast to thermodynamic stability, dynamical stability concerns itself with dynamical features of the QBH rather than equilibrium ones. A *dynamically stable* QBH is one in which the expectation value of any observable in any physical state remains bounded for in time. The simplest example, again, is the harmonic

[9] Thermodynamic stability is traditionally defined by requiring that the Hamiltonians bounded from below. Our slight generalization is to address the trivial fact that, if one has a Hamiltonian H that is bounded from above, then $-H$ is bounded from below. Thus, H is trivially associated to a thermodynamically stable system with indistinguishable physical characteristics. In particular, thermal states of $-H$ can be thought of as negative temperature thermal states of H.

[10] The qualifier "absolute" here is meant to explicitly address thermodynamically unstable Hamiltonians that conserve total boson number, like the example $H = \omega_1 a_1^\dagger a_1 - \omega_2 a_2^\dagger a_2$. In a fixed number sector, a ground state can exist (in particular, it is a BEC, where each bosonic particle occupies the lowest energy state). However, this ground state is not absolute in the sense that adding an additional boson will always lower the ground state energy. In particular, such Hamiltonians are not bounded as the number of particles diverges (e.g., in the thermodynamic limit).

oscillator $H = \omega a^\dagger a$, where all observables exhibit, at worst, quasi-periodic motion. If there is even one observable-state pair where the expectation value diverges, we say the system is dynamically unstable. In finite-dimensional systems (such as a finite collection of fermionic modes), such instabilities are impossible since all observables necessarily correspond to bounded operators. Even with a single bosonic mode, dynamical instabilities can arise as exemplified by the inverted harmonic oscillator $H = \lambda(p^2 - x^2)/2$, with $\lambda > 0$. A quick calculation reveals $\langle x(t)\rangle = \langle x(0)\rangle \cosh(\lambda t) + \langle p(0)\rangle \sinh(t)$, which diverges exponentially in time, save for special choices of the initial mean position and momentum. Generally speaking, dynamical instabilities reveal themselves (both classically and quantum mechanically) when using mean-field quadratic approximations to model unstable equilibria. Less obviously, polynomial instabilities may also arise. The most elementary example is that of a free quantum particle, $H = p^2/2m$, whereby $\langle x(t)\rangle = (\langle p(0)\rangle /m)t + \langle x(0)\rangle$ diverges linearly in t. As we will soon see, polynomial instabilities like this originate from so-called *exceptional points* (EPs), or points of degeneracy in the Hamiltonian parameter space where the dynamical matrix forgoes diagonalizability.

All examples thus far have revealed that the two above notions of stability are generally independent. That is, each of the four possible combinations of thermodynamic stability and dynamical stability exist. The harmonic oscillator is stable in both senses. The free particle is dynamically unstable but thermodynamically stable. The two-mode Hamiltonian $H = \omega_1 a_1^\dagger a_1 - \omega_2 a_2^\dagger a_2$ is dynamically stable but thermodynamically unstable. Finally, the inverted oscillator is unstable in both senses.[11]

While these notions of stability are defined in a physically natural way, we do not have, a priori, a natural way for assessing them. After all, it is unreasonable to expect that one will always be able to check that all observables have bounded expectation values in all states, or to verify that the expectations of H in arbitrary states are bounded in one direction or the other. Since the QBH is uniquely defined by its dynamical matrix $\mathbf{G}$, it is natural to expect the two types of stability to be directly diagnosable from it alone. Indeed, this is the case. Firstly, a QBH is thermodynamically stable if and only if $\mathbf{H} = \boldsymbol{\tau}_3\mathbf{G}$ is positive-semidefinite or negative-semidefinite, i.e., $\mathbf{H} \geq 0$ or $\mathbf{H} \leq 0$. The "if" direction follows from heavy restrictions placed on the spectra and Jordan block structure of $\mathbf{G}$ when $\mathbf{H} \geq 0$ or $\mathbf{H} \leq 0$. We will revisit these restrictions and their implications in the next section. The proof of the "only-if" direction is considerably more difficult, but can be found, e.g., in Ref. [4].

Dynamical stability is considerably easier to diagnose. A QBH is dynamically stable if and only if the dynamical matrix $\mathbf{G}$ has an entirely real spectrum, and is diagonalizable. For the if-direction, it is immediately apparent that these conditions

[11] Thermodynamic instability of the inverted oscillator can be verified by noting that the average energy of a coherent state is $\langle\alpha|H|\alpha\rangle = \mathrm{Im}(\alpha)^2 - \mathrm{Re}(\alpha)^2$, which is unbounded in both directions.

on **G** imply that the LTI system defined in Eq. (2.43) exhibits bounded evolution.[12] Thus, the creation and annihilation operators, and all operators built from them, exhibit bounded motion for all time. The only if-direction is best approached by proving the contrapositive. If **G** does not satisfy the stated conditions, then it either has a non-real eigenvalue, a non-trivial Jordan block, or both. In Sect. 2.4, we will see how either one of these occurrences will imply the existence of an operator that is (1) a linear combination of creation and annihilation operators, and (2) exhibits unbounded motion in generic states.

2.3.3 *The Indefinite Inner-Product Structure of Nambu Space and Reformulating Dynamics*

We have thus far seen that the LTI system Eq. (2.43) entirely characterizes the dynamics generated by a given QBH. In particular, the pseudo-Hermitian generator **G** plays a crucial role. This intrinsic pseudo-Hermiticity suggests that there is an indefinite inner-product structure embedded within bosonic systems. Making this structure explicit is most easily accomplished by formalizing the structure of Nambu space. Namely, consider the $2N$-dimensional complex vector space consisting of all complex linear combinations of the creation and annihilation operators Φ_j. We call elements of this space *linear forms*. Finite-dimensionality of this space allows us to put it in 1-to-1 correspondence with $\mathbb{C}^{2N}$. We do so in an antilinear fashion:

$$\mathbb{C}^{2N} \ni \vec{\alpha} \mapsto \widehat{\vec{\alpha}} = \vec{\alpha}^{\dagger}\boldsymbol{\tau}_3\Phi = \sum_{j=1}^{N}\left(\alpha_{2j-1}^{*}a_j - \alpha_{2j}^{*}a_j^{\dagger}\right). \tag{2.46}$$

We have essentially introduced an antilinear map $\widehat{\cdot}$ that takes $2N$-dimensional complex vectors $\vec{\alpha}$ to elements of Nambu space. This mapping, while clearly 1-to-1, is constructed in this antilinear way in conjunction with $\boldsymbol{\tau}_3$ entirely for utility in later calculations. We will call the association $\alpha \leftrightarrow \vec{\alpha}$ the *Nambu representation* of the linear form α.

So far, the indefinite inner-product structure is not obvious. A hint comes from the CCR, which explicitly include $\boldsymbol{\tau}_3$ in Eq. (2.1). In particular, they imply that the commutator of any two elements in Nambu space is always a constant multiple of

---

[12] In the language of dynamical systems theory, this is equivalent to saying that the LTI system Eq. (2.43), or more generally, the Heisenberg EOM for any operators, is "marginally" stable—bounded expectation values remain bounded for all times. Note that this is stronger than the more standard "asymptotic", or "Hurwitz" stability, which requires that every eigenvalue of the generator $-i\mathbf{G}$ have strictly negative real part. Such a requirement would mean $\langle\Phi\rangle(t) \to 0$ (in fact, $\langle O\rangle(t) \to 0$ for any traceless operator O) as $t \to \infty$, regardless of the initial condition. However, the fourfold symmetry of the spectrum of **G** means that the dynamical matrix of a QBH can never be Hurwitz stable.

the identity. That is, we have a product (the commutator) that produces a scalar in a well-defined way. Following this thread, we find for $\vec{\alpha}, \vec{\beta} \in \mathbb{C}^{2N}$,

$$[\widehat{\vec{\alpha}}, \widehat{\vec{\beta}}^\dagger] = \vec{\alpha}^\dagger \boldsymbol{\tau}_3 \vec{\beta}\, 1_{\mathcal{F}} = (\vec{\alpha}, \vec{\beta})_{\tau_3}\, 1_{\mathcal{F}}. \quad (2.47)$$

That is, the commutator of two Nambu operators naturally maps to the $\boldsymbol{\tau}_3$ indefinite inner-product of the associated vectors. The CCRs are naturally recovered by taking $\vec{\alpha}$ and $\vec{\beta}$ to be canonical basis vectors of $\mathbb{C}^{2N}$. Ultimately, we have identified Nambu space, paired with the commutator, to a $2N$-dimensional complex vector space, paired with an indefinite inner-product with metric $\boldsymbol{\tau}_3$. We also note that the fermionic realization of Nambu space (paired with the anticommutator) naturally maps to $\mathbf{C}^{2N}$, with the usual inner product.

We can now start importing concepts from one space to the other. For example, indefinite inner-product spaces have a natural indefinite "norm", i.e., $(\vec{\alpha}, \vec{\alpha})_\eta \equiv \vec{\alpha}^\dagger \eta \vec{\alpha}$. In the case where $\eta = \boldsymbol{\tau}_3$, this "norm" can be positive negative and zero. The sign of this norm is called the *Krein signature* (with the sign of zero taken to be zero). Examples of $+1$, -1, and 0 Krein-signature vectors are respectively $\vec{e}_{2j-1}$, $\vec{e}_{2j}$, and $\vec{x}_j = (\vec{e}_{2j-1} - \vec{e}_{2j})/\sqrt{2}$, with $\vec{e}_n$ a canonical basis vector of $\mathbb{C}^{2N}$. Applying our mapping to these vectors reveals

$$\widehat{\vec{e}}_{2j-1} = a_j, \quad \widehat{\vec{e}}_{2j} = a_j^\dagger, \quad \widehat{\vec{x}}_j = \frac{a_j + a_j^\dagger}{\sqrt{2}} = x_j. \quad (2.48)$$

In other words, positive and negative Krein signature vectors can be interpreted as annihilation and creation operators, respectively. Meanwhile, zero Krein signature vectors are equal mixtures of the two. In fact, to remain consistent with Eq. (2.47), all Hermitian linear forms correspond to zero Krein signature vectors. We can verify this by considering the vector corresponding to the adjoint of a given linear form:

$$\widehat{\vec{\alpha}}^\dagger = \widehat{\vec{\beta}}, \quad \vec{\beta} = -\boldsymbol{\tau}_1 \vec{\alpha}^*. \quad (2.49)$$

Thus, Hermitian linear forms must satisfy $\vec{\alpha} = -\boldsymbol{\tau}_1 \vec{\alpha}^*$. More generally, if $\vec{\beta} = -\boldsymbol{\tau}_1 \vec{\alpha}$, then

$$\vec{\beta}^\dagger \boldsymbol{\tau}_3 \vec{\beta} = \vec{\alpha}^T \boldsymbol{\tau}_1 \boldsymbol{\tau}_3 \boldsymbol{\tau}_1 \vec{\alpha}^* = -\vec{\alpha}^T \boldsymbol{\tau}_3 \vec{\alpha}^* = -\vec{\alpha}^\dagger \boldsymbol{\tau}_3 \vec{\alpha}. \quad (2.50)$$

This reveals that, to each positive (negative) Krein signature vector $\vec{\alpha}$, we have an associated negative (positive) Krein signature $\vec{\beta} = -\boldsymbol{\tau}_1 \vec{\alpha}^*$. This associated vector corresponds to the adjoint linear form, consistent with Eq. (2.47). Thus, when $\vec{\alpha} = \vec{\beta}$, the Krein signature is zero.

This mapping further provides utility in reformulating the Heisenberg equations of motion (EOMs) of Eq. (2.43). Specifically, Eq. (2.43) implies that if α is a linear form at time t_0, then it will remain a linear form for all $t \geq t_0$. Thus, if $\alpha(t_0) = \widehat{\vec{\alpha}}$

for some vector $\vec{\alpha}$, then $\alpha(t) = \widehat{\vec{\alpha}(t)}$ for some vector $\vec{\alpha}(t)$, with $\vec{\alpha}(t_0) = \vec{\alpha}$. In this sense, we have "contravariant" dynamics according to

$$\hat{\vec{\alpha}}(t) = \sum_{j=1}^{2N} \alpha_{2j-1}^* a_j(t) - \alpha_{2j}^* a_j^\dagger(t) \equiv \sum_{j=1}^{2N} \alpha_{2j-1}^*(t) a_j(0) - \alpha_{2j}^*(t) a_j^\dagger(0) \equiv \widehat{\vec{\alpha}(t)}, \tag{2.51}$$

where the final two equalities *define* the time-dependence of the vector $\vec{\alpha}(t)$. In order to be consistent with the Heisenberg EOMs, the contravariant EOM for $\vec{\alpha}(t)$ is found to be[13]

$$\frac{d}{dt}\vec{\alpha}(t) = i\mathbf{G}\vec{\alpha}(t), \quad \vec{\alpha}(t) = e^{i\mathbf{G}t}\vec{\alpha}. \tag{2.52}$$

We have successfully mapped the Heisenberg EOM of the Nambu array to a pseudo-Hermitian LTI system on an indefinite inner-product space. One may then naturally ask what role the second property of $\mathbf{G}$ plays, i.e., $\boldsymbol{\tau}_1\mathbf{G}^*\boldsymbol{\tau}_1 = -\mathbf{G}$. Recalling Eq. (2.49), one can verify that the second property of $\mathbf{G}$ ensures that if $\vec{\beta} = -\boldsymbol{\tau}_1\vec{\alpha}^*$ at $t = 0$, then $\vec{\beta}(t) = -\boldsymbol{\tau}_1\vec{\alpha}^*(t)$ for all t. This reflects the Hermitian conjugate-preserving nature of the Heisenberg EOMs.

Having placed the indefinite inner-product space in 1-to-1 correspondence with the Nambu space, one may ask further what linear transformations on this inner product space naturally correspond to. Our first hint is via the Heisenberg commutator:

$$i[H, \hat{\vec{\alpha}}] = \widehat{i\mathbf{G}\vec{\alpha}}. \tag{2.53}$$

In this way, the commutator $i[H, \cdot]$ implements the pseudo-Hermitian transformation $\mathbf{G}$. Thus we may identify H with the matrix $\mathbf{G}$. In fact, for any $2N \times 2N$ matrix satisfying $\boldsymbol{\tau}_1\mathbf{A}^*\boldsymbol{\tau}_1 = -\mathbf{A}$, we can unambiguously[14] define a quadratic form[15]

[13] This equation motivates the use of the term "contravariant". The coefficients α_j undergo a (time-dependent) transformation that is the inverse of that of the basis vectors Φ_j, cf. Eq. (2.43). That is, $\hat{\vec{\alpha}}(t)$ may be computed by evolving $\Phi(0)$ to $\Phi(t) = e^{-\mathbf{G}t}\Phi(0)$ and projecting with $\boldsymbol{\tau}_3\vec{\alpha}(0)$ (i.e., the covariant approach, whereby the basis vectors Φ_j evolve in time), or by evolving $\vec{\alpha}(0)$ to $\vec{\alpha}(t) = e^{i\mathbf{G}t}\vec{\alpha}(0)$ and then projecting onto $\boldsymbol{\tau}_3\Phi(0)$ (i.e., the contravariant approach, whereby the coefficients α_j evolve in time).

[14] Fixing the condition $\boldsymbol{\tau}_1\mathbf{A}^*\boldsymbol{\tau}_1 = -\mathbf{A}$ ensures that the corresponding quadratic form is expressed with the creation and annihilation operators symmetrically ordered. Without this condition, the corresponding quadratic form would be equal to the symmetrically ordered one modulo a constant shift. Fixing this condition thus allows us to focus on the equivalence class of quadratic forms that differ by a constant.

[15] Notational comment: when the coefficient vector of a given linear form is not particularly relevant, we will drop the hat-vector notation. That is, we will write α for $\hat{\vec{\alpha}}$. Similarly, we will

$$\widehat{\mathbf{A}} = \frac{1}{2}\Phi^\dagger \boldsymbol{\tau}_3 \mathbf{A} \Phi. \tag{2.54}$$

We have effectively lifted our $\widehat{\cdot}$-map to quadratic forms and matrices on the indefinite inner product space. From this definition, $H = \widehat{\mathbf{G}}$ in a natural way. Three nontrivial properties follow:

$$\text{(i) } \widehat{\mathbf{A}}^\dagger = \widehat{\boldsymbol{\tau}_3 \mathbf{A} \boldsymbol{\tau}_3}, \quad \text{(ii) } i[\widehat{\mathbf{A}}, \widehat{\vec{\alpha}}] = \widehat{i\mathbf{A}\vec{\alpha}}, \quad \text{(iii) } [\widehat{\mathbf{A}}, \widehat{\mathbf{B}}] = \widehat{[\mathbf{A}, \mathbf{B}]}. \tag{2.55}$$

The first says that QBHs are associated to pseudo-Hermitian transformations. The second says that a commutator between a quadratic and a linear form implements a linear transformation on the indefinite inner-product space. The final says that the commutator of two quadratic forms maps directly to the commutator of the corresponding transformations on the indefinite inner product space. It follows that the set of quadratic forms form a Lie algebra that maps directly to the Lie algebra of pseudo-Hermitian matrices.

As an immediate application of this formalism, we can formulate conditions for a given QBH to have a homogeneous Gaussian symmetry. Specifically, consider a many-body unitary or antiunitary operation S that effects the transformation $\Phi \mapsto \mathbf{S}\Phi$, with $\mathbf{S}$ satisfying Eqs. (2.18). This operator is a symmetry of the QBH $H = \widehat{\mathbf{G}}$ if and only if $[H, S] = 0$. A quick calculation reveals that this is equivalent to $[\mathbf{G}, \mathbf{S}] = 0$. In other words, a QBH possesses a Gaussian symmetry if and only if the Nambu space representation of the QBH commutes with the Nambu space representation of the symmetry.

We conclude this digression by introducing two operations defined on $2N \times 2N$ dimensional complex matrices that will provide utility in future sections. Given such a matrix $\mathbf{X}$, we define the fermionic and bosonic projectors $\mathcal{F}$ and $\mathcal{B}$ according to

$$\mathcal{F}(\mathbf{X}) \equiv \frac{1}{2}\left(\mathbf{X} - \boldsymbol{\tau}_1 \mathbf{X}^T \boldsymbol{\tau}_1\right), \quad \mathcal{B}(\mathbf{X}) \equiv \frac{1}{2}\left(\mathbf{X} + \boldsymbol{\tau}_1 \mathbf{X}^T \boldsymbol{\tau}_1\right). \tag{2.56}$$

These provide orthogonal projectors in the sense that $\mathcal{F}^2 = \mathcal{F}$, $\mathcal{B}^2 = \mathcal{B}$, $\mathcal{F} \circ \mathcal{B} = \mathcal{B} \circ \mathcal{F} = 0$, and $\mathcal{F} + \mathcal{B} = \mathcal{I}$, with $\mathcal{I}(\mathbf{X}) = \mathbf{X}$. Thus, any matrix is uniquely defined by its fermionic and bosonic projections. The nomenclature comes from the fact that a purely quadratic fermionic (bosonic) Hamiltonian is defined uniquely modulo constant shifts by a fermionic (bosonic) Hermitian matrix $\mathbf{H}$, i.e., a Hermitian matrix that satisfies $\mathcal{F}(\mathbf{H}) = \mathbf{H}$ ($\mathcal{B}(\mathbf{H}) = \mathbf{H}$). If one changes to a Majorana or quadrature basis (in the fermionic and bosonic cases, respectively), these projections transform to the antisymmetric and symmetric projections, respectively.

---

stick to notation like H for QBHs when the dynamical matrix $\mathbf{G}$ is understood, rather than using to the hat-notation.

2.4 Diagonalization of Quadratic Bosonic Hamiltonians

When faced with a mean-field Hamiltonian in the traditional condensed matter setting, the first instinct is to diagonalize it via a Bogoliubov transformation. That is, one maps the fundamental degrees of freedom (creation and annihilation operators) to a set of decoupled quasiparticles that excite modes with a fixed energy. The eigenstates of the Hamiltonian are then the transformed Fock states, i.e., the occupation number states of these quasiparticle excitations. A related concept in dynamical systems is that of normal mode analysis. Namely, to solve a (linearized) dynamical system, one will first compute the set of normal modes, which are defined by a simple dynamical Ansatz (usually having all the time-dependence absorbed in an exponential prefactor). The full solution is then constructed out of these normal modes in a linear fashion, subject to the set initial conditions. Remarkably, these two problems are solved simultaneously for quadratic Hamiltonians (both fermionic and bosonic). The aforementioned quasiparticles correspond in a natural way to the normal modes of the LTI system Eq. (2.52) (and its fermionic counterpart[16]). The quasiparticle energies then map on to the normal mode frequencies. However, in the fermionic case, the normal mode frequencies are always guaranteed to be real thanks to Hermiticity of the corresponding generator. So, the interpretation of them as quasiparticle energies is straightforward. In what sense does a non-real normal-mode frequency in a bosonic system correspond to a quasiparticle energy? The QBH must have real eigenvalues, after all.

The first step to seeing the correspondence between the diagonalization problem of a QBH and the normal mode analysis of Eq. (2.52) is to precisely state the diagonalization problem. Given a QBH $H = \widehat{\mathbf{G}}$, we seek a Bogoliubov transformation (unitary, homogeneous, Gaussian transformation) $U\Phi U^\dagger = \Psi = \mathbf{T}\Phi$, such that $H = \frac{1}{2}\Psi^\dagger \mathbf{D}\Psi$, with $\mathbf{D}$ diagonal. Explicitly, the quasiparticles $\Psi = [\psi_1, \psi_1^\dagger, \ldots, \psi_N, \psi_N^\dagger]^T$ satisfy the CCRs and

$$H = \frac{1}{2}\sum_{n=1}^{N} \omega_n \left(\psi_n^\dagger \psi_n + \psi_n \psi_n^\dagger\right) = \sum_{n=1}^{N} \omega_n \left(\psi_n^\dagger \psi_n + \frac{1}{2}\right), \tag{2.57}$$

with the quasiparticle energies $\omega_n \in \mathbb{R}$. Once we have accomplished this, the eigenstates are built up from the quasiparticle vacuum, i.e., the state $|\widetilde{0}\rangle$ satisfying $\psi_n|\widetilde{0}\rangle = 0$ for all n, in the usual way (e.g., Eq. (2.4)). We can then conclude that, if

[16] To be concrete, a quadratic fermionic Hamiltonian is an operator of the form $H = \sum_{i,j=1}^{N}\left(c_i^\dagger \mathbf{K}_{ij} c_j - c_i \mathbf{K}_{ij}^* c_j^\dagger + c_i^\dagger \boldsymbol{\Delta}_{ij} c_j^\dagger + c_j \boldsymbol{\Delta}_{ij}^* c_i\right)$, with the c_j ($c_j^\dagger$) the fermionic annihilation (creation) operator satisfying the canonical anticommutation relations $\{c_i, c_j^\dagger\} = \delta_{ij} 1_{\mathcal{F}}, c_i, c_j = 0$, $\mathbf{K}$ a complex Hermitian matrix, and $\boldsymbol{\Delta}$ a complex anti-symmetric matrix. It follows that $\partial_t \Psi = i[H, \Psi] = -i\mathbf{H}\Psi$, with $\Psi = [c_1, c_1^\dagger, \ldots]^T$ the fermionic Nambu array and $\mathbf{H}$ a *Hermitian* matrix whose elements are defined by $\mathbf{K}$ and $\boldsymbol{\Delta}$. This matrix, known as the Bogoliubov-de Gennes Hamiltonian, is the fermionic analogue of $\mathbf{G}$.

such a transformation is possible, the Hamiltonian will be thermodynamically stable if and only if the ω_n are either all non-negative (H bounded from below) or all non-positive (H bounded from above). If this is the case then the quasiparticle vacuum is precisely the ground state (or highest energy state when H is bounded from above), with ground state energy

$$H|\widetilde{0}\rangle = E_{\text{GS}}|\widetilde{0}\rangle, \quad E_{\text{GS}} = \frac{1}{2}\sum_{n=1}^{N}\omega_n. \tag{2.58}$$

Regardless of the signs of the energies, a representation of the form Eq. (2.57) implies dynamical stability. Namely, in the Heisenberg picture, the quasiparticles are normal modes: $\psi_n(t) = e^{-i\omega_n t}\psi_n(0)$. We can then invert the transformation $\mathbf{T}$ to find the dynamics of $a_j(t)$, which are linear combinations of $\psi_n(t)$, and thus, dynamically stable. Ultimately, if diagonalization via Bogoliubov transformation is possible, the system must be dynamically stable. Further, thermodynamic stability can be assessed via the sign distribution of the single particle energies. This reveals something surprising: dynamically unstable Hamiltonians (such as the inverted harmonic oscillator or the free particle) *cannot* be diagonalized via Bogoliubov transformation. In fact, the set of all dynamically stable QBHs is *precisely* the set of QBHs that can be diagonalized via a Bogoliubov transformation. Let us explore this fact in more detail.

Suppose we have a dynamically stable QBH. In Sect. 2.3.2, we explained how $\mathbf{G}$ being diagonalizable with a real spectrum is sufficient for $H = \widehat{\mathbf{G}}$ to be dynamically stable. With the tools of Sect. 2.3.3, we can prove necessity. We do so via the contrapositive: suppose $\mathbf{G}$ either has a non-real eigenvalue or is non-diagonalizable. In the first case, let $\vec{\xi}$ be the corresponding eigenvector, i.e., $\mathbf{G}\vec{\xi} = \omega\vec{\xi}$, with $\omega = x - iy$ and $y > 0$. Note that if $y > 0$, we can use symmetry of the spectrum to find an eigenvector with eigenvalue $\omega = x + iy$. One may immediately verify that $\widehat{\vec{\xi}}$ is a normal mode of the system with $\widehat{\vec{\xi}}(t) = e^{-i\omega^* t}\widehat{\vec{\xi}}(0) = e^{-ixt}e^{yt}\widehat{\vec{\xi}}(0)$ unbounded in time. If $\mathbf{G}$ is non-diagonalizable, then there is a Jordan chain of length $r > 1$ at a particular eigenvalue ω_0. That is, we have the system of equations

$$(\mathbf{G} - \omega_0 \mathbb{1}_{2N})\vec{\chi}_{0,1} = 0, \quad (\mathbf{G} - \omega_0 \mathbb{1}_{2N})\vec{\chi}_{0,k} = \vec{\chi}_{0,k-1}, \quad 2 \le k \le r, \tag{2.59}$$

for some vectors $\vec{\chi}_{0,k}$. If ω_0 is non-real, then the previous argument suffices to prove dynamical instability. If ω_0 is real, then we find that

$$\widehat{\vec{\chi}}_{0,k}(t) = e^{-i\omega_0 t}\sum_{\ell=0}^{k-1}\frac{(-it)^\ell}{\ell!}\widehat{\vec{\chi}}_{0,k-\ell}(0), \tag{2.60}$$

which generically exhibit polynomial divergence as $t \to \infty$. Since the vectors $\vec{\chi}_{0,k}$ are called generalized eigenvectors of rank k, we henceforth refer to normal modes of the form Eq. (2.60) as generalized normal modes of rank k. As it turns out, such generalized normal modes are sharply constrained in thermodynamically stable systems. Positive (or negative) semidefiniteness of $\mathbf{H}$ guarantee the longest Jordan chain is at most length 2 and that it must occur at eigenvalue $\omega = 0$ [22]. In this case, the $\omega = 0$ normal mode and its associated generalized mode can be interpreted as free particle terms in the Hamiltonian [2].

Now we are in a position to identify dynamically stable QBHs and those QBHs whose dynamical matrices are diagonalizable with a real spectrum. What would it mean for the system to be diagonalizable via a Bogoliubov transformation? If such a transformation $\Psi = \mathbf{T}\Phi$ were to exist, then we would have

$$H = \frac{1}{2}\Phi^\dagger \boldsymbol{\tau}_3 \mathbf{G}\Phi = \frac{1}{2}\Psi^\dagger \mathbf{T}^{-1\dagger} \boldsymbol{\tau}_3 \mathbf{G}\mathbf{T}^{-1}\Psi = \frac{1}{2}\Psi^\dagger \mathbf{D}\Psi, \tag{2.61}$$

with $\mathbf{D}$ diagonal. Since this transformation is canonical, the first equation in (2.18) yields

$$\mathbf{T}\mathbf{G}\mathbf{T}^{-1} = \mathbf{D}',$$

with $\mathbf{D}' = \boldsymbol{\tau}_3\mathbf{D}$ still diagonal. We further recognize $\mathbf{D}' = \text{diag}(\omega_1, -\omega_1, \ldots, \omega_N, -\omega_N)$. In short, the hunt for a Bogoliubov transformation is equivalent to the hunt for a matrix $\mathbf{T}$ satisfying Eqs. (2.18) that diagonalizes $\mathbf{G}$. Moreover, we immediately see that, if such a transformation is found, the eigenvalues of $\mathbf{G}$ (i.e., the normal mode frequencies) constitute the quasiparticle energies, as advertised earlier.

Let us denote the n'th column of $\mathbf{T}^{-1}$ by $\vec{v}_n$. Since $\mathbf{T}$ diagonalizes $\mathbf{G}$, we have $\mathbf{G}\vec{v}_{2n-1} = \omega_j \vec{v}_{2n}$ and $\mathbf{G}\vec{v}_{2n} = -\omega_j \vec{v}_{2n}$. Equations (2.18) provide two further requirements:

$$\vec{v}_n \boldsymbol{\tau}_3 \vec{v}_m = (\boldsymbol{\tau}_3)_{nm}, \quad \vec{v}_{2n} = \boldsymbol{\tau}_1 \vec{v}^{\,*}_{2n-1}. \tag{2.62}$$

These requirements motivate a more refined notation, namely, let $\vec{\psi}_n^+ = \vec{v}_{2n-1}$ and $\vec{\psi}_n^- = \vec{v}_{2n}$. We can reformulate the problem once more: given a dynamically stable QBH, can one find a basis of $\mathbb{C}^{2N}$ consisting of eigenvectors of $\mathbf{G}$ denoted by $\{\vec{\psi}_n^\pm\}_{n=1}^N$ satisfying

$$\vec{\psi}_n^{s\dagger} \boldsymbol{\tau}_3 \vec{\psi}_m^{s'} = s\delta_{ss'}\delta_{nm}, \quad \vec{\psi}_n^- = \boldsymbol{\tau}_1 \vec{\psi}_n^{+*}, \quad \mathbf{G}\vec{\psi}_n^s = s\omega_n \vec{\psi}_n^s, \quad s, s' \in \{+, -\}? \tag{2.63}$$

The answer is *yes*. In Appendix B.1, we prove this in the case where ω_n are non-zero. For a more comprehensive account, we refer the reader to Refs. [1–4].

Equipped with an eigenbasis satisfying Eqs. (2.63), we obtain a useful resolution of the identity, and consequentially, a sort-of $\boldsymbol{\tau}_3$ spectral decomposition,

$$\mathbb{1}_{2N} = \sum_{n=1}^{N} \left(\vec{\psi}_n^+ \vec{\psi}_n^{+\dagger} - \vec{\psi}_n^- \vec{\psi}_n^{-\dagger}\right) \boldsymbol{\tau}_3, \quad \mathbf{G} = \sum_{n=1}^{N} \omega_n \left(\vec{\psi}_n^+ \vec{\psi}_n^{+\dagger} + \vec{\psi}_n^- \vec{\psi}_n^{-\dagger}\right) \boldsymbol{\tau}_3, \tag{2.64}$$

which can be verified, for example, by applying the above expressions to the eigenbasis of **G**. Plugging this decomposition directly into H reveals

$$H = \frac{1}{2}\Phi^\dagger \boldsymbol{\tau}_3 \mathbf{G}\Phi = \frac{1}{2}\sum_{n=1}^{N} \omega_n \left(\psi_n^\dagger \psi_n + \psi_n \psi_n^\dagger\right), \tag{2.65}$$

where we have introduced the simplified notation $\psi_n = \widehat{\vec{\psi}}_n^{+} = -\widehat{\vec{\psi}}_n^{-\dagger}$. Ultimately, we see that for dynamically stable QBHs, the positive Krein signature eigenvectors $\vec{\psi}_n^+$ and their associated N eigenvalues define the quasiparticle annihilation operators via $\psi_n = \widehat{\vec{\psi}_n^+} = \vec{\psi}_n^{+\dagger}\boldsymbol{\tau}_3\Phi$ and their quasiparticle energies, respectively. While it may seem that we have lost the negative Krein signature eigenvectors, we note that the second relation in Eq. (2.63) implies $\widehat{\vec{\psi}_n^-} = -\widehat{\vec{\psi}_n^+}^\dagger$, and moreover, the eigenvalues are simply the negative of the eigenvalues of the positive Krein signature eigenvectors. Importantly, we have not, and need not, specify the sign of the N eigenvalues ω_n associated to these positive Krein signature eigenvectors. We see that it is precisely the sign distribution of these particular eigenvalues that define the thermodynamic stability of the system. In order for a dynamically stable system to be thermodynamically unstable all positive Krein signature eigenvectors must correspond to either all nonnegative or nonpositive eigenvalues. This is guaranteed precisely when $\mathbf{H} = \boldsymbol{\tau}_3\mathbf{G}$ is positive or negative semidefinite [22].

The identification between the Bogoliubov transformation diagonalizing the QBH and the matrix that diagonalizes **G** provides one immediate application. Namely, observe that we can write the corresponding matrix **T** as

$$\mathbf{T} = \sum_{j,n=1}^{N} \vec{e}_j \vec{e}_n^{\,\dagger} \otimes \begin{bmatrix} \mathbf{X}_{jn} & \mathbf{Y}_{jn}^* \\ \mathbf{Y}_{jn} & \mathbf{X}_{jn}^* \end{bmatrix}, \tag{2.66}$$

where $\mathbf{X}_{jn} = (\vec{\psi}_n^+)_{2j-1}$ and $\mathbf{Y}_{jn} = (\vec{\psi}_n^+)_{2j}$ are matrices that, roughly speaking, encode the weight of each quasiparticle on the annihilation operator and creation operator sectors, respectively. In terms of these matrices, we obtain a closed form expression for the unitary U taking Φ to $\Psi = U\Phi U^\dagger$, i.e.,

$$U = \det(\mathbf{X}^\dagger\mathbf{X})^{-1/4} : \exp\left[\frac{1}{2}\sum_{j,k=1}^{N}\left(-\mathbf{X}^{-1}\mathbf{Y}^*\right)_{jk} a_j^\dagger a_k^\dagger + (\mathbf{Y}\mathbf{X}^{-1})_{jk} a_j a_k \right.$$
$$\left. + \left(\mathbf{X}_{jk}^{-1} - \delta_{jk}\right) a_j^\dagger a_k \right] :, \tag{2.67}$$

where the normal ordering symbol $: \cdot :$ ensures all creation operators are to the left of the annihilation operators. One may verify that this is well-defined, i.e., that **X** is invertible, and that the associated quasiparticle vacuum is then

$$|\widetilde{0}\rangle = U|0\rangle = \det(\mathbf{X}^\dagger\mathbf{X})^{-1/4}\exp\left[\frac{1}{2}\sum_{j,k=1}^{N}\left(-\mathbf{X}^{-1}\mathbf{Y}^*\right)_{jk}a_j^\dagger a_k^\dagger\right]|0\rangle. \tag{2.68}$$

The quasiparticle vacuum is evidently Gaussian, with vanishing mean vector and covariance matrix given by

$$\mathbf{C}_{|\widetilde{0}\rangle} = \frac{1}{2}\boldsymbol{\tau}_3\mathbf{T}^\dagger\mathbf{T}\boldsymbol{\tau}_3. \tag{2.69}$$

Following this extensive account of the diagonalization procedure associated to dynamically stable QBHs, we are led to ask to what extent a normal form along the lines of Eq. (2.65) can be obtained in the absence of dynamical stability. For simplicity, let us focus on the case where $\mathbf{G}$ is diagonalizable. The eigenvalues and eigenvectors can then be partitioned into three sectors satisfying various properties [3, 6]:

1. Let $\pm\omega_n$, $n = 1, \ldots N_R$, denote the real eigenvalues. Just as in the dynamically stable case, we may arrange for these eigenvalues and their eigenvectors to satisfy

$$\mathbf{G}\vec{\psi}_n^{\pm} = \pm\omega_n\vec{\psi}_n^{\pm}, \quad \vec{\psi}_n^{-} = \boldsymbol{\tau}_1(\vec{\psi}_n^{+})^*, \tag{2.70}$$

in addition to the $\boldsymbol{\tau}_3$-orthonormality conditions $\vec{\psi}_n^{\pm\dagger}\boldsymbol{\tau}_3\vec{\psi}_m^{\pm} = \pm\delta_{nm}$ and $\vec{\psi}_n^{\pm\dagger}\boldsymbol{\tau}_3\vec{\psi}_m^{\mp} = 0$. In Eq. (2.70), the relationship between $\vec{\psi}^{+}$ and $\vec{\psi}_n^{-}$ ensures that $\widehat{\vec{\psi}_n^{+}}^{\dagger} = -\widehat{\vec{\psi}_n^{-}}$.

2. Let $\pm i\lambda_n$, $n = 1, \ldots, N_I$ denote the purely imaginary eigenvalues with $\lambda_n > 0$. For these, we denote the eigenvectors according to

$$\mathbf{G}\vec{z}_n^{\pm} = \pm i\lambda_n\vec{z}_n^{\pm}. \tag{2.71}$$

Since these purely imaginary eigenspaces are invariant under the antilinear map $\vec{z} \mapsto \boldsymbol{\tau}_1\vec{z}^*$ (which takes $i\lambda_n$ to $-(i\lambda_n)^* = i\lambda_n$), we can always ensure $\boldsymbol{\tau}_1\vec{z}_n^{\pm *} = -\vec{z}_n^{\pm}$. This consequentially guarantees $\widehat{\vec{z}}^{\pm}$ are Hermitian. Furthermore, we can arrange so that $\vec{z}_n^{\pm\dagger}\boldsymbol{\tau}_3\vec{z}_m^{\mp} = \pm i\delta_{nm}$ and $\vec{z}_n^{\pm\dagger}\boldsymbol{\tau}_3\vec{z}_m^{\pm} = 0$.

3. Let $\pm\mu_n$, $\pm\mu_n^*$, $n = 1, \ldots, N_C$, denote the remaining complex eigenvalues with μ_n in the upper right quadrant, i.e., $\mathrm{Re}(\mu_n), \mathrm{Im}(\mu_n) > 0$. We denote the eigenvectors according to

$$\mathbf{G}\vec{\xi}_n^{+} = \mu_n\vec{\xi}_n^{+}, \quad \mathbf{G}\vec{\xi}_{n*}^{+} = \mu_n^*\vec{\xi}_*^{+}, \quad \mathbf{G}\vec{\xi}_n^{-} = -\mu_n^*\vec{\xi}_n^{-}, \quad \mathbf{G}\vec{\xi}_{n*}^{-} = -\mu_n\vec{\xi}_*^{-}, \tag{2.72}$$

with $\vec{\xi}_n^{-} = \boldsymbol{\tau}_1(\vec{\xi}_n^{+})^*$ and $\vec{\xi}_{n*}^{-} = \boldsymbol{\tau}_1(\vec{\xi}_{n*}^{+})^*$. Finally, we can ensure $\vec{\xi}_n^{\pm\dagger}\boldsymbol{\tau}_3\vec{\xi}_{m*}^{\pm} = \pm\delta_{nm}$ and that all other $\boldsymbol{\tau}_3-$inner products vanish.

Note that we have the counting constraint $2N_R+2N_I+4N_C = 2N$. Furthermore, eigenvectors in distinct sectors are $\boldsymbol{\tau}_3$-orthogonal thanks to pseudo-Hermiticity. Altogether, we obtain a decomposition of the form

$$\begin{aligned}\mathbf{G} &= \sum_{n=1}^{N_R} \omega_n \left(\vec{\psi}_n^+ \vec{\psi}_n^{+\dagger} + \vec{\psi}_n^- \vec{\psi}_n^{-\dagger}\right)\boldsymbol{\tau}_3 - \sum_{n=1}^{N_I} \lambda_n \left(\vec{z}_n^+ \vec{z}_n^{-\dagger} + \vec{z}_n^- \vec{z}_n^{+\dagger}\right)\boldsymbol{\tau}_3 \\ &+ \sum_{n=1}^{N_C} \mu_n \left(\vec{\xi}_n^+ \vec{\xi}_{n*}^{+\dagger} + \vec{\xi}_{n*}^- \vec{\xi}_n^{-\dagger}\right)\boldsymbol{\tau}_3 + \mu_n^* \left(\vec{\xi}_{n*}^+ \vec{\xi}_n^{+\dagger} + \vec{\xi}_{n*}^- \vec{\xi}_n^{-\dagger}\right)\boldsymbol{\tau}_3.\end{aligned} \tag{2.73}$$

The resultant normal form of the Hamiltonian is then

$$\begin{aligned}H &= \frac{1}{2}\sum_{n=1}^{N_R} \omega_n \left(\psi_n^\dagger \psi_n + \psi_n \psi_n^\dagger\right) - \frac{1}{2}\sum_{n=1}^{N_I} \lambda_n \left(z_n^+ z_n^- + z_n^- z_n^+\right) \\ &+ \frac{1}{2}\sum_{n=1}^{N_C}\left[\mu_n \left(\xi_n^\dagger \xi_{n*} + \xi_{n*}\xi_n^\dagger\right) + \mu_n^* \left(\xi_{n*}^\dagger \xi_n + \xi_n \xi_{n*}^\dagger\right)\right],\end{aligned} \tag{2.74}$$

with the simplified notation $\psi_n = \widehat{\vec{\psi}_n^+}$, $z_n^\pm = \widehat{\vec{z}_n^\pm}$, $\xi_n = \widehat{\vec{\xi}_n}$, and $\xi_{n*} = \widehat{\vec{\xi}_{n*}}$ implemented. The first term is analogous to Eq. (2.65), that is, it represents the stable normal modes ψ_n which satisfy the CCRs. The second term consists of operators $z_n^\pm$ which satisfy the HWRs $[z_n^+, z_n^-] = i\delta_{nm}1_{\mathcal{F}}$. This is best understood by introducing the bosonic degrees of freedom $b_n = (z_n^+ + iz_n^-)/\sqrt{2}$, in which case

$$\frac{1}{2}\sum_{n=1}^{N_I} \lambda_n \left(z_n^+ z_n^- + z_n^- z_n^+\right) = \frac{1}{2}\sum_{n=1}^{N_I} i\lambda_n \left(b_n^2 - b_n^{\dagger 2}\right) \tag{2.75}$$

is a sum of degenerate parametric amplifiers (DPAs) (equivalently, single-mode squeezing Hamiltonians). The final term consists of so-called *pseudo-bosonic modes*, i.e., pairs of modes $(\xi_n, \xi_{n*}^\dagger)$ satisfying $[\xi_n, \xi_{m*}^\dagger] = \delta_{nm}1_{\mathcal{F}}$. Since $\xi_{n*} \neq \xi_n$, these are not canonical bosonic degrees of freedom.

The unstable modes in Eq. (2.74) present an obstruction to the usual construction of the many-body eigenstates. In particular, the traditional quasiparticle picture, in which one identifies a vacuum and builds up the eigenstates by exciting bosonic normal modes, is lost. The most useful perspective is instead to stick to the dynamical point of view and interpret $z_n^\pm$, ξ_n and ξ_{n*} as normal modes of the Heisenberg EOMs, with non-real normal mode frequencies. Such an interpretation does not require any additional algebraic specification. It also lends itself naturally to the non-diagonalizable case, whereby algebraic relationships between normal modes and generalized normal modes are far more ambiguous.

2.5 Bulk-Translationally Invariant QBHs

2.5.1 The Translationally Invariant Case

We will now characterize those QBHs on lattices that possess discrete translational symmetry. In particular, we will explicitly focus on the 1D setting, and examine both finite systems on a ring and infinite systems on a bi-infinite lattice.

Let us consider a QBH defined on a finite ring with d_{int} degrees of freedom on each of the N sites. Imposing discrete translation invariance engenders the general form

$$H = \frac{1}{2}\sum_{j=1}^{N}\sum_{r=-R}^{R} \phi_j^\dagger \mathbf{h}_r \phi_{j+r}, \tag{2.76}$$

with $\mathbf{h}_r$ the $2d_{\text{int}} \times 2d_{\text{int}}$ matrix of coupling constants, $0 \leq R < N$ the coupling range, and periodic BCs (PBCs) encoded via $\phi_{N+\ell} = \phi_\ell$ for any $\ell \geq 1$. Since the coupling between sites j and $j + r$ is only a function of r, discrete translation invariance is evident. The coupling matrices are required to satisfy $\mathbf{h}_r^\dagger = \mathbf{h}_{-r}$ and, without loss of generality,[17] additionally obey $\boldsymbol{\tau}_1\mathbf{h}_r^*\boldsymbol{\tau}_1 = \mathbf{h}_r$. In this context, $\boldsymbol{\tau}_j = \mathbb{1}_{d_{\text{int}}} \otimes \boldsymbol{\sigma}_j$ is $2d_{\text{int}} \times 2d_{\text{int}}$.

With the general structure specified, we identify the dynamical matrix as

$$\mathbf{G}_N^{\text{PBC}} = \mathbb{1}_N \otimes \mathbf{g}_0 + \sum_{r=1}^{R}\left(\mathbf{V}_N^r \otimes \mathbf{g}_r + \mathbf{V}_N^{\dagger\, r} \otimes \mathbf{g}_{-r}\right), \tag{2.77}$$

with $\mathbf{g}_r = \boldsymbol{\tau}_3\mathbf{h}_r$. The notational choices here indicate that there are a finite number of sites N and PBCs. The conditions that $\mathbf{G}_N^{\text{PBC}}$ must satisfy (see Eq. (2.39)) follow from $\boldsymbol{\tau}_3\mathbf{g}_r^\dagger\boldsymbol{\tau}_3 = \mathbf{g}_{-r}$ and $\boldsymbol{\tau}_1\mathbf{g}_r^*\boldsymbol{\tau}_r = \mathbf{g}_r$. Furthermore, we observe that $\mathbf{G}_N^{(P)}$ is a *block-circulant matrix*, i.e., a circulant matrix whose elements are matrices themselves.

It is immediate to see that the dynamical matrix Eq. (2.77) commutes with the Nambu space representation of discrete translations Eq. (2.35). Thus, we can block diagonalize $\mathbf{G}_N^{\text{PBC}}$ utilizing the known basis of eigenvectors for $\mathbf{V}_N$. What we have just described is precisely Bloch's theorem: a translation-invariant system can be block-diagonalized in terms of eigenstates of the translation operator. At the many-body level, this block-diagonalization is accomplished by a discrete Fourier transform:

$$b_{m,k} = \frac{1}{\sqrt{N}}\sum_{j=1}^{N} e^{-ijk} a_{m,j}, \quad k \in \mathcal{K}_N, \quad a_{m,k} = \frac{1}{\sqrt{N}}\sum_{k\in\mathcal{K}_N} e^{ijk} b_{m,k}. \tag{2.78}$$

[17] This condition is equivalent to requiring that the matrix $\mathbf{H}$ is bosonic, i.e., $\mathcal{B}(\mathbf{H}) = \mathbf{H}$.

Accordingly, the Fourier transform of the local Nambu array is

$$\widetilde{\phi}_k = \frac{1}{\sqrt{N}} \sum_{j=1}^{N} e^{-ijk} \phi_j = [b_{1,k}, b^\dagger_{1,-k}, \ldots, b_{d_{\text{int}},k}, b^\dagger_{d_{\text{int}},-k}]^T. \tag{2.79}$$

In k-space, the equation of motion Eq. (2.43) takes the form

$$\frac{d}{dt}\widetilde{\phi}_k(t) = -i\mathbf{g}(k)\widetilde{\phi}_k(t), \quad \mathbf{g}(k) = \sum_{r=-R}^{R} \mathbf{g}_r e^{ikr}. \tag{2.80}$$

The matrix-valued function $\mathbf{g}(k)$ constitutes the bosonic equivalent of the Bloch Hamiltonian so often encountered in standard condensed matter settings. For this reason, we refer to $\mathbf{g}(k)$ as the *Bloch dynamical matrix*. In fact, $\mathbf{g}(k)$ provides exactly the blocks of $\mathbf{G}_N^{(P)}$ after performing the Fourier transform. Hence, the normal mode frequencies are precisely the normal mode frequencies of each $\mathbf{g}(k)$, with $k \in \mathcal{K}_N$. We call these eigenvalues $\omega_n(k)$, $n = 1, \ldots 2d_{\text{int}}$, the frequency bands. In the case where the system is dynamically stable, these are exactly the usual energy bands. However, $\omega_n(k)$ may become complex at one, or multiple k-values, yielding dynamical instabilities. Similarly, $\mathbf{g}(k)$ may lose diagonalizability at any given k value, engendering further instabilities.

The generalization of this discussion to a bi-infinite chain or, equivalently, bi-infinite BCs (BIBCs), follows in a straightforward manner. Firstly, the finite sum over j in Eq. (2.76) is replaced with a sum over all integers. The dynamical matrix, which we will denote as $\mathbf{G}^{\text{BIBC}}$, is then exactly as in Eq. (2.77), but with $\mathbf{V}_N$ replaced with $\mathbf{V}$. Mathematically, $\mathbf{G}^{\text{BIBC}}$ is called a *block-Laurent operator*. We must further take the continuum limit of the Fourier transform

$$\widetilde{\phi}_j = \sum_{j=-\infty}^{\infty} e^{-ijk} \phi_j, \quad k \in [-\pi, \pi], \quad \phi_j = \int_{-\pi}^{\pi} \frac{dk}{2\pi} e^{ijk} \widetilde{\phi}_k. \tag{2.81}$$

The key distinction from the finite case is that the Bloch dynamical matrix $\mathbf{g}(k)$ is now considered for all values of $k \in [-\pi, \pi]$, rather than just the discrete subset $\mathcal{K}_N \subset [-\pi, \pi]$. The frequency bands thus provide a continuum of normal mode frequencies.

2.5.2 Boundary Conditions and the Generalization of Bloch's Theorem

We have thus far characterized QBHs that possess translation symmetry. It is often the case that one is confronted with a system that is translationally invariant in the bulk, but possesses non-trivial BCs. The simplest example is a finite chain with isotropic couplings, up to hardwall boundaries. A more complicated example would be one where the link between sites 1 and N on a finite ring interact with slightly different couplings than the rest of the sites. In these cases we say that translational-symmetry is broken by BCs and that the system is instead just bulk translation-invariant.

In 1D QBHs, bulk-translation invariance can be characterized in terms of the dynamical matrix. One dimensional, bulk translation-invariant QBHs take the general form

$$H = \frac{1}{2}\sum_{r=1}^{R}\sum_{j=0}^{N-r} \phi_j^\dagger \mathbf{h}_r \phi_{j+r} + \sum_{b,b' \in B} \phi_b^\dagger \mathbf{w}_{bb'} \phi_{b'} + \text{H.c.} \tag{2.82}$$

where $B = \{1, \ldots, R, N - R + 1, \ldots, N\}$ are the boundary lattice sites, $\mathbf{h}_r$ are the bulk coupling matrices as before, and $\mathbf{w}_{bb'}$ are $2d_{\text{int}} \times 2d_{\text{int}}$ matrices specifying the BCs. In order for the boundary to be well-defined, we must have $R < N/2$. However, in practice, one often has $R \ll N$ (e.g., $R = 1$ for nearest neighbor couplings).

The dynamical matrix of such a system is explicitly given by

$$\mathbf{G} = \mathbf{G}_N^{\text{OBC}} + \mathbf{B}, \tag{2.83}$$

where $\mathbf{G}_N^{\text{OBC}}$ is the dynamical matrix of the system subject to open BCs (OBCs) $\mathbf{w}_{bb'} = 0$ and $\mathbf{B}$ encodes the BCs.

Let us first consider the OBC case $\mathbf{B} = 0$. Explicitly,

$$\mathbf{G} = \mathbf{G}_N^{\text{OBC}} = \mathbb{1}_N \otimes \mathbf{g}_0 + \sum_{r=1}^{R} \left(\mathbf{T}_N^r \otimes \mathbf{g}_r + \mathbf{T}_N^{\dagger\, r} \otimes \mathbf{g}_{-r} \right), \tag{2.84}$$

with $\mathbf{g}_r$ again given by $\boldsymbol{\tau}_3 \mathbf{h}_r$. Mathematically, the OBC dynamical matrix is a *block-Toeplitz matrix*. We note that by taking the semi-infinite limit (replacing $\mathbf{T}_N$ with $\mathbf{T}$), one obtains a *block-Toeplitz operator*, which we will denote generally by $\mathbf{G}^{\text{SIBC}}$. We summarize the dynamical matrix structure for the four most commonly encountered BCs (OBCs, PBCs, SIBCs, and BIBCs) in Table 2.1.

Table 2.1 The correspondence between the four most commonly encountered BCs and the structure of the dynamical matrix

Boundary conditions	Dynamical matrix type
Open	Block-Toeplitz matrix
Periodic	Block-circulant matrix
Semi-infinite	Block-Toeplitz operator
Bi-infinite	Block-Laurent operator

Allowing for non-trivial BCs, one finds that the contribution $\mathbf{B}$ is given by

$$\mathbf{B} = \begin{bmatrix} \mathbf{b}^{(l)}_{11} & \cdots & \mathbf{b}^{(l)}_{1R} & 0 & \mathbf{b}_{11} & \cdots & \mathbf{b}_{1R} \\ \vdots & \ddots & \vdots & \vdots & \vdots & \ddots & \vdots \\ \mathbf{b}^{(l)}_{R1} & \cdots & \mathbf{b}^{(l)}_{RR} & \vdots & \mathbf{b}_{R1} & \cdots & \mathbf{b}_{RR} \\ 0 & \cdots & \cdots & 0 & \cdots & \cdots & 0 \\ \widetilde{\mathbf{b}}_{11} & \cdots & \widetilde{\mathbf{b}}_{1R} & 0 & \mathbf{b}^{(r)}_{11} & \cdots & \mathbf{b}^{(r)}_{1R} \\ \vdots & \ddots & \vdots & \vdots & \vdots & \ddots & \vdots \\ \widetilde{\mathbf{b}}_{R1} & \cdots & \widetilde{\mathbf{b}}_{RR} & 0 & \mathbf{b}^{(r)}_{R1} & \cdots & \mathbf{b}^{(r)}_{RR} \end{bmatrix}, \tag{2.85}$$

with $\mathbf{b}^{(l)}_{bb'} = \boldsymbol{\tau}_3 \mathbf{w}_{bb'}$, $\mathbf{b}^{(r)}_{bb'} = \boldsymbol{\tau}_3 \mathbf{w}_{N-b+1,N-b'+1}$, $\mathbf{b}_{bb'} = \boldsymbol{\tau}_3 \mathbf{w}_{b,N-b'+1}$, and $\widetilde{\mathbf{b}}_{bb'} = \boldsymbol{\tau}_3 \mathbf{b}^{\dagger}_{bb'} \boldsymbol{\tau}_3$. Such a matrix is called a *corner modification* and transforms the block-Toeplitz matrix $\mathbf{G}^{\text{OBC}}_N$ into the more general corner-modified block-Toeplitz matrix.[18]

The main challenge of introducing arbitrary BCs is that there is no longer an obvious diagonalization or block-diagonalization scheme like provided by the Fourier transform in the translation-invariant case. In other words, Bloch's theorem is no longer valid. For finite systems, even the simplest case of block-Toeplitz matrices can have wildly unpredictable spectra when compared to their circulant counterparts. To this end, a procedure for diagonalizing corner-modified block-Toeplitz matrices has been developed [7] and used to generalize Bloch's theorem to arbitrary BCs in free fermionic systems [9]. This procedure will ultimately allow us to study the dependence of stability on BCs in later sections. We refer the reader to Appendix A.3 for an account of the basic aspects of this procedure.

We conclude with a remark on the special case of SIBCs. As previously mentioned, the relevant matrices are block-Toeplitz operators. Unlike the block-Toeplitz matrices, the spectra of these operators have been completely characterized [24, 25]. In fact, the SIBC spectra is determined entirely by the so-called *symbol*, namely, the analytic extension of the Bloch dynamical matrix into the complex plane. While the necessary tools will be introduced as necessary in later sections, we refer the reader to Appendix A for a detailed summary.

[18] More specifically, it is a *corner-modified banded block-Toeplitz matrix*. The "banded" here refers to the fact that the constant diagonals are banded around the diagonal with thickness $2R < N$. We will drop 'banded' henceforth, since it is the only case that arises in this thesis.

References

1. J.H.P. Colpa, Diagonalization of the quadratic bosonic hamiltonian. Physica A **93**, 327 (1978); "with zero modes: I. Mathematical" *ibid.* **134**, 377 (1986); "with zero modes: II. Physical" *ibid.* 417 (1986)
2. J.-P. Blaizot, G. Ripka, *Quantum Theory of Finite Systems* (MIT Press Cambridge, Cambridge, 1986), Chap. 2–3
3. R. Rossignoli, A.M. Kowalski, Phys. Rev. A **72**, 032101 (2005)
4. P.T. Nam, M. Napiórkowski, J.P. Solovej, J. Func. Anal. **270**, 4340 (2016)
5. J. Dereziński, J. Math. Phys. **58**, 121101 (2017)
6. K. Kustura, C.C. Rusconi, O. Romero-Isart, Phys. Rev. A **99**, 022130 (2019)
7. E. Cobanera, A. Alase, G. Ortiz, L. Viola, J. Phys. A **50**, 195204 (2017)
8. A. Alase, E. Cobanera, G. Ortiz, L. Viola, Phys. Rev. Lett. **117**, 076804 (2016)
9. A. Alase, E. Cobanera, G. Ortiz, Ly Viola, Phys. Rev. B **96**, 195133 (2017)
10. V.P. Flynn, E. Cobanera, L. Viola, New J. Phys. **22**, 083004 (2020)
11. V.P. Flynn, E. Cobanera, L. Viola, Phys. Rev. Lett. **127**, 245701 (2021)
12. V.P. Flynn, E. Cobanera, L. Viola, Phys. Rev. B **108**, 214312 (2023)
13. Hy Schulz-Baldes, Integr. Equ. Oper. Theory **78**, 323 (2014)
14. H. Schulz-Baldes, C. Villegas-Blas, J. Phys. A **49**, 405201 (2016)
15. V. Peano, Hy Schulz-Baldes, J. Math. Phys. **59**, 031901 (2018)
16. A. Royer, Phys. Rev. A **15**, 449 (1977)
17. G. Adesso, S. Ragy, A.R. Lee, Open Syst. Inf. Dyn. **21**, 1440001 (2014)
18. Jy Rosenberg, Contemp. Math. **365**, 331 (2004)
19. A.S. Wightman, Il Nuovo Cimento B **110**, 751 (1995)
20. C. Gerry, P. Knight, P.L. Knight, *Introductory Quantum Optics* (Cambridge University Press, 2005)
21. A. Mostafazadeh, J. Math. Phys. **43**, 205; *ibid.* 2814; *ibid.* 3944 (2002)
22. I. Gohberg, P. Lancaster, L. Rodman, *Indefinite Linear Algebra and Applications* (Springer, 2005)
23. Y. Kawaguchi, M. Ueda, Phys. Rep. **520**, 253 (2012)
24. L.N. Trefethen, M. Embree, *Spectra and Pseudospectra: The Behavior of Nonormal Matrices and Operators* (Princeton University Press, 2005)
25. A. Böttcher, B. Silbermann, *Introduction to Large Truncated Toeplitz Matrices* (Springer Science & Business Media, 2012)

Chapter 3
Dynamical Stability Phase Transitions

Abstract We present an in-depth analysis of the dynamical stability phase diagrams of QBHs. The first key observation is that boundaries separating dynamically stable and dynamically unstable phases are defined via two distinct flavors of spectral degeneracies: *EPs* and *Krein collisions* (KCs). These two flavors of degeneracy are distinguished by whether or not the relevant dynamical matrix is diagonalizable. This difference of origin will entail tangible differences in the dynamical and algebraic features of the normal modes at the transition point. Despite their differences, we are able to unify them through the lens of spontaneous *generalized PT-symmetry (GPT) breaking*. We connect this unifying perspective to the long-established, and rapidly growing, field of PT-symmetric quantum mechanics. Moreover, we prove equivalence between GPT-symmetry and pseudo-Hermiticity, a property intrinsic to the dynamical matrices of QBHs. Our dissection of dynamical stability phase boundaries utilizes the mathematical techniques of Krein stability theory, which we describe along the way as necessary. Putting these tools to use, we introduce a numerical indicator for dynamical stability phase transition known as *Krein phase rigidity* (KPR). Our development of this indicator, along with the understanding of its behavior, is inspired by the previously established algebraic features of bosonic normal modes in the vicinity of stability phase transition. Despite this boson-centric perspective, we find that it is, in fact, an extension of *phase rigidity* (a quantity relevant to the study of EPs in certain non-Hermitian systems) to the pseudo-Hermitian realm. Three example Hamiltonians are studied in detail: a single-mode toy model, a two-mode cavity QED model, and a bosonic Kitaev chain (BKC) under a family of BCs. In all cases, the stability phase diagrams are computed and boundaries analyzed. The GPT-symmetry breaking and behavior of the KPR are studied in detail. Further general features regarding phase-dependent transport in QBHs are elucidated by means of the BKC.

In this chapter, we develop a theory of dynamical stability phase transitions (henceforth, stability phase transitions) in closed, non-interacting bosonic systems (i.e., QBHs) and provide an in-depth analysis of the roles played by symmetries,

V. P. Flynn, *Effective Non-Hermiticity and Topology in Markovian Quadratic Bosonic Dynamics*, Springer Theses, https://doi.org/10.1007/978-3-031-52045-7_3

degeneracies, BCs, and system size.[1] The question that this chapter concerns itself with is the following: "How does one characterize, both mathematically and physically, the stability phase boundaries of a given QBH?". To answer this question, we introduce two key concepts. The first is the notion of *generalized parity-time* (PT) *symmetry*. This notion is inspired by the vast literature on non-Hermitian, but PT-symmetric, Hamiltonians that have been long-known to exhibit entirely real spectra [4, 5]. By suitably generalizing what is meant by "PT", we uncover that the dynamical matrices of QBHs (in fact, all pseudo-Hermitian matrices) possess such a symmetry, independently of the details of the system. Moreover, the spontaneous breaking of this symmetry coincides precisely with stability phase transitions. This observation allows us to characterize stability phase transitions in a novel, symmetry-breaking-oriented manner.

The second is the numerical indicator that we call *Krein phase rigidity*. This indicator, which is inspired by the notions of *phase rigidity* used in non-Hermitian quantum mechanics [6–8], is developed and introduced by bringing forth the mathematical infrastructure of *Krein stability theory*, i.e., the stability theory of linear time-invariant dynamical systems with pseudo-Hermitian generators. These tools provide us with a correspondence between stability phase boundaries and two types of spectral degeneracies in bosonic dynamical matrices: *exceptional points* and *Krein collisions*. With this correspondence in hand, we craft the Krein phase rigidity by demanding physically motivated algebraic relationships (namely, canonical, or suitability generalized, commutations relations) between the normal modes of our system and studying the behavior of the associated Nambu space vectors. This culminates in the demonstration that the Krein phase rigidity should always vanish at stability phase transitions. We summarize the key components of this theory of stability phase transitions as follows:

(i) All QBHs possess an underlying generalized PT symmetry;
(ii) Stability phase transitions are associated to the breaking of this symmetry; and
(iii) Stability phase transitions are detected by the vanishing of a new type of "phase rigidity" that we call Krein phase rigidity.

We put this general theory to the test in three models. The first model consists of a single mode exhibiting both harmonic and inverted oscillator stability phases. We identify the generalized PT symmetry as time-reversal and proceed to compute the Krein phase rigidity analytically throughout the 2-parameter phase diagram. The second model is derived from the cavity QED Hamiltonian introduced in Ref. [9] and consists of a single magnonic mode coupled to a photonic degree of freedom. We again identify the generalized PT symmetry as time-reversal and completely characterize the stability phase boundaries. We confirm that the Krein phase rigidity vanishes as expected. Finally, we conclude with an extensive analysis of the bosonic Kitaev chain (introduced in Ref. [10]) generalized to allow for a two-parameter

[1] *The vast majority of content in this chapter originates in Ref. [1]. The pseudospectral perspective on the BKC presented at the conclusion of the chapter stems from the later works [2] and [3]. Each of these works were jointly co-authored with Emilio Cobanera & Lorenza Viola.*

family of non-trivial BCs. Following a detailed normal mode analysis (which constitutes the first application of the generalization of Bloch's theorem to a bosonic system), we completely characterize both the bulk and boundary parameter stability phase diagrams. One remarkable consequence of this analysis is the following: the BCs for which the BKC is dynamically stable are *precisely* those that host Majorana fermion edge modes in the fermionic Kitaev chain. Along the way, we will further uncover a previously unknown connection between phase dependent transport, time-reversal symmetry, squeezing invariance, and unstable equilibria. As with the previous models, we successfully demonstrate the utility of the Krein phase rigidity as an indicator of stability phase transitions. We are lead to two conjectures regarding the generic features of stability phase transitions in many-mode systems, including an intriguing connection to the emergence of zero modes.

The outline of this chapter is as follows. In Sect. 3.1.1 we introduce the notion of generalized PT symmetry, establish its equivalence with pseudo-Hermiticity, and discuss the physical implications of its presence and spontaneous breaking in QBHs. In Sect. 3.2 we introduce the necessary mathematical tools of Krein stability theory and leverage them, in addition to the concept of spontaneous generalized PT symmetry-breaking, to classify the stability phase boundaries of QBHs in terms of spectral degeneracies (namely, exceptional points and Krein collisions) in the dynamical matrix. This leads to our introduction of Krein phase rigidity as a physically motivated numerical indicator of stability phase transitions. Finally, we conclude the chapter with a detailed analysis of a single-mode model, a two-mode model derived from a cavity QED system, and a generalization of the many-mode bosonic Kitaev chain. The bosonic Kitaev chain analysis, in particular, contains several general results regarding phase-dependent transport and the roles of time reversal symmetry and topology in QBHs.

3.1 Spontaneous Generalized PT Symmetry-Breaking

3.1.1 The Equivalence Between Pseudo-Hermiticity and Generalized PT Symmetry

In recent decades there has been a considerable interest in reexamining the self-adjointness constraint enforced on quantum Hamiltonians. One such reexamination has considered replacing the self-adjointness assumption with one of parity-time (PT) symmetry which originates from the observation that the non-Hermitian Hamiltonian $H = p^2 + x^2 + ix^3$ has an entirely real and positive spectrum [4]. This fact is attributed to the presence of PT symmetry, i.e., invariance under the simultaneous action of parity P taking $(x, p) \mapsto (-x, -p)$ and the antilinear time-reversal $\mathcal{T}$ taking $(zx, zp) \mapsto (z^*x, -z^*p)$, with $z \in \mathbb{C}$. In particular, it is simple to show that the eigenvalues of a non-Hermitian Hamiltonian that commutes with $P\mathcal{T}$ necessarily come in complex-conjugate pairs. These observations sparked a multitude of investigations into the role PT symmetry plays in ensuring an entirely real spectrum [5, 11]. Ultimately, reality of the spectrum depends on whether PT symmetry is broken or unbroken. A Hamiltonian has *unbroken* PT symmetry if there exists a basis of simultaneous eigenstates of H and $P\mathcal{T}$. PT symmetry is *broken*, otherwise. Immediately, we see that unbroken PT symmetry implies that each eigenvalue is real, i.e., if $H\psi = E\psi$ and $P\mathcal{T}\psi = \lambda\psi$, then $0 = [H, P\mathcal{T}]\psi = \lambda(E - E^*)\psi$. Since invertibility of $P\mathcal{T}$ ensures $\lambda \neq 0$, we conclude that $E = E^*$. Hence, a broken PT symmetry implies that either (1) there exists a non-real eigenvalue of H; or (2) there is no basis of eigenvectors to begin with (i.e., H is not diagonalizable).

In a similar vein, a comparable degree of interest in pseudo-Hermitian systems began to emerge due to a similar spectral constraint. Eigenvalues of pseudo-Hermitian operators come in complex conjugate pairs with entirely real spectra being a particular instance of this. Naturally, a slew of connections between the two types of non-Hermitian operators was established. In the following, PT symmetry is defined for finite dimensional systems (i.e., matrices) in a natural way: an $n \times n$ complex matrix $\mathbf{M}$ is said to be PT-symmetric if it commutes with an antilinear operator of the form $\mathbf{P}\mathcal{T} : \mathbb{C}^n \to \mathbb{C}^n$, with $\mathbf{P}$ a linear involution ($\mathbf{P}^2 = \mathbb{1}_n$) and $\mathcal{T}$ complex conjugation with respect to the canonical basis of $\mathbb{C}^n$. Equivalently, $\mathbf{PMP}^{-1} = \mathbf{M}^*$. For example, every real matrix is PT-symmetric with $\mathbf{P} = \mathbb{1}_n$.

- Every PT-symmetric matrix is pseudo-Hermitian [12].
- A diagonalizable operator is pseudo-Hermitian if and only if it commutes with an antilinear invertible mapping [13]. If, in addition, the spectrum of the operator is real and discrete, then there is a basis in which this anti-linear mapping can be decomposed as a product of an involutory linear operator (i.e., a linear operator that squares to the identity) and complex conjugation with respect to this basis.
- An operator commutes with an involutory antilinear operator if and only if it is weakly pseudo-Hermitian (a generalization of pseudo-Hermiticity that coincides with standard pseudo-Hermiticity when the spectrum is discrete) [14, 15].

While this web of relations is appealing, it evades the simplicity of stating a precise equivalence between PT-symmetric and pseudo-Hermitian systems. The key to establishing such an equivalence is by expanding the basis-dependent definition of PT-symmetric matrices to a more general notion of PT symmetry that applies to linear operators.

Definition 3.1.1 (GPT Symmetry) A linear transformation $\mathbf{M}$ has a *generalized PT symmetry*, or equivalently, is *GPT-symmetric* if there exists (1) an invertible antilinear map Θ such that $[\mathbf{M}, \Theta] = 0$; and (2) a basis relative to which $\Theta = \mathbf{P}\mathcal{T}$, with $\mathbf{P}$ an involutory linear map and $\mathcal{T}$ complex conjugation with respect to this basis.

Clearly, every PT-symmetric matrix is automatically GPT-symmetric with respect to the canonical basis. The notion of broken and unbroken PT symmetry extend naturally as well: a GPT-symmetric operator **M** has unbroken GPT symmetry Θ if there is a simultaneous eigenbasis of **M** and Θ. As in the PT-symmetric case, those GPT-symmetric operators with an unbroken GPT symmetry are precisely those that are diagonalizable with entirely real spectra. Ultimately, this generalization allows for the following proposition.

Proposition 3.1.2 *Let* **M** *denote a linear transformation on a finite-dimensional Hilbert space. Then* **M** *is pseudo-Hermitian if and only if* **M** *is GPT-symmetric.*

Proof One direction of the proof, namely that GPT symmetry implies pseudo-Hermiticity, follows from the fact that any linear operator that commutes with an invertible antilinear operator is necessarily pseudo-Hermitian [15].

For the other direction, suppose that **M** is pseudo-Hermitian. We begin by constructing a basis of eigenvectors and generalized eigenvectors of **M**. We build up this basis in the following way:

- For each real eigenvalue λ_j, with $j = 1, \ldots, \alpha$ of **M**, we denote the generalized eigenvector of rank k by $\vec{v}_{jk}$, with $k = 1, \ldots, p_j$ and p_j the length of the associated Jordan chain.
- For each non-real eigenvalue μ_j, with $\mathrm{Im}(\mu_j) > 0$ and $j = 1, \ldots, \beta$ of **M**, we denote the generalized eigenvectors of rank k by $\vec{w}_{jk}$, with $k = 1, \ldots, r_j$ and r_j the length of the associated Jordan chain.
- For each pair (μ_j, r_j) in the above, there is an eigenvalue μ_j^* with Jordan chain of equal length r_j (this follows from pseudo-Hermiticity, see e.g., Prop. 4.2.3 in [16]). We denote these generalized eigenvectors of rank k by $\vec{w}'_{jk}$, with $k = 1, \ldots, r_j$.

The set $B = \{\vec{v}_{jk}, \vec{w}_{jk}, \vec{w}'_{jk}\}$ is then a basis of $\mathbb{C}^n$ consisting of generalized eigenvectors of **M**. As we will show, **M** is GPT-symmetric with respect to this basis.

First, define an operator **P** on B according to

$$\mathbf{P}\vec{v}_{jk} \equiv \vec{v}_{jk}, \quad \mathbf{P}\vec{w}_{jk} \equiv \vec{w}'_{jk}, \quad \mathbf{P}\vec{w}'_{jk} = \vec{w}_{jk}, \tag{3.1}$$

which is evidently involutory. We then take $\Theta \equiv \mathbf{P}\mathcal{T}$, with $\mathcal{T}$ complex-conjugation with respect to B. Immediately, we have $[\mathbf{M}, \Theta]\vec{v}_{jk} = 0$ for all j and k. Furthermore,

$$\mathbf{M}\Theta\vec{w}_{jk} = \mathbf{M}\vec{w}'_{jk} = \mu_j^*\vec{w}'_{jk} + \vec{w}'_{j(k-1)} = \Theta\mathbf{M}\vec{w}_{jk}, \tag{3.2}$$

where we have used the Jordan chain identity $(\mathbf{M} - \mu_j^*)\vec{w}'_{jk} = \vec{w}'_{j(k-1)}$, with $\vec{w}'_{j(k-1)} = 0$ for $k = 1$. A similar calculation additionally shows $[\mathbf{M}, \Theta]\vec{w}'_{jk} = 0$. Altogether, we have demonstrated that **M** and Θ commute on all of B, and thus commute as operators. $\square$

Notably, in the second direction of the proof, we have presented a constructive method for realizing a particular instance of a GPT symmetry given an arbitrary pseudo-Hermitian operator. In light of this, three remarks are in order.

(i) The operator Θ explicitly constructed in the proof is generally dependent on system parameters. This follows from the fact that the definition of $\mathbf{P}$ directly depends on the basis of generalized eigenvectors B. Thus, if system parameters change, Θ may change accordingly.
(ii) This Θ need not be the only GPT symmetry of the system. That is, there may exist multiple antilinear operators with the appropriate properties, that commute with $\mathbf{M}$. We will see an example of a (parameter-independent) GPT symmetry that is distinct from the above constructed Θ in Sect. 3.3.2.
(iii) It is possible to prove this direction in a shorter, but non-constructive manner. Theorem 3 in Ref. [15] asserts that every pseudo-Hermitian operator has an associated antilinear operator that commutes with it. Such an antilinear operator can always be expressed as conjugation with respect to some basis [17]. This conjugation (paired with $\mathbf{P} = \mathbb{1}_n$) then constitutes a GPT symmetry. While this proof does not specify the basis, it does reveal that we may always find at least one GPT symmetry with $\mathbf{P} = \mathbb{1}_n$. Again, the GPT symmetry in Sect. 3.3.2 will exemplify this point.

3.1.2 Generalized PT Symmetry in Quadratic Hamiltonians

We will now apply the preceding results to QBHs. Recalling that the dynamical matrices of QBHs are always pseudo-Hermitian, i.e., $\boldsymbol{\tau}_3\mathbf{G}^\dagger\boldsymbol{\tau}_3 = \mathbf{G}$, we have the following theorem.

Theorem 3.1.3 *The dynamical matrices of QBHs are always GPT-symmetric. Furthermore, a given QBH is dynamically stable if and only if there is an unbroken GPT symmetry.*

These facts follow as straightforward consequences of the previous section. We further remark that this theorem holds also for fermions. Each fermionic dynamical matrix (which coincides with the Bogoliubov-de Gennes matrix) is pseudo-Hermitian with metric $\boldsymbol{\eta} = \mathbb{1}_{2N}$, i.e., they are simply Hermitian. The GPT symmetry is then just conjugation with respect to the normal mode basis. In comparison, bosonic systems offer a larger degree of complexity. Since both complex eigenvalues and nontrivial Jordan chains may arise, unbroken GPT symmetry is a possibility. Furthermore, we see that all of the properties of dynamically stable systems (e.g., the existence of a quasiparticle vacuum) are equivalent to the presence of an unbroken GPT symmetry.

3.2 Classification and Detection of Stability Phase Transitions

The introduction of GPT symmetry allows us to draw close analogy between stability phase transitions of QBHs and traditional phase transitions in statistical mechanics. Namely, the transitions between two phases are marked by breaking of a certain symmetry. In the traditional statistical mechanics case, it is a physical symmetry of the Hamiltonian (e.g., rotational symmetry breaking in ferromagnetic phase transitions), while in our case, it is a symmetry of the dynamical matrix of a given QBH. Along this line of thinking, it is natural to pursue indicators of GPT symmetry breaking just as is typical in traditional phase transitions. To accomplish this, let us reflect on exactly what happens when GPT symmetry is broken. GPT symmetry is broken by either (1) the loss of diagonalizability (equivalently, the system is at an *exceptional point* (EP) in phase space) or (2) the presence of a non-real eigenvalue, regardless of diagonalizability. Any indicator we propose must be sensitive to either of these occurrences.

In the study of symmetric non-Hermitian matrices, i.e., those non-Hermitian matrices that satisfy $\mathbf{M}^T = \mathbf{M}$, an indicator known as *phase rigidity* (PR) has been developed and extensively used for detecting EPs [6–8]. To understand phase rigidity, we must introduce the concept of *biorthogonal bases*. Given an arbitrary diagonalizable matrix $\mathbf{M}$, one can always produce two bases $\{\vec{v}_j^R\}$ and $\{\vec{v}_j^L\}$, where $\vec{v}_j^{R(L)}$ are right (left) eigenvectors of $\mathbf{M}$ with eigenvalues λ_j. Importantly, it is possible to ensure biorthonormality, i.e., $\vec{v}_j^{L\dagger}\vec{v}_k^R = \delta_{jk}$. This allows for a decomposition of the form $\mathbf{M} = \sum_j \lambda_j \vec{v}_j^R \vec{v}_j^L$. In the case of symmetric complex matrices, it always is possible to construct a biorthonormal basis with $\vec{v}_j^L = (\vec{v}_j^R)^*$ by virtue of $(\vec{v}_j^R)^T\mathbf{M} = (\mathbf{M}\vec{v}_j^R)^T = \lambda_j \vec{v}_j^{RT}\mathbf{M}$. With this, the PR of $\vec{v}_j^R$ is defined as

$$\rho_j = \frac{(\vec{v}_j^R)^T \vec{v}_j^R}{\vec{v}_j^{R\dagger}\vec{v}_j^R} = \frac{1}{||\vec{v}_j^R||^2}, \quad \rho_j \in [0, 1], \tag{3.3}$$

where $||\vec{v}||^2 = \vec{v}^\dagger \vec{v}$ is the usual 2-norm. Importantly, this quantity vanishes smoothly as $\mathbf{M}$ approaches an EP. We cannot immediately lift PR to bosonic matrices because they are generally not symmetric. Even if we had a dynamical matrix that was symmetric, the PR need not detect those stability phase transitions that retain diagonalizability throughout. In order to address these issues, we aim to extend PR to the realm of dynamical matrices and assess its general utility in detecting stability phase transitions. This will be accomplished by first studying more closely those LTI systems whose generators are pseudo-Hermitian and, specifically, by dissecting the precise mechanisms for GPT symmetry-breaking.

3.2.1 Tools from Krein Stability Theory and Their Implications for QBHs

The stability theory of LTI systems with pseudo-Hermitian generators is known as *Krein stability theory*. Krein stability theory[2] concerns itself, in particular, with equations of the form

$$\frac{d}{dt}\vec{v}(t) = i\mathbf{M}\vec{v}(t), \tag{3.4}$$

with $\mathbf{M}$ an $n \times n$ pseudo-Hermitian matrix with metric $\boldsymbol{\eta}$. Of course, this is precisely the type of equation describing the Nambu space dynamics generated by a given QBH, see e.g., Eq. (2.52). Such systems are dynamically stable (in the sense of bounded evolution) if and only if $\mathbf{M}$ is diagonalizable and has an entirely real spectrum. Krein stability theory asks (and answers) the question "how far away is a given dynamically stable system from becoming dynamically unstable?". Equivalently, how small of a perturbation is needed to destabilize the system. To answer, we must examine more closely the Krein signature distribution of the eigensystem of $\mathbf{M}$.

Recalling that the *Krein signature* of a vector $\vec{v}$ is given by $\text{sgn}(\vec{v}^\dagger\boldsymbol{\eta}\vec{v})$ (with $\text{sgn}(0) = 0$ by convention), we can classify the eigenspaces $\mathcal{E}_\lambda$ of $\mathbf{M}$, $\lambda \in \sigma(\mathbf{M})$, as follows. If every vector $\vec{v}$ in a given eigenspace $\mathcal{E}_\lambda$ has a positive (negative) Krein signature, then we say $\mathcal{E}_\lambda$ is *$\boldsymbol{\eta}$-definite*. Otherwise, it is called *$\boldsymbol{\eta}$-indefinite*. Firstly, it is clear that $\mathcal{E}_\lambda$ can only be $\boldsymbol{\eta}$-definite if $\lambda \in \mathbb{R}$. To see this, note that if $\mathbf{M}\vec{v} = \lambda\vec{v}$, with $\text{Im}(\lambda) \neq 0$, then

$$\lambda\vec{v}^\dagger\boldsymbol{\eta}\vec{v} = \vec{v}^\dagger\boldsymbol{\eta}\mathbf{M}\vec{v} = \vec{v}^\dagger\mathbf{M}^\dagger\boldsymbol{\eta}\vec{v} = \lambda^*\vec{v}^\dagger\boldsymbol{\tau}_3\vec{v}. \tag{3.5}$$

Since $\lambda \neq \lambda^*$, we must have $\vec{v}^\dagger\boldsymbol{\tau}_3\vec{v} = 0$. This calculation also shows that nondegenerate real eigenvalues engender $\boldsymbol{\eta}$-definite eigenspaces.

Eigenspaces corresponding to real eigenvalues can still be $\boldsymbol{\eta}$-indefinite, however. For instance, if λ becomes degenerate with (at least) two eigenvectors consisting of opposite Krein signatures. This motivates the following definition.

Definition 3.2.1 (Krein Collision) Let $\mathcal{E}_\lambda$ denote an $\boldsymbol{\eta}$-indefinite eigenspace of a given pseudo-Hermitian matrix. We say there is a *Krein collision* (KC) at λ if there exists both an eigenvector for λ with a Krein signature of $+1$ and an eigenvector for λ with a Krein signature of -1.

[2] Krein stability theory deals with more than just pseudo-Hermitian LTI systems. For example, it can be applied even in the presence of time-periodic coefficients [18]. These advanced techniques are beyond the scope of our needs.

Less obvious is the fact that EPs also lead to η-indefiniteness of the associated eigenspace. This follows as a particular case of the following cornerstone result in Krein stability theory (adapted from Ch. III of Ref. [18]).

Theorem 3.2.2 (Krein-Gel'fand-Lidskii) *Let* $\mathbf{M}_0$ *denote a pseudo-Hermitian matrix with metric* η *and let* λ_0 *denote a real eigenvalue of* $\mathbf{M}_0$. *Further, let* $\|\cdot\|$ *be any matrix norm.*

(a) If $\mathcal{E}_{\lambda_0}$ *is* η*-definite, then all Jordan chains associated to* λ_0 *are of length one. In addition, there exists* $\epsilon, \delta > 0$ *such that if* $\mathbf{M}$ *is pseudo-Hermitian with metric* η *and* $\|\mathbf{M} - \mathbf{M}_0\| < \delta$, *then all the eigenvalues* λ *of* $\mathbf{M}$ *such that* $|\lambda - \lambda_0| < \epsilon$ *are real and correspond to Jordan chains of length one.*

(b) If $\mathcal{E}_{\lambda_0}$ *is* η*-indefinite and all the Jordan chains associated to* λ_0 *are of length one, then for every* $\epsilon > 0$ *there exists a pseudo-Hermitian matrix* $\mathbf{M}$ *with metric* η *possessing non-real eigenvalues in an open neighborhood of* λ_0 *and such that* $\|\mathbf{M} - \mathbf{M}_0\| < \epsilon$.

These results reveal the fact that spectral degeneracies are necessary, but not sufficient for a normal mode to become unstable. In particular, if the degeneracy preserves η-definiteness of the eigenspace, (a) tells us that any sufficiently small perturbation will not destabilize the system. On the contrary, if η-definiteness is lost either through a KC or an EP, then the associated normal modes can either be destabilized by arbitrarily small perturbations (in the case of a KC) or they are already unstable (in the case of an EP). These observations are recorded in Table 3.1. Let us now return our focus to QBHs. The relevance of Theorem 3.2.2 had been noticed in Refs. [19, 20] (see also Ref. [21] for earlier related, albeit less general, results). For us, its major power is allowing us to completely characterize the stability phase boundaries (equivalently, points of spontaneous GPT symmetry breaking). Consider a QBH $H(\vec{p})$ with dynamical matrix $\mathbf{G}(\vec{p})$ depending continuously on some parameters $\vec{p}$. Suppose we trace out a path in parameter space $\vec{p}(s)$, with $0 < s < 1$ that crosses a stability phase boundary. That is, there is an s_c such that the system is dynamically stable for $s < s_c$ and dynamically unstable for $s > s_c$. The system is either dynamically stable or unstable at $\vec{p}(s_c) = \vec{p}_c$. In the former case, it must be that an arbitrarily small perturbation (namely $s_c \mapsto s_c + \delta s$,

Table 3.1 The stability properties of eigenspaces corresponding to real eigenvalues of a pseudo-Hermitian matrix in light of Theorem 3.2.2. The first column is the type of degeneracy. The second column assesses the definiteness of the corresponding eigenspace. The third column specifies the dynamics of the associated eigenvectors, i.e., the normal modes associated to λ. The final column assesses the response to perturbations. In particular, No means that arbitrarily small perturbations can destabilize the normal modes

Real eigenvalue	Eigenspace	Normal modes	Stable to perturbations
Non-degenerate	η-definite	Stable	Yes (sufficiently small)
Degenerate, neither KC or EP	η-definite	Stable	Yes (sufficiently small)
Degenerate, KC	η-indefinite	Stable	No
Degenerate, EP	η-indefinite	Unstable	No

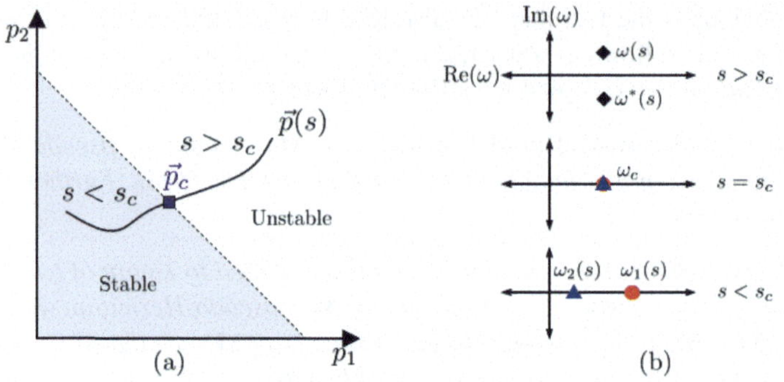

Fig. 3.1 (**a**) An example stability phase transition in a QBH that depends on two real parameters p_1 and p_2. (**b**) The spectral flow around a phase boundary that hosts a KC. The red circle, blue triangle, and black diamond indicate eigenvectors of Krein signature 1, −1, and 0 respectively. Earlier versions of these figures first appeared in Ref. [1]

with δs arbitrarily small) can render our stable system unstable. Hence, $\vec{p}_c$ must be a KC. A typical instance of this phenomena is illustrated in Fig. 3.1. In the latter case, it must be that we are at a real eigenvalue EP.[3] If this were not the case, and the system were still unstable at $s = s_c$, then there would an eigenvalue $\lambda(s)$ that discontinuously develops a non-zero imaginary part going from $s_c - \delta s$ to $s = s_c$, i.e., $\lim_{\delta s \to 0} |\lambda(s_c) - \lambda(s - \delta s)| > |\mathrm{Im}(\lambda(s_c))| > 0$. This violates continuity. The culmination of this analysis is as follows.

Theorem 3.2.3 *A QBH undergoes a stability phase transition only if, at the phase boundary, the associated dynamical matrix hosts a KC or an EP at a real eigenvalue.*

Equivalently, we have obtained the necessary spectral conditions for the spontaneous breaking of GPT symmetry. We can now leverage this to develop a proper extention of PR that we will ultimately use to *numerically* detect stability phase transitions in QBHs.

3.2.2 *Krein Phase Rigidity*

With this concrete understanding of the nature of stability phase transitions in hand, we introduce our proposed indicator as follows.

Definition 3.2.4 Consider a dynamical matrix **G** and let $\vec{\psi}$ denote an eigenvector corresponding to an eigenvalue ω, such that $\vec{\psi}$ is not in the range of $\mathbf{G} - \omega \mathbb{1}_{2N}$.

[3] Here, and henceforth, when we say that the eigenvalue itself is, or hosts, an EP, we mean that the matrix is at an EP and this particular eigenvalue hosts a non-trivial Jordan chain.

Then there exists an eigenvector $\vec{\psi}_*$ corresponding to an eigenvalue ω^*, that may be normalized to satisfy $\vec{\psi}_*^\dagger \boldsymbol{\tau}_3 \vec{\psi} = 1$. The KPR (KPR) of $\vec{\psi}$ is then the quantity

$$r \equiv \frac{\vec{\psi}_*^\dagger \boldsymbol{\tau}_3 \vec{\psi}}{\|\vec{\psi}_*\| \, \|\vec{\psi}\|} = \frac{1}{\|\vec{\psi}_*\| \, \|\vec{\psi}\|}, \quad r \in [0, 1]. \tag{3.6}$$

Firstly, we must establish that KPR is well-defined; namely, that such a vector $\vec{\psi}_*$ always exists. The conditions set on $\vec{\psi}$ imply that one can construct a biorthogonal basis[4] containing $\vec{\psi}$, with a biorthogonal partner $\vec{\phi}$ satisfying $\mathbf{G}^\dagger \vec{\phi} = \omega^* \vec{\phi}$ and $\vec{\phi}^\dagger \vec{\psi} = 1$. Pseudo-Hermiticity then implies that $\vec{\psi}_* \equiv \boldsymbol{\tau}_3 \vec{\phi}$ satisfies all the required properties. Finally, the fact that $r \in [0, 1]$ follows immediately from the Cauchy-Schwarz inequality.

Several remarks are then in order.

(i) If $\psi \equiv \widehat{\vec{\psi}}$ and $\psi_* \equiv \widehat{\vec{\psi}}_*$ are the associated normal modes of H, then the normalization condition is equivalent to the algebraic requirement $[\psi_*, \psi^\dagger] = 1$. Depending on whether ω is real, purely imaginary, or fully complex, this condition is akin to demanding a bosonic, Heisenberg-Weyl, or pseudo-bosonic commutation relations, respectively (see Sect. 2.4). Specifically, if ω is real, we have $\vec{\psi}_* = \kappa \vec{\psi}$, with κ the Krein signature of $\vec{\psi}$. While this interpretation is physically appealing, we note that r is, in fact, invariant under any scaling transformation $\vec{\psi} \mapsto z\vec{\psi}$, $\vec{\psi}_* \mapsto w\vec{\psi}_*$. By taking $|z|^2 = \|\vec{\psi}_*\|/\|\vec{\psi}\|$ and $w = 1/z^*$, we can compute r as $r = 1/\|\vec{\psi}\|^2$, which is a formula identical to the standard PR.

(ii) The definition of KPR can be immediately generalized to any pseudo-Hermitian system by replacing $\boldsymbol{\tau}_3$ with the general indefinite metric $\boldsymbol{\eta}$. In fact, it may be generalized to any non-Hermitian matrix by replacing $\boldsymbol{\tau}_3 \vec{\psi}_*$ with the biorthogonal partner of $\vec{\psi}$. However, like in the case of the standard PR, the specialization to a particular class of non-Hermitian matrices (for KPR, pseudo-Hermitian matrices, and for PR, complex symmetric matrices) allows for a direct relationship between $\vec{\psi}$ and its biorthogonal partner to be made (for KPR, $\vec{\phi} = \boldsymbol{\tau}_3 \vec{\psi}_*$, and for PR, $\vec{\phi} = \vec{\psi}^*$).

(iii) If $\mathbf{G}$ happens to also be complex symmetric, the KPR generically reduces to the PR. To see this, consider the simplest case of a non-degenerate, real eigenvalue ω. For the KPR, we have that $\vec{\psi}_* = \kappa \vec{\psi}$, and so $r = 1/\|\vec{\psi}\|^2$. We also see that the symmetric condition along with pseudo-Hermiticity ensure that $\boldsymbol{\tau}_3 \vec{\psi}^*$ is also an eigenvector corresponding to ω, and so $\vec{\psi}^* = \mu \boldsymbol{\tau}_3 \vec{\psi}$. Now,

$$|\mu|^2 \kappa = (\vec{\psi}^*)^\dagger \boldsymbol{\tau}_3 \vec{\psi}^* = \left(\vec{\psi}^\dagger \boldsymbol{\tau}_3 \vec{\psi}\right)^* = \kappa, \tag{3.7}$$

[4] While we have not specified diagonalizability of $\mathbf{G}$, the condition that $\vec{\psi}$ lie outside of the range of $\mathbf{G} - \omega \mathbb{1}_{2N}$ ensures that ω hosts only length one Jordan chains.

so that $|\mu| = 1$. In conclusion, $\vec{\psi}$ satisfies both the KPR normalization condition $\vec{\psi}_*^\dagger \boldsymbol{\tau}_3 \vec{\psi} = \kappa^2 = 1$ and the PR normalization condition $(\vec{\psi}^*)^\dagger \vec{\psi} = 1$. Hence $\rho = r$.

The KPR, thus far, only supplies us with a particular numerical quantity to associate to a given eigenvector of $\mathbf{G}$. Its utility as a stability phase transition indicator comes from the following claim.

Claim 3.2.5 *Let* $\mathbf{G}(\vec{p})$ *be a bosonic dynamical matrix depending smoothly on a set of parameters* $\vec{p}$. *If* $\mathbf{G}(\vec{p})$ *undergoes a stability phase transition along a smooth path* $\vec{p}(s)$, *then there exists an eigenvector of* $\mathbf{G}$ *whose KPR vanishes. Equivalently,* $\mathbf{G}(\vec{p})$ *hosts eigenvectors of arbitrarily small KPR in any open neighborhood of* $\vec{p}_c$.

Heuristic Justification Using the same notation as in the context of Fig. 3.1, let $\vec{p}_c = \vec{p}(s_c)$ denote the stability transition point separating the stable phase $s < s_c$ from the unstable phase $s > s_c$. The results of Sect. 3.2.1 allow us to conclude that $\mathbf{G}(\vec{p}_c)$ hosts either a KC or an EP. For simplicitly, let us assume that the degeneracy (algebraic multiplicity) of the KC or EP is 2. We will handle the case of a KC first.

For $s < s_c$, there must be two real eigenvalues of $\mathbf{G}(\vec{p}(s))$, denoted by $\omega_1(s)$ and $\omega_2(s)$, that deform smoothly into the degenerate eigenvalue $\omega_c = \omega_j(s_c)$, $j = 1, 2$. They then split smoothly into a complex conjugate pair $\omega_1(s) = \omega_2(s)^*$ for $s > s_c$ (this spectral flow is depicted heuristically in Fig. 3.1b). Denoting the associated eigenvectors by $\vec{\psi}_j(s)$, $j = 1, 2$, we have that $\kappa_1 = -\kappa_2$, with κ_j the Krein signature of $\vec{\psi}_j(s)$ for $s \leq s_c$. Moreover, we can assume $\boldsymbol{\tau}_3$-orthogonality, $\vec{\psi}_1^\dagger(s)\boldsymbol{\tau}_3\vec{\psi}_2(s) = 0$ for $s < s_c$. The KPR in the stable phase can then be written as

$$r_j(s) = \frac{\vec{\psi}_j^\dagger(s)\boldsymbol{\tau}_3\vec{\psi}_j(s)}{\|\vec{\psi}_j(s)\|^2} = \hat{\psi}_j^\dagger(s)\boldsymbol{\tau}_3\hat{\psi}_j(s), \quad \hat{\psi}_j(s) \equiv \frac{1}{\|\vec{\psi}_j(s)\|}\vec{\psi}_j(s), \quad s < s_c. \tag{3.8}$$

In the above, we have introduced the conventionally normalized eigenvectors $\hat{\psi}_j(s)$. Recalling that eigenvectors corresponding to complex eigenvalues always have vanishing $\boldsymbol{\tau}_3$-norms, smoothness ultimately dictates that $\vec{\psi}_j(s)$, and hence $\hat{\psi}_j(s)$, evolve into Krein signature 0 eigenvectors for $s > s_c$. Thus we have

$$\lim_{s \to s_c^-} r_j(s) = 0. \tag{3.9}$$

For $s > s_c$, we have that $\vec{\psi}_1(s)$ must be the biorthogonal partner to $\vec{\psi}_2(s)$ for $s > s_c$. That is, $\vec{\psi}_1^\dagger(s)\boldsymbol{\tau}_3\vec{\psi}_2(s)$ may be normalized to 1 for $s > s_c$. Thus,

$$r_1(s) = r_2(s) = \frac{\vec{\psi}_1^\dagger(s)\boldsymbol{\tau}_3\vec{\psi}_2(s)}{\|\vec{\psi}_1(s)\|\,\|\vec{\psi}_2(s)\|} = \hat{\psi}_1^\dagger(s)\boldsymbol{\tau}_3\hat{\psi}_2(s),$$

$$\hat{\psi}_j(s) \equiv \frac{1}{\|\vec{\psi}_j(s)\|}\vec{\psi}_j(s), \quad s > s_c, \tag{3.10}$$

where we have again used the conventionally normalized eigenvectors. Since $\vec{\psi}_1(s)$ and $\vec{\psi}_2(s)$, and hence $\hat{\psi}_1(s)$ and $\hat{\psi}_2(s)$, evolve into mutually $\boldsymbol{\tau}_3$-orthogonal eigenvectors, we conclude

$$\lim_{s \to s_c^+} r_j(s) = 0. \tag{3.11}$$

Altogether, the KPR will vanish at the KC.

The argument for the EP is similar, except now we have that $\vec{\psi}_j(s)$ converge to (a multiple of) the same eigenvector, $\vec{\chi} \propto \vec{\psi}_j(s)$ at $s = s_c$, regardless of the side of approach. By Theorem 3.2.2(a),[5] this eigenvector must have vanishing Krein signature. So, following nearly identical arguments as before, the KPR will again vanish from both sides.

We conclude by remarking that we omit a formal proof of this claim in order to avoid an unnecessarily over-technical discussion about the smoothness assumption and its consequences on the parameter dependence of eigenvectors and eigenvalues. It is known, for instance, that eigenvalues can depend nonanalytically on parameters in the neighborhood of EPs (see, e.g., Ch. 2, Sec. 1.5 of Ref. [22]). Despite this, we believe that the claim will hold generically even if one relaxes the assumption of smoothness to one of differentiability. Further technical complications could arise by considering higher order degeneracies. These technical aspects of a formal proof will be left to future research.

3.3 Examples

We will now explore a series of examples that best exemplify the rich structure of stability phase boundaries in QBHs. Among other things, we will demonstrate the utility of KPR as an indicator of stability phase transitions and call pointed attention to explicit examples of GPT symmetry-breaking. Beginning with the simplest possible model exhibiting a nontrivial stability phase diagram, we will gradually ramp up the complexity by increasing the number of modes, culminating in a flagship model with an arbitrarily large number $N \to \infty$. This final example, known as the *bosonic Kitaev chain* (BKC), will also serve as a playground for exploring the role bulk-translation invariance and BCs play in determining the stability of QBHs.

---

[5] Specifically, since this phase boundary spawns a Jordan chain of length greater than 1, the one-dimensional eigenspace must be $\boldsymbol{\tau}_3$-indefinite by the Krein-Gel'fand-Lidskii theorem.

3.3.1 A Single-Mode Model

Our first example is a single bosonic mode with Hamiltonian

$$H \equiv \frac{\alpha+\beta}{2}\left(a^\dagger a + a a^\dagger\right) - \frac{\alpha-\beta}{2}\left(a^{\dagger 2} + a^2\right) = \alpha p^2 + \beta x^2, \quad \alpha, \beta \in \mathbb{R}. \tag{3.12}$$

The landscape of dynamical and thermodynamical stability can be assessed via the dynamical matrix

$$\mathbf{G}(\alpha, \beta) = \begin{bmatrix} \alpha+\beta & \beta-\alpha \\ \alpha-\beta & -\alpha-\beta \end{bmatrix}. \tag{3.13}$$

Reality of $\mathbf{G}$ immediately lends a useful GPT symmetry, namely, $\Theta = \mathcal{T}$, with $\mathcal{T}$ complex conjugation on $\mathbb{C}^2$. Physically, this corresponds to time-reversal symmetry $(zx, zp) \mapsto (z^* x, -z^* p)$, with $z \in \mathbb{C}$. The spontaneous breaking of this symmetry, i.e., the stability phase boundaries, may be determined via the normal mode frequencies $\pm\omega = \pm 2\sqrt{\alpha\beta}$, while thermodynamic stability is assessed via the eigenvalues of $\mathbf{H} = \boldsymbol{\tau}_3 \mathbf{G}$, which are $\sigma(\mathbf{H}) = \{2\alpha, 2\beta\}$. In total, there are four distinct parameter regimes:

(i) If $\alpha, \beta > 0$ or $\alpha, \beta < 0$, then H is simply a quantum harmonic oscillator (QHO) of frequency ω with an overall negative sign in latter case. This is both dynamically and thermodynamically stable. Note that H is bounded below for $\alpha, \beta > 0$ and from above for $\alpha, \beta < 0$. Taking $\alpha, \beta > 0$ without loss of generality, there is one bosonic normal mode pair (ψ, ψ) satisfying

$$H = \frac{\omega}{2}(\psi^\dagger \psi + \psi \psi^\dagger), \quad \psi \equiv \cosh\theta\, a + \sinh\theta\, a^\dagger, \quad \tanh\theta \equiv \frac{\sqrt{\beta} - \sqrt{\alpha}}{\sqrt{\beta} + \sqrt{\alpha}}, \tag{3.14}$$

with $\psi(t) = e^{-i\omega t}\psi(0)$. An unbroken GPT symmetry manifests as invariance of this mode under time-reversal symmetry, i.e., $\mathcal{T}\psi\mathcal{T}^{-1} = \psi$.

(ii) If $\alpha > 0 > \beta$ or $\beta > 0 > \alpha$, then H is an inverted harmonic oscillator (IHO), again with an overall negative sign in the latter case. This is both dynamically and thermodynamically unstable. Taking $\alpha > 0 > \beta$ without loss of generality, there is a pair of canonically conjugate quadrature normal modes $z^\pm$ satisfying

$$H = -|\omega|(z^+ z^- + z^- z^+), \quad z^\pm = \frac{1}{\sqrt{|\omega|}}\left(\sqrt{|\beta|}x \mp \sqrt{\alpha}p\right), \tag{3.15}$$

with $z^\pm(t) = e^{\pm\lambda t} z^\pm(0)$. Broken GPT symmetry can be seen in $\mathcal{T} z^\pm \mathcal{T}^{-1} = z^\mp$. That is, the normal modes are no longer time-reversal invariant.

(iii) If only one of α or β vanishes, then H is either the free particle Hamiltonian αp^2 or a "generalized" free particle Hamiltonian[6] βx^2. This model is dynamically unstable, but thermodynamically stable. In the case where $\beta = 0$, p is a normal mode with $p(t) = p(0)$ and $x(t)$ is a generalized normal mode with $x(t) = 2\alpha p(0)t + x(0)$. The roles swap in the other case. Broken GPT symmetry now manifests in the lack of a normal mode basis (or equivalently, in the need to introduce generalized normal modes).

(iv) If $\alpha = \beta = 0$ the Hamiltonian vanishes and is (trivially) dynamically stable and thermodynamically stable. All linear forms are then normal modes with no time-dependence. For example, one may choose $(a, a^\dagger)$ as (time-reversal) invariant normal modes.

In the (α, β) parameter space, cases (iii) and (iv) constitute stability phase boundaries. Predictably, $\mathbf{G}(\alpha, \beta)$ is at an EP in case (iii) and a KC (albeit, trivial) in case (iv). Notably, the locus of the two EP phase boundaries ($\alpha = 0$ and $\beta = 0$) is the KC. We analytically obtain the KPR[7]

$$r(\alpha, \beta) = \frac{2\sqrt{|\alpha||\beta|}}{|\alpha| + |\beta|}. \tag{3.16}$$

The KPR is plotted throughout the stability phase diagram shown in Fig. 3.2a. Evidently, $r(\alpha, \beta)$ vanishes consistently at EPs (case (iii) above). More ambiguously, however, is the situation at the origin. Note that $\lim_{(\alpha,\beta)\to(0,0)} r(\alpha, \beta)$ is ill-defined, or more precisely, the limiting value is contour-dependent. In Fig. 3.2b, the KPR is evaluated over 6 contours $\beta = \alpha^n$ that pass through the origin. In all cases, other than $n = 1$, the KPR vanishes. The case $n = 1$ is not in contradiction to our heuristic argument, however. In particular, there is no stability transition along this contour. The same holds for any odd n but, in these cases, the contours flatten out along the EP boundary $\beta = 0$ (whereby the KPR vanishes unambiguously). Let us consider the more general case where the contour is defined by $\beta = f(\alpha)$, with $f(\alpha) = c_1\alpha + c_2\alpha^2 + \cdots$ analytic and $f(0) = 0$. For $|\alpha| \ll 1$ we have

$$r(\alpha, f(\alpha)) \simeq \frac{2\sqrt{|c_1|}}{1 + |c_1|}. \tag{3.17}$$

It may be surprising that r doesn't vanish at $\alpha = 0$ generally. However, this is to be expected. Contours that behave linearly near the origin (i.e., $c_1 \neq 0$) do not undergo a stability phase transition as they pass through the origin. More importantly, the

[6] The canonical transform $(x, p) \mapsto (-p, x)$ takes $H = \beta x^2$ to $H = \beta p^2$ and so the former is mathematically indistinguishable from a free particle.

[7] While KPR is a quantity tied to a particular eigenvector, the 2×2 nature of $\mathbf{G}(\alpha, \beta)$ makes the choice irrelevant. That is, the KPR is the same for both eigenvectors.

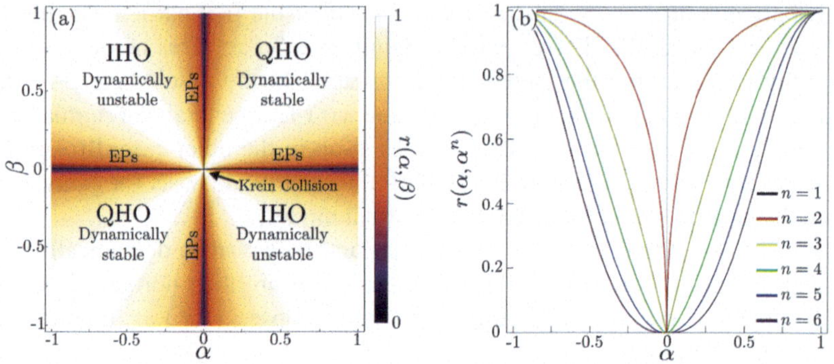

Fig. 3.2 (**a**) The KPR of the single-mode Hamiltonian Eq. (3.12) plotted throughout the stability phase diagram. Note that it vanishes along the boundaries separating the stable QHO phase and the unstable IHO phase. (**b**) The KPR evaluated over the diagonalizable contours $\beta = \alpha^n$, with $n = 1, \ldots, 6$ and $-1 < \alpha < 1$. Note that the KPR vanishes at the KC in all cases other than $n = 1$. Earlier versions of these figures first appeared in Ref. [1]

KPR vanishes along every analytic[8] contour for which a transition occurs at the origin.

The contour dependence of the KPR around the KC is easy to understand if one looks at the eigenvector of $\mathbf{G}(\alpha, f(\alpha))$ explicitly. Near the origin, the conventionally normalized eigenvectors are

$$\vec{\psi}_{\pm} \simeq \frac{1}{\sqrt{2(1+c_1)}} \begin{bmatrix} 1 \pm \sqrt{c_1} \\ 1 \mp \sqrt{c_1} \end{bmatrix}, \quad c_1 \geq 0. \tag{3.18}$$

Different choices of c_1 yield different limiting values of the eigenvectors as $\alpha \to 0$. The only reason this is possible is because $\mathbf{G}(0, 0) = 0$ has a diagonalizable degeneracy.

3.3.2 A Two-Mode Cavity QED Model

It is straightforward to check that if a single-mode model has KC, then it must be at a point in parameter space where the Hamiltonian vanishes. By increasing the number of modes, we are able to find more exotic KCs. The example we will consider for this is a two-mode model descendent from the cavity QED Hamiltonian of Ref. [9]. The full Hamiltonian describes N identical neutral spin-1/2 atoms interacting with a single-cavity mode:

[8] While real analytic functions are a subset of smooth functions, it is reasonable to expect the same conclusion holds if f is smooth non-analytic.

$$H = \omega_c a^\dagger a + \omega_s S_z + g(a^\dagger + a)(S_+ + S_-), \tag{3.19}$$

with $a^\dagger$ (a) the creation (annihilation) operator associated with the optical cavity mode, S_z the collective z-direction spin operator of the atoms, and S_+ (S_-) the collective spin raising (lowering) operator. The frequencies $\omega_c, \omega_s > 0$ correspond to the resonant frequency of the cavity and the transition frequency of the atoms, respectively, whereas the atom-cavity coupling strength is given by $g \in \mathbb{R}$. As in Ref. [9], we obtain a two-mode bosonic Hamiltonian by employing a large-spin (Holstein-Primakoff) approximation

$$S_z = \frac{N}{2} - b^\dagger b, \quad S_+ \simeq \sqrt{N} b, \quad S_- \simeq \sqrt{N} b^\dagger, \tag{3.20}$$

with $b^\dagger$ (b) denoting a canonical bosonic mode that creates (annihilates) a magnonic excitation, i.e., one lowering (raising) the collective spin by $1/2$. In this approximation we obtain the desired Hamiltonian

$$H \simeq H_0 \equiv \omega_c a^\dagger a - \omega_s b^\dagger b + \chi(a^\dagger + a)(b^\dagger + b), \quad \chi = g\sqrt{N}. \tag{3.21}$$

For any choice of parameters, H_0 is not thermodynamically stable. This can be seen by considering the Fock states $|n_a, n_b\rangle$, with $n_a, n_b \in \mathbb{Z}_{\geq 0}$, which represent a state of n_a photons and n_b magnons. The mean energy of these states is $\langle n_a, n_b | H_0 | n_a, n_b \rangle = n_a \omega_c - n_b \omega_s$, which is not bounded in either direction. This thermodynamical instability is not surprising, however. The large-spin approximation is only valid in the limit where there are a few magnonic excitations—in particular, if n_b (or the expectation value of $b^\dagger b$, more generally) exceeds N, the state $|n_a, n_b\rangle$ does not represent any physical state of the full system. Introducing the two-mode Nambu array $\Phi = [a, a^\dagger, b, b^\dagger]^T$ allows us to identify

$$H_0 = \frac{1}{2}\Phi^\dagger \boldsymbol{\tau}_3 \mathbf{G}_0 \Phi - \frac{\delta}{2} 1_{\mathcal{F}}, \quad \mathbf{G}_0 \equiv \begin{bmatrix} \omega_c & 0 & \chi & \chi \\ 0 & -\omega_c & -\chi & -\chi \\ \chi & \chi & -\omega_s & 0 \\ -\chi & -\chi & 0 & \omega_s \end{bmatrix}, \tag{3.22}$$

where we have introduced the detuning parameter $\delta \equiv \omega_c - \omega_s$. Immediately we see that $\mathbf{G}_0$ is a real matrix. As in the single-mode model, complex conjugation is then a valid GPT symmetry of the system with time-reversal being the physical realization. In Ref. [9] it was noted that spontaneous time-reversal symmetry breaking is responsible for the stability phase transitions in the model. This is entirely consistent with our more general analysis of GPT symmetry.

Spontaneous GPT symmetry breaking, which we may now consider to be the losing of time-reversal symmetric normal modes, occurs when the normal mode frequencies become complex or when diagonalizability is lost. Introducing the

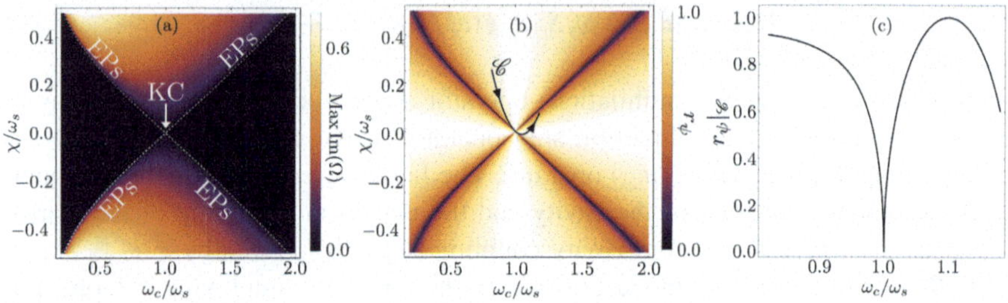

Fig. 3.3 (**a**) The stability phase diagram of the two-mode cavity QED Hamiltonian Eq. (3.21). The phase diagram is constructed by plotting the largest imaginary part of the normal mode spectrum. Recall that if any eigenvalue develops a nonzero imaginary part, then fourfold symmetry of the normal mode spectrum (Eq. (2.45)) ensures dynamical stability is lost. (**b**) The KPR of a representative eigenvector $\vec{\psi}$ computed throughout the stability phase diagram. As expected from Claim 3.2.5, it vanishes at phase boundaries. (**c**) The phase rigidity computed over a KC-crossing contour $\mathscr{C}$ (defined by $y = 5x^2 - x/2$, $x = \delta/\omega_s$, and $y = \chi/\omega_s$) depicted in (**b**). Note that, by construction, diagonalizability is retained along $\mathscr{C}$. Earlier versions of these figures first appeared in Ref. [1]

dimensionless parameters $x \equiv \delta/\omega_s \in (-1, \infty)$ and $y \equiv \chi/\omega_s \in (-\infty, \infty)$ and letting $f(x, y) \equiv x^2(x+2)^2 - 16y^2(x+1)$ allows us to write these frequencies as

$$\Omega_{1,\pm} = \pm\frac{\omega_s}{\sqrt{2}}\sqrt{x^2 + 2x + 2 + \sqrt{f(x, y)}},$$
$$\Omega_{2,\pm} = \pm\frac{\omega_s}{\sqrt{2}}\sqrt{x^2 + 2x + 2 - \sqrt{f(x, y)}}. \tag{3.23}$$

Dynamical instability occurs precisely when $f(x, y) < 0$ which defines the dynamical phase boundaries, $y = y_{\pm}(x) = \pm(x^2 + 2x)/(4\sqrt{x+1})$). The stability phase diagram is shown in Fig. 3.3a.

Similar to the single-mode example, the phase diagram hosts two EP-dominated boundaries that coalesce at a KC. In sharp contrast to the single-mode model, however, the KC does not correspond to the zero Hamiltonian. Instead, it occurs at the resonant ($\delta = 0$) decoupled ($\chi = 0$) limit:

$$H_0\big|_{\delta=\chi=0} = \omega_c(a^\dagger a - b^\dagger b). \tag{3.24}$$

The fact that an infinitesimal coupling ($\chi \neq 0$) can destabilize this (otherwise dynamically stable) limit is consistent with the Krein-Gelf'and-Lidskii theorem which, in particular, asserts that KCs are dynamically stable points that sit at the cusp of instability. If the KC is avoided via even the slightest detuning $\delta \neq 0$, a window of stable couplings $\chi \in (-|\chi_0|, |\chi_0|)$ with $\chi_0 \equiv \delta(\delta + 2\omega_s)/4\sqrt{\omega_s(\delta + \omega_s)} = (\omega_s^2 - \omega_c^2)/4\sqrt{\omega_c\omega_s}$ develops.

For further verification of Claim 3.2.5, we numerically compute the KPR of a representative eigenvector in Fig. 3.3b. As expected, the KPR vanishes at the EP phase boundaries while its behavior around the KC is more subtle. The KPR again behaves in a contour dependent fashion in the vicinity of the KC. However, in Fig. 3.3c we verify that it vanishes along the smooth contour $\mathscr{C}$ depicted in Fig. 3.3b.

3.3.3 A Bosonic Kitaev Chain

The preceding two models have offered simple arenas for testing the core features of our theory of stability phase transitions in QBHs. Namely, that (1) stability phase transitions are dictated by the spontaneous breaking of GPT symmetries and (2) the KPR provides a useful numerical indicator for stability phase boundaries. We will now explore the utility of this theory in a much richer model, namely, a BKC. This model is a generalization of the BKC introduced in Ref. [10], which was originally proposed as bosonic analogue to the well-known fermionic Kitaev chain (FKC) [23]. The connection to the original FKC was motivated by the potential for uncovering SPT-like (or more specifically, Majorana-like) physics in a bosonic setting. While our primary goal is to explore the interplay between system size, BCs, topology, and stability phases, our generalization will ultimately serve to further elucidate the deep connections between the BKC and its fermionic progenitor. More dramatically, it will play an instrumental part in eventually revealing that closed quadratic bosonic dynamics provide an insufficient platform for realizing tight bosonic analogues to fermionic SPT physics (see Chap. 5).

3.3.3.1 The Model, Its Symmetries, and Phase-Dependent Transport

Our generalization of the BKC (which we will henceforth refer to as 'the BKC') can be written as $H(s, \varphi) = H^{\text{OBC}} + sW(\varphi)$, with $s \in [0, 1]$ and

$$\begin{aligned} H^{\text{OBC}} &\equiv \frac{1}{2}\sum_{j=1}^{N-1}\left(iJa_{j+1}^\dagger a_j + i\Delta a_{j+1}^\dagger a_j^\dagger + \text{H.c.}\right), \quad J, \Delta > 0, \\ W(\varphi) &\equiv \frac{1}{2}\left(iJe^{i\varphi}a_1^\dagger a_N + i\Delta e^{i\varphi}a_1^\dagger a_N^\dagger + \text{H.c.}\right), \quad \varphi \in [0, \pi]. \end{aligned} \tag{3.25}$$

The first term H^{OBC} is the BKC of Ref. [10] subject to OBCs while the second term $sW(\varphi)$ encodes the BCs. Physically, J and Δ encode the hopping and non-degenerate parametric amplification (NDPA) amplitudes, respectively. We will more commonly call Δ the pairing amplitude in direct analogy to the fermionic counterpart. The two boundary parameters s and φ determine the coupling strength between ends of the chain and the joint hopping-pairing twisting angle, respectively. In particular, taking s from 0 to 1, with $\varphi = 0$ smoothly interpolates between open and periodic BCs. The dynamical matrix of $H(s, \varphi)$ is $\mathbf{G}(s, \varphi) \equiv \mathbf{G}^{\text{OBC}} + \mathbf{B}(s, \varphi)$

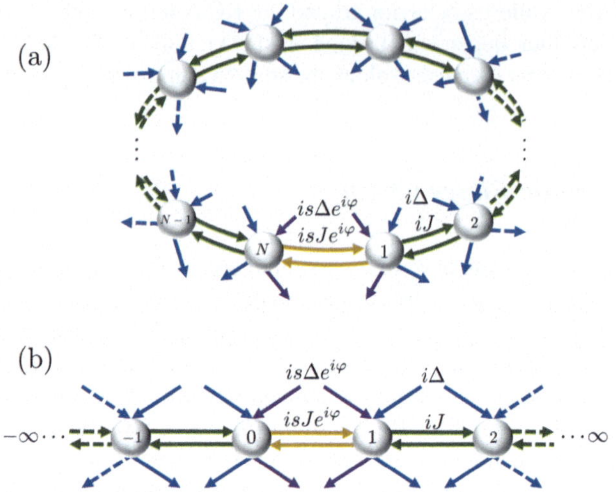

Fig. 3.4 A pictorial representation of the BKC (Eqs. (3.25)) along with the relevant parameters for (**a**) a finite lattice of N sites and (**b**) an infinite lattice. Earlier versions of these figures first appeared in Ref. [1]

with

$$\mathbf{G}^{\mathrm{OBC}} \equiv \mathbf{T}\otimes \mathbf{g}_1 + \mathbf{T}^\dagger \otimes \mathbf{g}_{-1}, \quad \mathbf{B}(s,\varphi) = \vec{e}_N\vec{e}_1^{\,\dagger} \otimes \mathbf{b}_1(s,\varphi) + \vec{e}_1\vec{e}_N^{\,\dagger} \otimes \mathbf{b}_{-1}(s,\varphi), \tag{3.26}$$

with the internal matrices given by

$$\mathbf{g}_1 \equiv -\frac{i}{2}\begin{bmatrix} J & -\Delta \\ -\Delta & J \end{bmatrix}, \quad \mathbf{b}_1(s,\varphi) \equiv -\frac{is}{2}\begin{bmatrix} Je^{-i\varphi} & -\Delta e^{i\varphi} \\ -\Delta e^{-i\varphi} & Je^{i\varphi} \end{bmatrix}, \tag{3.27}$$

and $\mathbf{g}_{-1} = \boldsymbol{\sigma}_3\mathbf{g}_1^\dagger\boldsymbol{\sigma}_3$, $\mathbf{b}_{-1}(s,\varphi) = \boldsymbol{\sigma}_3\mathbf{b}_1^\dagger(s,\varphi)\boldsymbol{\sigma}_3$.

The BKC, which is depicted pictorially in Fig. 3.4, enjoys a number of notable transformation properties with respect to physical symmetries. The first, and most relevant for our purposes, is the explicit breaking of number symmetry ushered in by the pairing terms. Non-Hermiticity of the dynamical matrix is thus inevitable. The second notable feature is the behavior under time-reversal. Explicitly,

$$\mathcal{T}(H^{\mathrm{OBC}} + sW(\varphi))\mathcal{T}^{-1} = -(H^{\mathrm{OBC}} + sW(-\varphi)). \tag{3.28}$$

In the special cases $\varphi = 0, \pi$, we observe that the Hamiltonian is odd under time-reversal. One immediate consequence is thermodynamic instability. If $|E\rangle$ is an eigenstate of the Hamiltonian with energy E, then $\mathcal{T}|E\rangle$ is an eigenstate with energy

$-E$, i.e., the spectrum has a "chiral"[9] $\pm$ symmetry. Since the spectrum of a (non-zero) QBH is always unbounded in at least one direction, it is unbounded in both. Using analytic diagonalization methods, we will see that thermodynamic instability persists, in particular, for $\varphi = \pi/2$.

Digression on Oddness Under Time-Reversal Oddness under a time-reversal is not a symmetry in the formal sense. However, in QBHs, it is equivalent to the existence of a particular unitary symmetry. Consider an arbitrary QBH with dynamical matrix $\mathbf{G}$. Then

$$\mathcal{T}H\mathcal{T}^{-1} = \frac{1}{2}\Phi^\dagger \boldsymbol{\tau}_3 \mathbf{G}^* \Phi, \quad \mathcal{T}H\mathcal{T}^{-1} = -H \iff \mathbf{G} = -\mathbf{G}^*. \tag{3.29}$$

Combining this with the fundamental property $\mathbf{G} = -\boldsymbol{\tau}_1 \mathbf{G}^* \boldsymbol{\tau}_1$ reveals that oddness under time-reversal is equivalent to $[\mathbf{G}, \boldsymbol{\tau}_1] = 0$. We may identify this as invariance under an isotropic local squeezing transformation[10] $U(\chi)$, $\chi \in \mathbb{R}$, implementing $a_j \mapsto \cosh(\chi)a_j - \sinh(\chi)a_j^\dagger$. This can be seen by noting $U(\chi)\Phi U(\chi)^\dagger = e^{\chi\boldsymbol{\tau}_1}\Phi$ and thus $[H, U(\chi)] = 0$ for all $\chi \in \mathbb{R}$ if and only if $[\mathbf{G}, \boldsymbol{\tau}_1] = 0$. Altogether, we are free to identify oddness under time-reversal with this particular squeezing symmetry. This symmetry manifests as a rather concrete feature of the quadrature dynamics. Computing the EOM for the quadrature array R (Eq. (2.2)), we have

$$i\frac{dR}{dt} = i\frac{d}{dt}\boldsymbol{\Sigma}\Phi = \boldsymbol{\Sigma}\left(i\frac{d}{dt}\Phi\right) = \boldsymbol{\Sigma}\mathbf{G}\Phi = \boldsymbol{\Sigma}\mathbf{G}\boldsymbol{\Sigma}^\dagger R, \tag{3.30}$$

or more explicitly,

$$\frac{d}{dt}\begin{bmatrix} x_j \\ p_j \end{bmatrix} = \sum_{k=1}^{N} \begin{bmatrix} \mathbf{C}_{kj}x_k + \mathbf{M}_{jk}p_k \\ -\mathbf{U}_{jk}x_k - \mathbf{C}_{jk}p_k \end{bmatrix}, \tag{3.31}$$

$$\mathbf{C} \equiv \mathrm{Im}(\boldsymbol{\Delta} - \mathbf{K}), \quad \mathbf{U} \equiv \mathrm{Re}(\mathbf{K} + \boldsymbol{\Delta}), \quad \mathbf{M} \equiv \mathrm{Re}(\mathbf{K} - \boldsymbol{\Delta}), \tag{3.32}$$

in terms of the matrices in Eq. (2.38). Oddness under time-reversal ensures that $\mathbf{G}^* = -\mathbf{G}$, or $\mathrm{Re}(\mathbf{K}) = \mathrm{Re}(\boldsymbol{\Delta}) = 0$. Ultimately, this means that the dynamics of the x (p) quadratures depend *only* on the other x (p) quadratures. This feature is known as "phase-dependent transport" and was identified in the OBC and PBC BKC in Ref. [10]. The terminology originates from the fact that the real and imaginary

[9] The term "chiral" is invoked to draw analogy with the consequences of chiral symmetry in non-interacting fermionic systems. In that context, a chiral symmetry manifests as an antiunitary operator that anticommutes with the single-particle or BdG Hamiltonian. As a consequence, the single-, or quasi-particle energies become symmetric about 0. However, it is important to note that the fermionic notion of a chiral symmetry is not perfectly transplanted into the bosonic context [24]. Our use of this term is simply to call attention to this $\pm$ quasi-particle energy symmetry.

[10] The $U(\chi)$ here is a special case of Eq. (2.21), with $\mathrm{Re}(z) = \chi$ and $\mathrm{Im}(z) = 0$.

components of any coherent state will propagate independently from one-another in such a system. In fact, the phase-dependent transport of the BKC is actually *chiral*: x and p quadratures propagate in opposite directions. This is a consequence of the non-symmetric nature of the coefficients $\mathbf{C}_{jk}$, as also noted in Ref. [10].

Let us summarize the results of the preceding analysis and further elucidate its dynamical consequences.

Theorem 3.3.1 *Let H be a QBH. Then the following statements are equivalent.*

(i) H is odd under time-reversal symmetry, i.e., $\mathcal{T}H\mathcal{T}^{-1} = -H$.
(ii) H has a local squeezing symmetry of the form $U(\chi)\Phi U(\chi)^\dagger = e^{\chi\tau_1}\Phi$, *with* $\chi \in \mathbb{R}$.
(iii) H supports phase-dependent transport.

Furthermore, any H satisfying (i)–(iii) is thermodynamically unstable, and either dynamically unstable, or at the cusp of dynamical instability (specifically, every real normal mode frequency ω *supports a KC).*

Proof Equivalence of (i)–(iii) has already been established in the preceding analysis. We have also already seen that such systems are thermodynamically unstable. All that is left to show is that systems satisfying (i)–(iii) are either dynamically stable, or at the cusp of instability.

If the system is dynamically unstable, we are done. If not, then consider a real eigenvalue $\omega \in \sigma(\mathbf{G})$ and the corresponding eigenvector $\vec{\psi}^+$ which we will assume, without loss of generality, to have Krein signature $+1$. Properties (i)–(iv) imply $[\mathbf{G}, \boldsymbol{\tau}_1] = 0$, from which we conclude that $\vec{\psi}^- \equiv \boldsymbol{\tau}_1\vec{\psi}^+$ is also an eigenvector of $\mathbf{G}$ with eigenvalue ω. The Krein signature of this eigenvector is

$$\vec{\psi}^{-\dagger}\boldsymbol{\tau}_3\vec{\psi}^- = \vec{\psi}^{+\dagger}\boldsymbol{\tau}_1\boldsymbol{\tau}_3\boldsymbol{\tau}_1\vec{\psi}^+ = -\vec{\psi}^{+\dagger}\boldsymbol{\tau}_3\vec{\psi}^+ = -1. \tag{3.33}$$

This allows us to conclude that $\vec{\psi}^- \not\propto \vec{\psi}^+$ and hence ω supports a KC. Since the choice of ω was irrelevant, each eigenvalue necessarily hosts a KC. □

Applying this result to the BKC allows us to make several predictions before engaging in detailed analysis. For BCs with $\varphi = 0, \pi$, the BKC will support phase-dependent transport and either be unstable, or at best, at the cusp of instability for all values of J, Δ, and s. We will now confirm this via exact diagonalization under several BCs of interest.

3.3.3.2 Exact Normal Mode Decomposition

Here we report the key results of the exact diagonalization undertaken in Appendix B.2. In particular, using the eigendecomposition of $\mathbf{G}(s, \varphi)$, we cast the BKC Hamiltonian into the normal form Eq. (2.74) when applicable. Various features of the normal modes and their frequencies are pictorially summarized in Fig. 3.5. We also note that exact diagonalization for OBCs and PBCs was

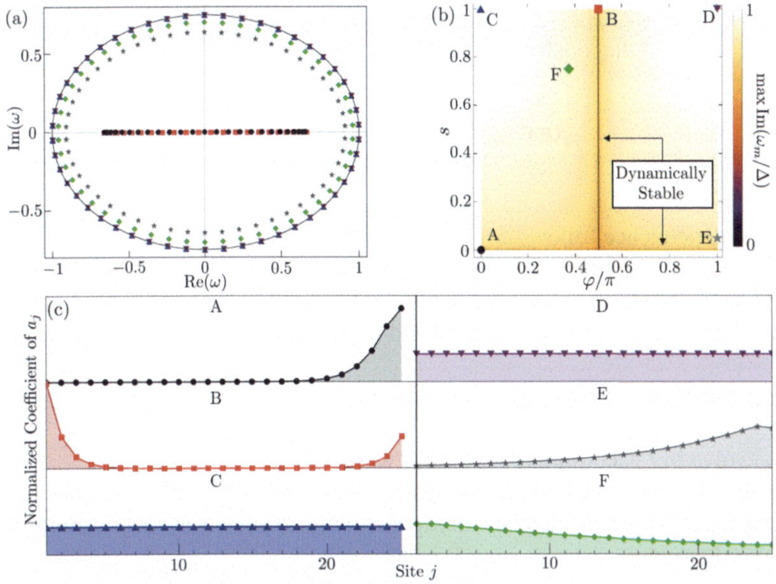

Fig. 3.5 Boundary dependence of the normal modes (eigenvectors of $\mathbf{G}(s, \varphi)$) and their frequencies (eigenvalues of $\mathbf{G}(s, \varphi)$) in the BKC with $J = 1$, $\Delta = 0.75$ and $N = 25$. (**a**) The normal mode frequencies for various choices of BCs. (**b**) The largest imaginary part for the normal mode frequencies numerically calculated on a grid of spacing $\Delta s = \Delta\varphi/\pi = 0.002$. Points A-F correspond to values of s and φ from which the eigenvalues and eigenvectors are sourced in (**a**) and (**c**), respectively. Points A–D correspond to points in parameter space where exact analytical solutions are found in the main text. (**c**) The (suitably normalized) coefficients of a_j for representative normal modes of the chain at various choices of s and φ. These normal modes are representative in the sense that their gross localization properties are independent of the particular eigenvector chosen. Earlier versions of these figures first appeared in Ref. [1]

accomplished in Ref. [10]. However, in these cases, we extend the analysis and derive new conclusions specifically oriented around the many-body features.

Open Boundary Conditions For OBCs ($s = 0$), the normal mode frequencies are doubly degenerate and given by $\omega_n = \sqrt{J^2 - \Delta^2}\cos(n\pi/(N+1))$, with $m = 1, \ldots, N$. The analysis then splits into three cases: $J < \Delta$, $J = \Delta$, and $J > \Delta$.

When $J < \Delta$, the normal mode spectrum is purely imaginary and hence the system is dynamically unstable. To each $\omega_n = i\lambda_n$, there are two localized Hermitian normal modes

$$z_n^+ \equiv \sqrt{\frac{2}{N+1}} \sum_{j=1}^{N} e^{-jr} \sin\left(\frac{n\pi j}{N+1}\right) x_j,$$

$$z_n^- \equiv \sqrt{\frac{2}{N+1}} \sum_{j=1}^{N} e^{jr} \sin\left(\frac{n\pi j}{N+1}\right) p_j, \tag{3.34}$$

with $2r \equiv \ln[(J+\Delta)/|J-\Delta|]$. These modes satisfy the HWRs $[z_n^+, z_m^-] = i\delta_{nm}1_{\mathcal{F}}$, $[z_n^\pm, z_m^\pm] = 0$, and diagonalize the BKC according to

$$H^{\text{OBC}} = \frac{1}{2}\sum_{n=1}^{N} \lambda_m \left(z_n^+ z_n^- + z_n^- z_n^+\right). \tag{3.35}$$

The time-dependence follows as $z_n^\pm(t) = e^{\pm\lambda_m t} z_n^\pm(0)$.

The point $J = \Delta$ is an EP for the dynamical matrix. In this case, there is one normal mode frequency $\omega = 0$ that hosts two Jordan chains of length N. A particular choice of (perfectly localized) generalized normal modes may be constructed as

$$\chi_{1k} \equiv (-iJ)^{-k} p_{N+1-k}, \quad \chi_{2k} \equiv (iJ)^{-k} x_k, \quad k = 1, \ldots, N, \tag{3.36}$$

which evolve in t according to Eq. (2.60) with $\omega_0 = 0$.

Finally, if $J > \Delta$, we find that the system is dynamically stable with quasiparticle energies ω_n. To each n there is an exponentially localized bosonic normal mode (e.g., quasiparticle)

$$\psi_n = \sqrt{\frac{2}{N+1}} \sum_{j=1}^{N} i^{-j} \sin\left(\frac{n\pi j}{N+1}\right)\left(\cosh(jr)a_j - \sinh(jr)a_j^\dagger\right), \tag{3.37}$$

with r defined exactly as before, and satisfying the CCRs $[\psi_n, \psi_m^\dagger] = \delta_{nm}1_{\mathcal{F}}$, $[\psi_n, \psi_m] = 0$. In terms of these,

$$H^{\text{OBC}} = \frac{1}{2}\sum_{n=1}^{N} \omega_n \left(\psi_n^\dagger \psi_n + \psi_n \psi_n^\dagger\right) = \sum_{n=1}^{\lfloor N/2 \rfloor} \omega_n \left(\psi_n^\dagger \psi_n - \psi_{\bar{n}}^\dagger \psi_{\bar{n}}\right), \tag{3.38}$$

where, in the second equality, we have introduced the notation $\bar{n} = N + 1 - n$ to highlight the chiral symmetry of the quasiparticle energy spectrum $\omega_{\bar{n}} = -\omega_n$. Note further that, in the case of N odd, there is an additional bosonic zero mode (ZM) ψ_n, with $n = (N+1)/2$. With this, we can explicitly construct the quasiparticle vacuum using Eq. (2.68). It is given by

$$|\widetilde{0}\rangle = \mathcal{M} \exp\left[\frac{1}{2}\sum_{j=1}^{M} \tanh(jr)(a_j^\dagger)^2\right]|0\rangle, \tag{3.39}$$

with $\mathcal{M}$ a normalization constant. The chiral nature of the quasiparticle spectrum reveals a, previously unnoticed, symmetry of the problem. Namely, consider the

Bogoliubov transformation

$$\begin{aligned}\psi_n &\mapsto \psi_n(s_n) \equiv \cosh(s_n)\psi_n + \sinh(s_n)\psi_{\bar{n}}^\dagger,\\ \psi_{\bar{n}} &\mapsto \psi_{\bar{n}}(s_n) \equiv \cosh(s_n)\psi_{\bar{n}} + \sinh(s_n)\psi_n^\dagger,\end{aligned} \tag{3.40}$$

where s_n, $n = 1, \ldots, \lfloor N/2 \rfloor$, are arbitrary real constants. Thanks to chiral symmetry, H^{OBC} is left invariant under this transformation. This follows directly from

$$\psi_n^\dagger(s_n)\psi_n(s_n) - \psi_{\bar{n}}^\dagger(s_n)\psi_{\bar{n}}(s_n) = \psi_n^\dagger\psi_n - \psi_{\bar{n}}^\dagger\psi_{\bar{n}}. \tag{3.41}$$

These new quasiparticles are given by

$$\psi_n(s_n) = \sqrt{\frac{2}{N+1}}\sum_{j=1}^{N} i^{-j}\sin\left(\frac{n\pi j}{N+1}\right)\left(\cosh(s_n + jr)a_j + \sinh(s_n + jr)a_j^\dagger\right). \tag{3.42}$$

Unlike the isotropic squeezing symmetry of the physical bosonic modes a_j discussed earlier, this transformation induces independent degrees of squeezing on each of the normal modes. Hence, s_n are free parameters that can be used to tune the localization properties of each normal mode. For the particular (isotropic) choice $s_n = -j_0 r$ for all n, we recover the parametric freedom identified in Ref. [10].

We conclude the OBC analysis by calling direct attention to the fact that, in each of the three cases, the normal modes of the system were spatially localized. As noted in Ref. [10], this is reminiscent of the so-called *non-Hermitian skin effect* (NHSE) [25], whereby a macroscopic number of eigenvectors of a matrix defined on a lattice localize around one of the edges. In fact, as will become immensely relevant in Part II, this is a manifestation of the NHSE that is descendant from topological features of the PBC spectrum.

Periodic and Anti-periodic Boundary Conditions In terms of boundary parameters, PBCs and anti-PBCs are characterized by $(s, \varphi) = (1, 0)$ and $(1, \pi)$ respectively. Dealing with PBCs first, we follow the procedure laid out in Sect. 2.5. Specifically, we perform a Fourier transform $b_k = N^{-1/2}\sum_{j=1}^{N} e^{-ijk}a_j$, with $k \in \mathcal{K}_N$ the discrete Brillouin zone. With this, we may block diagonalize $H(1,0) = \sum_{k\in\mathcal{K}_N} H_k$, with

$$H_k = \frac{1}{2}\widetilde{\phi}_k^\dagger \boldsymbol{\sigma}_3 \mathbf{g}(k)\widetilde{\phi}_k = \frac{J}{2}\sin(k)(b_k^\dagger b_k - b_{-k}b_{-k}^\dagger) + \frac{i\Delta}{2}\cos(k)(b_k^\dagger b_{-k}^\dagger - b_{-k}b_k), \tag{3.43}$$

where $\widetilde{\phi}_k = [b_k, b_{-k}^\dagger]^T$ is the Fourier Nambu array and

$$\mathbf{g}(k) = e^{ik}\mathbf{g}_1 + e^{-ik}\mathbf{g}_{-1} = \begin{bmatrix} J\sin(k) & i\Delta\cos(k) \\ i\Delta\cos(k) & J\sin(k) \end{bmatrix} = J\sin(k)\mathbb{1}_2 + \Delta\cos(k)\boldsymbol{\sigma}_1. \tag{3.44}$$

is the Bloch dynamical matrix. The two normal mode frequency bands follow as the eigenvalues of $\mathbf{g}(k)$, which are explicitly given by $\omega_\pm(k) = J\sin(k) \pm i\Delta\cos(k)$. Immediately we see that the system is dynamically unstable with the frequency bands forming counter-oriented ellipses in the complex plane. The eigenvectors of $\mathbf{g}(k)$ are given by $[1, \pm 1]/\sqrt{2}$, which allow us to construct (generally) pseudo-bosonic normal modes $\xi_\pm(k) = \xi_{\mp *}(k) = (b_k \mp b_{-k}^\dagger)/\sqrt{2}$ such that

$$H_k = \frac{1}{2}\left(\omega_+(k)\xi_+^\dagger\xi_- + \omega_-(k)\xi_-^\dagger\xi_+\right). \tag{3.45}$$

In the case where k is 0 or $-\pi$, $\xi_+(k) = ip_k$, and $\xi_-(k) = x_k$, with p_k and x_k the Fourier transforms of the quadrature p_j and x_j, respectively. Similarly, when $k = \pm\pi/2$ (which occurs if N is a multiple of 4), we actually have that $H_{\pm\pi/2} = J(b_{\pm\pi/2}^\dagger b_{\pm\pi/2} - b_{\mp\pi/2} b_{\mp\pi/2}^\dagger)/2$ is already diagonalized in terms of the bosonic normal modes $b_{\pm\pi/2}$.

The case of anti-PBCs can be accommodated without much modification to the preceding analysis. The first step is to observe that the dynamical matrix takes on a almost identical to Eq. (2.77), but with the translation operator $\mathbf{V}_N$ replaced with the anti-periodic translation operator $\mathbf{V}_N' = \mathbf{V}_N - 2\vec{e}_N\vec{e}_1^\dagger = \mathbf{T}_N - \vec{e}_N\vec{e}_1^\dagger$. As in the periodic case, this matrix is straightforwardly diagonalized with plane waves, however, with the discrete Brillouin zone slightly modified. Ultimately, the PBC analysis carries over except $\mathcal{K}_N$ is replaced with

$$\mathcal{K}_N' = \begin{cases} \{0, \pm\pi/N, \pm 3\pi/N, \ldots, \pm\pi(1-1/N)\}, & N \text{ odd}, \\ \{0, \pm\pi/N, \pm 3\pi/N, \ldots, \pm\pi(1-1/N), \pi\}, & N \text{ even}. \end{cases} \tag{3.46}$$

The normal mode spectrum is then that of PBCs (i.e., $\omega_\pm(k)$, $k \in \mathcal{K}_N$), translated by $\Delta k = \pi/N$. Importantly, dynamical instability is retained.

Twisted Boundary Conditions: $\varphi = \pi/2$ The emergence of Majorana edge modes in the FKC with twisted BCs is a well-studied phenomena [26]. Like in the case of the OBCs, the presence of a $\pi/2$ relative phase rotation between bulk and boundary hopping and pairing amplitudes is known to host such modes. For comparison, we obtain an exact diagonalization in the BKC. The normal mode spectrum is given by $\pm\omega_n = \pm\sqrt{J^2 - \Delta^2}\sin(k_n)$, with $k_n = (4n-1)\pi/2N$ and $n = 1, \ldots, N$. As in the OBC case, we have three distinct cases $J < \Delta$, $J = \Delta$, and $J > \Delta$.

If $J < \Delta$, the system is dynamically unstable with purely imaginary normal mode frequencies $\omega_n = i\lambda_n$. To each n, there are two localized Hermitian normal modes

$$
\begin{aligned}
z_n^+ &\equiv \frac{1}{\sqrt{N}} \sum_{j=1}^{N} \left(e^{-(j-(N+2)/2)r} \cos(jk_n) x_j + e^{(j-(N+2)/2)r} \sin(jk_n) p_j \right), \\
z_n^- &\equiv \frac{1}{\sqrt{N}} \sum_{j=1}^{N} \left(e^{(j-(N+2)/2)r} \cos(jk_n) p_j - e^{-(j-(N+2)/2)r} \sin(jk_n) x_j \right),
\end{aligned}
\tag{3.47}
$$

with $2r \equiv \ln[(J+\Delta)/|J-\Delta|]$ as in the OBC case. These modes satisfy the HWRs $[z_n^+, z_m^-] = i\delta_{nm} 1_{\mathcal{F}}$, $[z_n^\pm, z_m^\pm] = 0$, and diagonalize the Hamiltonian according to

$$
H(1, \pi/2) = \frac{1}{2} \sum_{n=1}^{N} \lambda_m \left(z_n^+ z_n^- + z_n^- z_n^+ \right). \tag{3.48}
$$

The time-dependence follows as $z_n^\pm(t) = e^{\pm\lambda_m t} z_n^\pm(0)$.

For $J = \Delta$, the dynamical matrix again hosts an EP. The only normal mode frequency is 0. For N even there are two Jordan chains of length N, while for N odd, there is a chain of length $N-1$ and one of length $N+1$. The generalized normal modes have a rather complex form which can be derived from Eqs. (B.16)–(B.19). We omit them here for simplicity.

For $J > \Delta$ the system is dynamically stable with quasiparticle energies ω_n. The associated bosonic normal modes (quasiparticles) are

$$
\psi_n = \frac{1}{\sqrt{N}} \sum_{j=1}^{N} e^{-ijk_n} \left(\cosh\left[\left(j - \frac{N+2}{2} \right) r \right] a_j - \sinh\left[\left(j - \frac{N+2}{2} \right) r \right] a_j^\dagger \right), \tag{3.49}
$$

satisfying the CCRs $[\psi_n, \psi_m^\dagger] = \delta_{nm} 1_{\mathcal{F}}$, $[\psi_n, \psi_m] = 0$. The Hamiltonian is then diagonalized as

$$
H(1, \pi/2) = \frac{1}{2} \sum_{n=1}^{N} \omega_n \left(\psi_n^\dagger \psi_n + \psi_n \psi_n^\dagger \right). \tag{3.50}
$$

We immediately see that the system is still thermodynamically unstable (specifically, $\omega_n < 0$ for $n > \lfloor N/2 \rfloor$). Unlike the case of OBCs, chiral symmetry of the quasiparticle spectrum that results from KCs throughout the spectrum is not always present. For N odd, there are no KCs while, if N is even, then $\omega_{N-n} = -\omega_n$ signals a KC at every normal mode frequency.

Notably, in all three cases the NHSE is retained despite the explicit lack of a hard-wall boundary. Instead, the modes localized around the boundary link with minimal amplitude near the center of the chain (see panel B of Fig. 3.5c, for instance).

Twisted Boundary Conditions: $J = \Delta$ and $\varphi \in (0, \pi)$ Under both open and $\pi/2$-twisted BCs, we have seen high order EPs when $J = \Delta$. On the contrary, periodic and anti-periodic BCs have yielded no such singularities. If we take $s = 1$ and twist φ from 0 to π, we can explore the way in which the EP emerges at $\varphi = \pi/2$. To this end, we diagonalize $\mathbf{G}(1, \varphi)$ and find the normal mode spectrum

$$\omega_n = iJ(\cos(\varphi))^{1/N}) \begin{cases} e^{-2in\pi/N}, & N \text{ even}, \\ e^{-\pi in/N}, & N \text{ odd}, \end{cases} \qquad n = 1, \ldots, 2N. \tag{3.51}$$

where we choose the branch of the N'th root whereby $(\cos(\varphi))^{1/N} = |\cos(\varphi)|^{1/N} e^{i\pi/N}$ for $\varphi \in (\pi/2, \pi]$. With this, we see that twisting scales the PBC (anti-PBC) spectrum by $|\cos(\varphi)|^{1/N}$ for $\varphi < \pi/2$ ($\varphi > \pi/2$). The N'th root behavior is characteristic of N'th order EPs [22]. Beyond the spectrum, we can also explicitly study the parametric dependence of the normal modes on the twisting angles. Focusing on the case of N odd for simplicity.[11] The pseudobosonic normal modes are found to be

$$\phi_n = \mathcal{N}_n \begin{cases} \sum_{j=1}^{N} z_n^{j-1} p_j - \tan(\varphi) x_1, & n \text{ even}, \\ \sum_{j=1}^{N} (-z_n)^{-j} x_j, & n \text{ odd}, \end{cases} \qquad z_n \equiv \frac{-iJ^*}{\omega_n}, \tag{3.52}$$

where $\mathcal{N}_n(\varphi)$ is a normalization constant that ensures $[\phi_n, \phi_{m*}^\dagger] = \delta_{nm} 1_{\mathcal{F}}$, with $\phi_{n*} = \phi_\ell$ and ℓ such that $\omega_\ell = \omega_n^*$. We verify that as $\varphi \to 0$, the pseudo-bosonic normal modes approach x_1 which is one of the two zero-frequency modes spawning Jordan chains.

3.3.3.3 Dynamical Stability Analysis

Dependence of Stability on BCs With a number of exact solutions found, we now undertake a detailed analysis of the BKC's stability phase diagram. As the general theory purports, stability phase transitions are mediated by one of two spontaneous GPT-symmetry breaking phenomena: (1) the emergence of EPs or (2) the emergence of KCs. As we have seen for open and $\pi/2$ twisted BCs, a transition of type (1) occurs at $J = \Delta$, whereby two long Jordan chains appear. For these transitions, there is a somewhat simple physical interpretation, namely, the NDPA amplitude Δ overcomes the lattice hopping J and, thus, amplification dominates. More surprising, however, is the fact that a stability phase transition

[11] The case of N even is discussed in detail in Ref. [1].

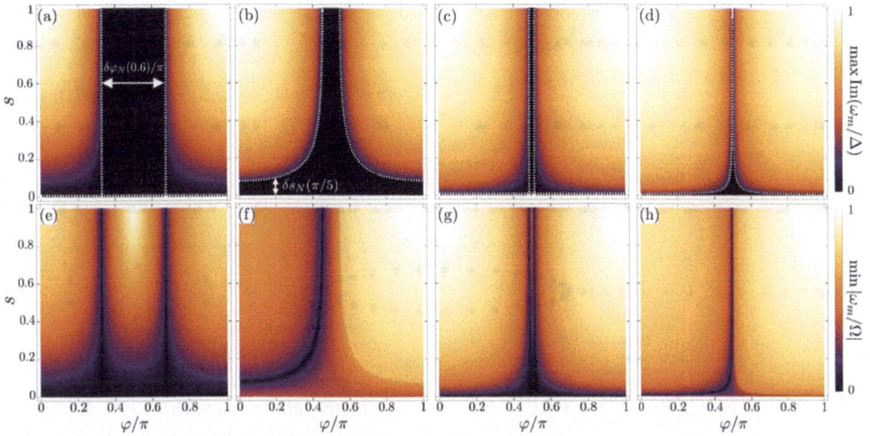

Fig. 3.6 (**a**)–(**d**): Numerical assessment of dynamical stability as a function of BCs, with $J = 1$ and $\Delta = 0.25$, and the boundary parameters sampled on a grid of spacing $\Delta s = \Delta\varphi/\pi = 0.002$. The systems size is (**a**) $N = 5$, (**b**) $N = 10$, (**c**) $N = 15$, and (**d**) $N = 20$. Phase boundaries are indicated by white, dashed lines (see Eq. (3.53)). The $s \neq 0$ phase boundaries host N (1) length-2 Jordan chains for N odd (even), while the $s = 0$ boundary hosts N Krein collisions. In (**a**) the stability width $\delta\varphi_N(s = 0.6)$ is shown, while in (**b**), the stability height $\delta s_N(\varphi = \pi/5)$ is shown. (**e**)–(**h**): Minimum-modulus eigenvalue of $\mathbf{G}(s, \varphi)$ sampled on the same grid as (**a**)–(**d**). The parameter values for (**e**), (**f**), (**g**), and (**h**) match those of (**a**), (**b**), (**c**), and (**d**), respectively. Here, Ω is the largest value of $\min|\omega_m/\Delta|$ over the whole sample grid. The lines of ZMs (along with their mirror-symmetric partners in (**f**) and (**h**)) appear to define the dynamical phase boundaries. Earlier versions of these figures first appeared in Ref. [1]

can occur simply as a function of boundary parameters. The most notable instance of this is the loss of dynamical stability when one passes from OBCs to PBCs. Instability under PBCs suggests, correctly, that the bulk dynamics of the BKC are amplifying (a feature first noted in Ref. [10]). Introducing boundaries may, or may not, stabilize this bulk amplification. Let us get a better hold on which choices of boundary parameters stabilize the bulk.

Along with Fig. 3.5, Fig. 3.6a–d shows numerically determined dynamical stability phase diagrams over the boundary parameter (s, φ)-space, for various choices of N. An immediate observation is that there is a *finite* region of stability around the line $\varphi = \pi/2$. More precisely, there is a width $\delta\varphi_N(s)$ such that the system is dynamically stable for $\varphi \in I_{\varphi_N}(s) \equiv (\pi/2 - \delta\varphi_N(s)/2, \pi/2 + \delta\varphi_N(s)/2)$. The length of this interval is minimized around $s = 1$, whereby we have the *minimal width* of stability, $\delta\varphi_N^{\min} \equiv \delta\varphi_N(1)$. Similarly, for a fixed N and φ, we define $\delta s_N(\varphi)$ to be the quantity such that the system is dynamically stable for $s \in I_{s_N}(\varphi) = [0, \delta s_N(\varphi))$. This is minimized at $\varphi = 0$ whereby we define the *minimum height* of stability, $\delta s_N^{\min} \equiv \delta s_N(0)$. The numerics in Fig. 3.6a–d suggest that both $\delta\varphi_N^{\min}$ and $\delta s_N^{\min}$ go to zero as $N \to \infty$, while our analytical solutions

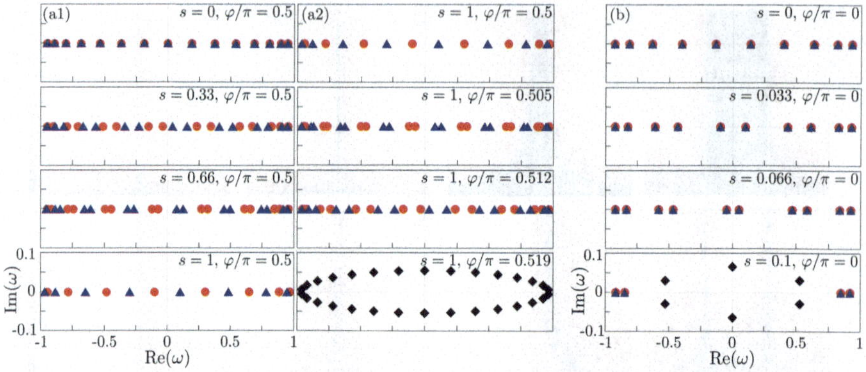

Fig. 3.7 The spectral flow of the BKC as boundary parameters are varied with $J = 1$, $\Delta = 0.25$. The Krein signatures for the corresponding eigenvectors are indicated by red circles (+1), blue triangles (−1), and black diamonds (0), respectively. (**a**) $N = 15$. Going down the left column (**a1**), we show how the eigenvalues evolve within the stable phase, as s goes from 0 to 1 at $\varphi = \pi/2$. Recall that, for N odd, $\varphi = \pi/2$, and $s = 1$, each eigenvalue except the extremal ones are twofold degenerate. Stability is seen to be preserved along this flow. Going down the right column (**a2**), we show how that eigenvalues evolves around the transition between $\pi/2$-twisted and periodic, as φ goes from $\pi/2$ to $1.03 \cdot (\pi/2)$ at $s = 1$. Note how eigenvalues are each split as φ increases, and eventually move symmetrically off the real axis. (**b**) $N = 10$. The spectral flow around the transition between open and periodic, as s goes from 0 to 0.1, with $\varphi = 0$. In this case, stability is retained for sufficiently small strength of the boundary perturbation due to non-zero s. Earlier versions of these figures first appeared in Ref. [1]

demonstrate OBCs ($s = 0$) and $\pi/2$-twisted BCs[12] ($\varphi = \pi/2$) are stable for all N. It then appears that these become isolated, measure zero regions of stability as $N \to \infty$. We can make this more concrete by studying the nature of the stability phase boundaries.

The stability phase boundaries can best be understood by studying the spectral flow of the dynamical matrix (see Fig. 3.7). Further numerical analysis suggests that zero eigenvalue plays a significant role in said transitions (see Fig. 3.6e–h). Our investigation culminates in four main observations.

(i) Figures 3.6a–d and 3.7a1 show that, despite the macroscopic number of KCs in the OBC spectrum, arbitrarily small boundary perturbations need not destabilize the system. For N even, we explicitly have $\delta s_N^{\min} \neq 0$. For N odd, whether or not arbitrarily small nonzero s can destabilize the system depends on φ. In particular, $H(s, \varphi)$ is unstable for all $s > 0$ when $\varphi \notin I_{\varphi N}(1)$.

(ii) For N odd, Fig. 3.7a2 demonstrates that the transition between $\pi/2$-twisted BCs and PBCs is mediated by non-degenerate eigenvalues of opposite Krein signature coming arbitrarily close together.

[12] The preceding analysis revealed that $H(1, \pi/2)$ is dynamically stable for $J > \Delta$. In Sect. 4.2.2 we will establish a duality transformation which further reveals that $H(s, \pi/2)$ is, in fact, dynamically stable for all $s \in [0, 1]$.

(iii) For N even, Fig. 3.7b shows that the OBC to PBC transition occurs when pairs of Krein collided eigenvalues themselves collide.
(iv) The left ($\varphi < \pi/2$) phase boundary is defined by the emergence of a zero frequency mode for arbitrary N. If N is odd, zero frequency also defines the right ($\varphi > 0$) boundary. Furthermore, the phase diagram is symmetric about $\varphi = \pi/2$.

In the context of Theorem 3.2.2, observation (i) demonstrates that, while the coalescence of eigenvalues of opposite Krein signatures is sufficient for stability phase transitions, it is not necessary. In other words, systems with KCs are not destabilized by *arbitrary* perturbations (in this case, for example, the weak linking of boundary sites preserves stability of OBCs for N even). Observations (ii) and (iii) quantify some of the notable even-odd effects evident in Fig. 3.6. Observation (iv) provides an analytical path for determining the phase boundaries. Specifically, if we determine those BCs which support ZMs, we have effectively determined the phase boundaries. Calculations detailed in Appendix B.2 reveal that phase boundaries are parameterized by

$$\cos(\varphi) = \pm\frac{1}{2}\begin{cases}(s+s^{-1})\text{sech}(Nr), & N \text{ even},\\ 2\text{sech}(Nr), & N \text{ odd},\end{cases} \tag{3.53}$$

which are plotted in Fig. 3.6a–d as white dashed lines. With this, we can more explicitly study the distribution of EPs and KCs along the boundaries. In Appendix B.2, we show that the ZMs host one or two Jordan chains of length two for N odd and even, respectively. Beyond zero frequency, we can numerically assess whether a particular eigenvalues hosts an EP or KC by computing the distance between coalescing eigenvectors (and correcting for arbitrary phases). For odd N between 5 and 55, and for various choices of $\Delta/J \in (0, 1)$, we find that two eigenvectors coalesce at degenerate eigenvalues along the phase boundaries. Thus, each eigenvalue hosts a Jordan chain of length 2. For N even, degeneracies in the spectrum make things more complicated. However, we conjecture that a similar behavior will hold in the case N odd. Altogether, the phase boundaries for $s \neq 0$ are EPs while the $s = 0$ (N odd) phase boundary hosts KCs.

Equation (3.53) further allows us to determine the minimal height[13] and width of the stability regions:

$$\delta s_N^{\min} = \begin{cases}e^{-Nr}, & N \text{ odd},\\ 0, & N \text{ even},\end{cases} \qquad \sin\left(\frac{\delta\varphi_N^{\min}}{2}\right) = \text{sech}(Nr). \tag{3.54}$$

[13] We note that in Ref. [27] an analogous quantity to $\delta s_N^{\min}$ was computed for the Hatano-Nelson chain. As shown in Ref. [10], the dynamical matrix of the BKC (under OBCs and PBCs) is unitarily equivalent to two-copies of the Hatano-Nelson chain. However, generic values of s and φ explicitly break this correspondence.

The exponential decrease of these two quantities as $N \to \infty$ indicates that the stable region of boundary phase space quickly approaches a set of measure zero, i.e., the left and right boundaries rapidly coalesce. Thus, in the case of N even, the right boundary hosts arbitrarily small frequency modes. This motivates the following conjecture:

Conjecture 3.3.2 Generically, ZMs indicate stability phase transitions in QBHs.

While we will examine the physical and mathematical validity of the conjecture more closely in Sect. 5.2.3, both the single-mode example and the current analysis of the BKC provide two pieces of immediate supporting evidence.

An additional point of contact between the BKC with $J > \Delta$ and the single-mode model is the interplay between EP and KC boundaries. For the BKC, the only KC-dominated phase boundary is $s = 0$ (for N odd) which, physically, represents one BC: OBCs. Thus, $s = 0$ can be safely interpreted as one point in parameters space. Similarly, the KC in the single-mode model occurs at one point in parameter space, namely $\alpha = \beta = 0$. Meanwhile, in both models, the EP phase boundaries are one-dimensional lines in the two-dimensional phase space. Thus, we present a second conjecture:

Conjecture 3.3.3 Generically, the $(d - 1)$-dimensional phase boundaries of d-dimensional stability phase diagram of a QBH are characterized by EPs, whereas the $(d - 2)$-dimensional boundaries are characterized by KCs.

Krein Phase Rigidity and Sensitivity to System Size The rich structure of the BKC stability phase diagram offers a non-trivial venue for assessing the utility of the KPR in detecting stability phase transitions. Explicitly, we assess the validity of Claim 3.2.5 by computing the KPR of a representative eigenvector of the BKC throughout the boundary phase space, see Fig. 3.8. We confirm that the KPR

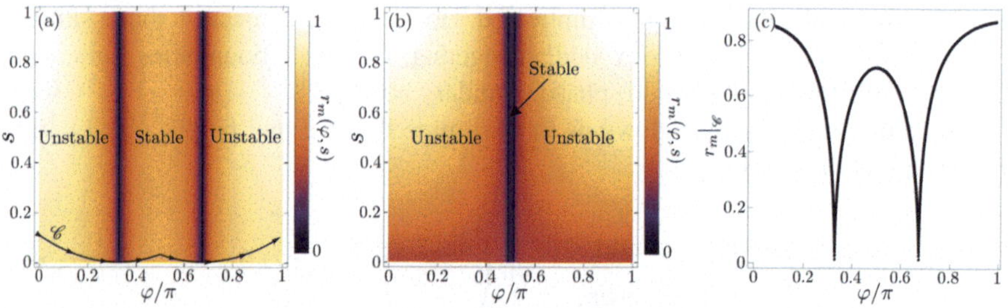

Fig. 3.8 The KPR evaluated numerically as a function of boundary parameters s, φ, on a grid of spacing $\Delta s = \Delta\Phi/\pi = 0.002$, for $J = 1$ and $\Delta = 0.25$. **(a)** $N = 5$ **(b)** $N = 15$ (we restrict to N odd to avoid difficulties in maintaining continuous eigenvector-tracking in the presence of the doubly degenerate spectrum for N even). **(c)** The KPR evaluated along the contour $\mathscr{C}$ in **(a)**, defined by $s(\varphi) \equiv (\varphi - \varphi^-)^2$, for $\varphi < \pi/2$, and $(\varphi - (\pi - \varphi^-))^2$ otherwise, with φ^- being the angle defining the left dynamical phase boundary. Earlier versions of these figures first appeared in Ref. [1]

vanishes unambiguously at the EP phase boundaries ($s > 0$). As in the single- and two-mode examples, the behavior around the KCs at $s = 0$ (N odd) is more subtle. However, we expect, and find, that the KPR should vanish on smooth paths through the KC boundary. Explicitly, Fig. 3.8c shows that the KPR vanishes as we pass from the stable region $\varphi \in I_{\varphi_N}(s)$ to the unstable $\varphi \in I_{\varphi_N}(s)$ through the point $s = 0$. As before, diagonalizability is retained over the contour.

We have thus far seen that the stability phase diagram responds to changes in system size in a rather dramatic way. Specifically, the regions of stability shrink exponentially fast in system size. We can refine this analysis by looking at the response of both the normal modes and their associated frequencies to system size. For the normal modes, the KPR is a useful tool to this end. Consider the KPR $r_n(\varphi)$ of a representative eigenvector $\vec{\psi}_n(\varphi)$, with $s = 1$ and φ arbitrary. With reference to Fig. 3.9, we see that as N increases, the KPR converges to a continuous curve that sharply vanishes at $\varphi = \pi/2$, akin to a 'stability order parameter'. We can contextualize the numerical results for $J = 1$, $\Delta = 0.5$ (Fig. 3.9a) by computing the KPR analytically for $J = \Delta$ using eigenvectors Eqs. (B.16)–(B.19). With N odd, we have

$$r_n(\varphi) = \frac{1}{\|\vec{\psi}(\varphi)_n\|^2} = \frac{N|\cos(\varphi)|}{|\cos\varphi|^2 - 1}\left(|\cos\varphi|^{2/N} - 1\right), \quad J = \Delta, \quad \varphi \in [0, \pi], \tag{3.55}$$

which is plotted for varies N in Fig. 3.9b. In the limit $N \to \infty$, we have

$$\lim_{N\to\infty} r_n(\varphi) = \frac{|\cos\varphi| \ln(|\cos\varphi|)}{|\cos\varphi|^2 - 1}. \tag{3.56}$$

Ultimately, we see that the normal modes, which are understood through the non-singular behavior KPR of the associated eigenvectors, respond in a rather tame fashion.

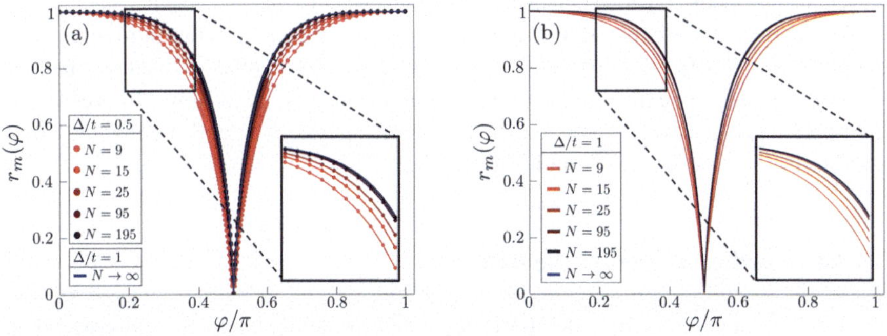

Fig. 3.9 The response of the KPR to system size (**a**) around the phase boundary near $\varphi = \pi/2$ for $J = 1$, $\Delta = 0.5$ and (**b**) around the EP near $\varphi = \pi/2$ for $J = \Delta = 1$. In (**a**), the blue line follows from Eq. (3.56). Earlier versions of these figures first appeared in Ref. [1]

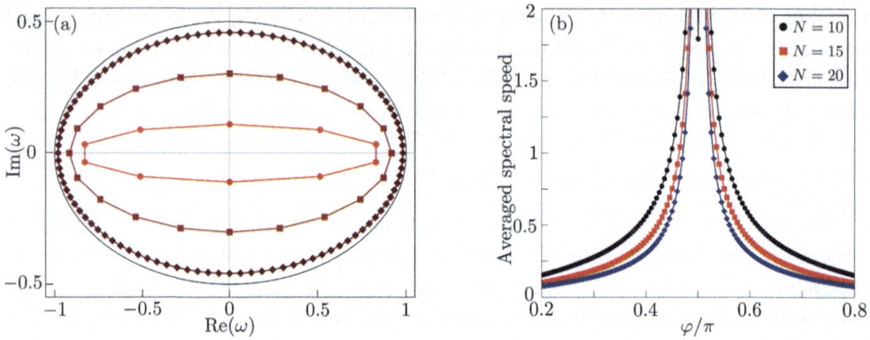

Fig. 3.10 (**a**) The spectrum with $J = 1$, $\Delta = 0.5$, $s = 1$, and $\varphi = 0.99 \cdot (\pi/2)$ for system sizes, $N = 10, 20$, and 100, from inner to outer. The solid outermost ellipse traces out the periodic spectrum for $N \to \infty$. (**b**) The spectral speed $d|\omega_n|/d\varphi$ for same J and Δ for various N, averaged over all eigenvalues. Earlier versions of these figures first appeared in Ref. [1]

The normal mode spectrum, on the other hand, behaves in a much more dramatic way as $N \to \infty$. In Fig. 3.10a, we show the normal mode frequencies as a function of N, with $s = 1$ and $\varphi = 0.99 \cdot (\pi/2)$. The choice of s and φ places the system extremely close to the stability phase transition near $\varphi = \pi/2$. As N increases, we see that the spectrum clings closer and closer to the PBC ellipse. The spectra must then must collapse down at an increasingly high speed, which we may quantify in terms of the *spectral speed* $d|\omega_n|/d\varphi$. We conjecture that as $N \to \infty$, the spectral speed behaves in a singular fashion according to $d|\omega_n|/d\varphi \to \delta(\varphi - \pi/2)$. The spectral speed is plotted for $J = 1$ and $\Delta = 0.5$ in Fig. 3.10b. As before with the KPR, we may leverage the analytical solutions (Eq. (3.51)) at $J = \Delta$ to test our conjecture in this particular case. We find

$$\frac{d|\omega_n|}{d\varphi} = \frac{1}{N} |\cos\varphi|^{1/N-1}, \tag{3.57}$$

which, as conjectured more generally, is proportional to $\delta(\varphi - \pi/2)$ as $N \to \infty$.

Altogether, the system-size dependence of the normal modes and their frequencies are drastically different. The normal modes evolve in a tame manner while the spectrum exhibits an extreme non-analyticity in its flow. Interestingly, a similar discrepancy between the behavior of eigenvectors and eigenvalues in a non-Hermitian system was found in Ref. [28].

3.3.3.4 Further Aspects of the BKC

We have demonstrated that the BKC responds dramatically to the presence of a family of non-trivial BCs. This extends the observations regarding the sharp change in stability phase between OBCs and PBCs made in Ref. [10], and in particular, has allowed us to put our theory of stability phase boundaries of QBHs to the test. As pointed out in Ref. [10], one physical implication of this sensitivity comes in the form of *signal amplification*. For concreteness, let us consider the BKC under OBCs[14] and suppose we have a state initialized in such a way that $\langle x_j\rangle(0) = \delta_{j,j_0}$ with j_0 somewhere near the center of the chain. As the state evolves, locality demands that the signal, i.e., the distribution of quadrature expectation values, remains 'unaware' of the presence of the boundaries for some time t_N. Here, t_N should be upper bounded by $d_N/v_{\rm LR}$ where d is the minimum distance from j_0 to the closest boundary and $v_{\rm LR}$ is the Lieb-Robinson velocity of the chain. For times $t < t_N$, the signal should evolve as though there were no boundaries at all. Since the boundaries are the stabilizing feature of the BKC, the signal should amplify.

This phenomenon of "transient amplification" will play a major role in the second half of this thesis. In particular, we will see that this behavior (in the more general open-system context) is a consequence of the dynamical matrix possessing non-trivial "*pseudospectra*" of the dynamical matrix. While we defer the detailed discussion to Sect. 8.1.2, we may intuitively describe pseudospectra as the set of *approximate* normal mode frequencies. Concretely, consider the bulk (PBC) normal mode $\xi_+(k)$ whose complex normal mode frequency is $\omega_+(k)$. While this is *not* a normal mode under OBCs, we do have the peculiar identity

$$[H^{\rm OBC}, \xi_+(k)] = -\omega_+^*(k)\xi_+(k) + \frac{1}{\sqrt{N}}\left(e^{i(N+1)k}\frac{(J-\Delta)}{2}p_N - \frac{(J+\Delta)}{2}p_1\right) \tag{3.58}$$

If we were to ignore the boundary term (which vanishes like $N^{-1/2}$ as $N \to \infty$), then we would conclude that $\xi_+(k)$ *is* a normal mode under OBCs with frequency $\omega_+(k)$. In this sense, $\xi_+(k)$ is an approximate *unstable* normal mode: it behaves more and more like a normal mode as N grows. Thus, if the initial wavepacket is a simple linear combination of these approximate unstable modes, it too will behave initially unstable. We will later recognize $\xi_+(k)$ and $\omega_+(k)$ as a pseudonormal mode, pseudoeigenvalue pair.

Finally, we briefly mention that an experimental proposal for realizing the BKC is given in Ref. [10]. We discuss further experimental capabilities that may be useful for additional modifications to the original Hamiltonian in Chap. 10.

References

1. V.P. Flynn, E. Cobanera, L. Viola, New J. Phys. **22**, 083004 (2020)
2. V.P. Flynn, E. Cobanera, L. Viola, Phys. Rev. Lett. **127**, 245701 (2021)
3. V.P. Flynn, E. Cobanera, L. Viola, Phys. Rev. B **108**, 214312 (2023)

[14] This argument is equally valid for any stabilizing BC, i.e., $\pi/2$-twisted BCs.

4. C.M. Bender, S. Boettcher, Phys. Rev. Lett. **80**, 5243 (1998)
5. C.M. Bender, Rep. Prog. Phys. **70**, 947 (2007)
6. S. Van Langen, P. Brouwer, C. Beenakker, Phys. Rev. E **55**, R1 (1997)
7. P.W. Brouwer, Phys. Rev. E **68**, 046205 (2003)
8. E.N. Bulgakov, I. Rotter, A.F. Sadreev, Phys. Rev. E **74**, 056204 (2006)
9. Y.-K. Lu, P. Peng, Q.-T. Cao, D. Xu, J. Wiersig, Q. Gong, Y.-F. Xiao, Sci. Bull. **63**, 1096 (2018)
10. A. McDonald, T. Pereg-Barnea, A.A. Clerk, Phys. Rev. X **8**, 041031 (2018)
11. C.M. Bender, P.D. Mannheim, Phys. Lett. A **374**, 1616 (2010)
12. R. Zhang, H. Qin, J. Xiao, J. Math. Phys. **61**, 012101 (2019)
13. A. Mostafazadeh, J. Math. Phys. **43**, 205; *ibid.* 2814; *ibid.* 3944 (2002)
14. L. Solombrino, J. Math. Phys. **43**, 5439 (2002)
15. G. Scolarici, L. Solombrino, J. Math. Phys. **44**, 4450 (2003)
16. I. Gohberg, P. Lancaster, L. Rodman, *Indefinite Linear Algebra and Applications* (Springer, 2005)
17. R. Ascoli, G. Teppati, C. Garola, L. Solombrino, *Physical Reality and Mathematical Description*, ed. by C.P. Enz, J. Mehra (Springer-Netherlands, Dordrecht, 1974), p. 239
18. V.A. Yakubovich, V.M. Starzhinskiĭ, *Linear Differential Equations with Periodic Coefficients* (Wiley, New York, 1975)
19. V. Peano, H. Schulz-Baldes, J. Math. Phys. **59**, 031901 (2018)
20. H. Schulz-Baldes, C. Villegas-Blas, J. Phys. A **49**, 405201 (2016)
21. Y. Nakamura, M. Mine, M. Okumura, Y. Yamanaka, Phys. Rev. A **77**, 043601 (2008)
22. T. Kato, *Perturbation Theory for Linear Operators*, vol. 132 (Springer Science & Business Media, 2013)
23. A. Kitaev, Phys. Usp. **44**, 131 (2001)
24. Q.-R. Xu, V.P. Flynn, A. Alase, E. Cobanera, L. Viola, G. Ortiz, Phys. Rev. B **102**, 125127 (2020)
25. S. Yao, Z. Wang, Phys. Rev. Lett. **121**, 086803 (2018)
26. K. Kawabata, R. Kobayashi, N. Wu, H. Katsura, Phys. Rev. B **95**, 195140 (2017)
27. R. Koch, J.C. Budich, Eur. Phys. J. D **74** (2020)
28. S. Longhi, Ann. Phys. **531**, 1900054 (2019)

Chapter 4
The Role of Pairing in Dynamically Stable QBHs

Abstract This chapter is dedicated to a set of results specialized to dynamically stable QBHs. In particular, we import mathematical techniques developed for pseudo-Hermitian quantum systems in order to construct a duality transformation that restores number symmetry in dynamically stable QBHs with pairing. We find that the duality transformation is intrinsically related to the covariance matrix of the quasi-particle vacuum of the QBH. Implications for analogue simulation of non-Hermitian quantum dynamics, as well as topological invariants and edge states are discussed. We specialize our results to a general dynamically stable single-band models, exemplified by a gapped harmonic chain, as well as the BKC considered previously.

In this chapter, we uncover the existence of a duality between number-non-conserving, but dynamically stable, QBHs and number-conserving ones.[1] Ultimately, we characterize the role of pairing in dynamically stable QBHs. Thus far, we have seen that bosonic pairing, or equivalently, parametric amplification, terms in QBHs play a key role in inducing dynamical instabilities. From a dynamical perspective, nonzero pairing terms are necessary and sufficient for non-Hermiticity of the dynamical matrix. From the many-body perspective, pairing terms explicitly break total bosonic number conservation, and thus, are ubiquitous for Hamiltonians describing massless particles such as phonons and photons. Beyond this, just as they do in fermionic systems (e.g., superconductors), pairing may arise for massive bosons in a mean-field context. There is an intuitive link between these two observations: namely, the loss of number conservation allows for the indefinite (coherent, as it is mediated by the Hamiltonian) pumping of particles into the system, and hence, instabilities in the form of unbounded particle number become relevant. We have argued that the crucial ingredient for this phenomenon is the infinite-

[1] With the exception of the connection between the duality and the covariance matrix of the vacuum state, the vast majority of content in this chapter originates in Ref. [1]. The aforementioned connection is instead established in Ref. [2]. Both of these works were jointly co-authored with Emilio Cobanera & Lorenza Viola.

V. P. Flynn, *Effective Non-Hermiticity and Topology in Markovian Quadratic Bosonic Dynamics*, Springer Theses, https://doi.org/10.1007/978-3-031-52045-7_4

dimensional nature of the single-mode Hilbert space. Loosely speaking, there is 'room' for infinitely many bosons. Fermions stand in sharp contrast whereby the finite-dimensional single-mode space prevents any mechanism, including pairing, from driving these types of instabilities. While this intuitive link is pleasing, we know from previous examples that bosonic pairing *need not* elicit instabilities. Rather, there is typically a competition between the stabilizing (e.g., hopping, on-site potentials) and these destabilizing mechanisms. So then, what is the precise role of pairing in dynamically stable systems?

The key observation for answering this question is the following: a dynamically stable QBH is entirely characterized by a dynamical matrix that is (i) pseudo-Hermitian, (ii) diagonalizable, and (ii) possesses an entirely real spectrum. Leveraging known mathematical results on the properties of pseudo-Hermitian matrices, we are are able to uncover a canonical mapping from this pseudo-Hermitian matrix, which acts on an indefinite inner-product space, to a *Hermitian* matrix that acts on a true Hilbert space. We ultimately show that this new matrix uniquely defines a number-conserving QBH and that the mapping itself lifts to a unitary one on the many-body Fock space. In fact, this mapping can be thought of as a *duality transformation*. In contrast to general unitary mappings, dualities are typically constrained to transform the microscopic degrees of freedom in such a way that preserves certain desirable features of the original Hamiltonian (e.g., locality, spectrum), while reducing the 'complexity' of the system in some predefined way [3–5]. For example, there a famous duality that connects the high- and low-temperature physics of the Ising model defined on a square lattice allows one to analytically determine the Curie temperature [6]. In our case, the duality transformation will remove the number-non-conserving terms in a natural way so that the dynamical stability of the system becomes evident. We also discuss a promising application of our dualities to the problem of simulating (in an analog fashion) both number non-conserving QBHs, as well as genuinely non-Hermitian PT-symmetric Hamiltonians. In particular, our duality provides us, under suitable conditions, a method for simulating such systems without the need for any destabilizing mechanisms such as pairing or incoherent driving.

Generally speaking, dualities can drastically transform symmetries of the Hamiltonian (e.g., a broken number symmetry of one system being mapped to the broken translational symmetry of the other [7, 8]). This feature, known as *symmetry transmutation*, can be useful for identifying and understanding symmetries via the dual system. In our case, the preimage of the total number symmetry turns out to be the total quasiparticle number symmetry—something that any dynamically stable QBH must posses (recall Sect. 2.4). Spatial symmetries are more subtle, however. While translational symmetry survives the duality, the locality structure of the couplings may change. However, by establishing a direct link between the duality transformation and the covariance matrix of the original vacuum state, we can infer locality properties of the dual system by means of vacuum correlations.

We explicitly construct the duality transformation in two models of interest: (i) A gapped harmonic chain; and (ii) the bosonic Kitaev chain under open and $\pi/2$-twisted BCs. In (i), we show that the number-conserving dual is a hopping model with real hopping amplitudes that decay exponentially with distance. In this sense, the dual Hamiltonian is still "short-range"—a hypothesis we put to the test by explicitly comparing the band structure of the original QBH and the dual with couplings truncated at a fixed, finite range. We further conclude that the analysis of the gapped harmonic chain extends naturally to any translationally invariant single-band model. In (ii), we show that the BKC may be mapped, by means of a perfectly local duality, to a number conserving chain with nearest neighbor imaginary hopping. This mapping allows us to nontrivially extend the analytical stability analysis of Sect. 3.3.3.

The outline for this chapter is as follows. In Sect. 4.1, we establish the existence of the duality and provide an explicit construction for arbitrary, dynamically stable QBHs. We further discuss various physical features of the duality including symmetry transmutation properties and its connection to the vacuum covariance matrix. In Sect. 4.2, we present the two example models: a gapped Harmonic chain, and the BKC. In both cases, we explicitly construct the duality and discuss important locality properties.

4.1 A Number Conservation-Restoring Duality Transformation

4.1.1 Existence and Construction of the Duality

The Hamiltonians of interest are now those QBHs that are dynamically stable. We place no restrictions on thermodynamic instability. However, being a unitary map, our duality transformation will preserve the boundedness properties of the Hamiltonian, and hence, the thermodynamic features. From Sect. 2.4, we know such Hamiltonians can be characterized in terms of their dynamical matrices. Specifically, $H = \Phi^\dagger \tau_3 \mathbf{G} \Phi/2$ is dynamically stable if and only if the dynamical matrix $\mathbf{G}$ is diagonalizable and possesses an entirely real spectrum.

Non-Hermitian matrices and operators possessing entirely real spectra are well-studied for, among other reasons, their potential for serving as appropriate generalizations of Hamiltonians modeling physical systems [9–12] (see also, Sect. 3.1.1). The most important result for this chapter is the following (adapted and specialized from Refs. [12, 13]).

Lemma 4.1.1 *Let* $\mathbf{M}$ *be an* $n \times n$ *complex matrix and let* $(\vec{v}, \vec{w}) = \vec{v}^\dagger \vec{w}$, *with* $\vec{v}, \vec{w} \in \mathbb{C}^n$, *denote the standard inner-product on* $\mathbb{C}^n$. *Then the following are equivalent.*

(i) $\mathbf{M}$ *is diagonalizable and possesses an entirely real spectrum;*
(ii) $\mathbf{M}$ *is pseudo-Hermitian with respect to a positive-definite metric* $\mathbf{S}$*;*
(iii) There exists a positive-definite metric $\mathbf{S}$ *such that* $\mathbf{M}$ *is a self-adjoint operator when viewed as acting on the Hilbert space* $\mathbb{C}^n$ *paired with the inner-product* $(\vec{v}, \mathbf{S}\vec{w}) = \vec{v}^\dagger \mathbf{S}\vec{w}$, *where* $\vec{v}, \vec{w} \in \mathbb{C}^n$.

We remark that, in (iii), the positive-definite nature of $\mathbf{S}$ ensures that $(\cdot, \mathbf{S}\cdot)$ is a proper inner-product (rather than just an indefinite inner-product). We have an immediate corollary for dynamically stable QBHs.

Corollary 4.1.2 *Let H be a dynamically stable N-mode QBH with dynamical matrix $\mathbf{G}$. Then there exists a positive-definite metric $\mathbf{S}$ with respect to which $\mathbf{G}$ is pseudo-Hermitian.*

A priori, it may seem that we haven not gained much information. We already know that $\mathbf{G}$ is pseudo-Hermitian. However, general QBHs are pseudo-Hermitian with respect to $\boldsymbol{\tau}_3$, which is *not* a positive-definite metric. Thus, when dynamical stability is present, at least one such $\mathbf{S}$ exists, is distinct from $\boldsymbol{\tau}_3$. In fact, it is possible to explicitly construct such a metric.

Proposition 4.1.3 *Let $\mathbf{G}$ be the dynamical matrix of an N-mode dynamically stable QBH. If $\{\vec{\psi}_n^{\pm}\}_{n=1}^N$ is a basis of $\mathbb{C}^{2N}$ consisting of eigenvectors of $\mathbf{G}$ satisfying Eq. (2.63), then*

$$\mathbf{S} \equiv \sum_{n=1}^{N} \boldsymbol{\tau}_3 \left(\vec{\psi}_n^{+} \vec{\psi}_n^{+\dagger} + \vec{\psi}_n^{-} \vec{\psi}_n^{-\dagger} \right) \boldsymbol{\tau}_3 \tag{4.1}$$

is positive-definite and satisfies (i) $\mathbf{S}^{-1}\mathbf{G}^\dagger\mathbf{S} = \mathbf{G}$, (ii) $\boldsymbol{\tau}_3\mathbf{S}\boldsymbol{\tau}_3 = \mathbf{S}^{-1}$, and (iii) $\boldsymbol{\tau}_1\mathbf{S}\boldsymbol{\tau}_1 = \mathbf{S}^*$.

Proof Positive definiteness of $\mathbf{S}$ follows from the fact that it is a sum of rank-one (unnormalized) projectors $\boldsymbol{\tau}_3\vec{\psi}_n^{s}\vec{\psi}_n^{s\dagger}\boldsymbol{\tau}_3 = \vec{v}_n^{s}\vec{v}_n^{s\dagger}$, with $\vec{v}_n^{s} = \boldsymbol{\tau}_3\vec{\psi}_n^{s}$ and $s \in \{+, -\}$.

To prove (i), we first note that it is equivalent to $\mathbf{G}^\dagger\mathbf{S} = \mathbf{SG}$. By direct computation,

$$\begin{aligned}
\mathbf{G}^\dagger\mathbf{S} &= \sum_{n=1}^{N} \mathbf{G}^\dagger\boldsymbol{\tau}_3 \left(\vec{\psi}_n^{+} \vec{\psi}_n^{+\dagger} + \vec{\psi}_n^{-} \vec{\psi}_n^{-\dagger} \right) \boldsymbol{\tau}_3 = \sum_{n=1}^{N} \boldsymbol{\tau}_3\mathbf{G} \left(\vec{\psi}_n^{+} \vec{\psi}_n^{+\dagger} + \vec{\psi}_n^{-} \vec{\psi}_n^{-\dagger} \right) \boldsymbol{\tau}_3 \\
&= \sum_{n=1}^{N} \boldsymbol{\tau}_3 \left(\omega_n \vec{\psi}_n^{+} \vec{\psi}_n^{+\dagger} - \omega_n \vec{\psi}_n^{-} \vec{\psi}_n^{-\dagger} \right) \boldsymbol{\tau}_3 = \sum_{n=1}^{N} \boldsymbol{\tau}_3 \left(\vec{\psi}_n^{+} \vec{\psi}_n^{+\dagger} - \vec{\psi}_n^{-} \vec{\psi}_n^{-\dagger} \right) \mathbf{G}^\dagger\boldsymbol{\tau}_3 \\
&= \sum_{n=1}^{N} \boldsymbol{\tau}_3 \left(\vec{\psi}_n^{+} \vec{\psi}_n^{+\dagger} - \vec{\psi}_n^{-} \vec{\psi}_n^{-\dagger} \right) \boldsymbol{\tau}_3\mathbf{G} = \mathbf{SG},
\end{aligned}$$

as desired. Meanwhile, (ii) is equivalent to $\boldsymbol{\tau}_3\mathbf{S}\boldsymbol{\tau}_3\mathbf{S} = \mathbb{1}_{2N}$. We can establish this by leveraging the mutual $\boldsymbol{\tau}_3$ inner-products of the basis elements in addition to the resolution of the identity in Eq. (2.64). Explicitly,

$$\boldsymbol{\tau}_3\mathbf{S}\boldsymbol{\tau}_3\mathbf{S} = \sum_{n,m=1}^{N} \left(\vec{\psi}_n^{+} \vec{\psi}_n^{+\dagger} + \vec{\psi}_n^{-} \vec{\psi}_n^{-\dagger} \right) \boldsymbol{\tau}_3 \left(\vec{\psi}_m^{+} \vec{\psi}_m^{+\dagger} + \vec{\psi}_m^{-} \vec{\psi}_m^{-\dagger} \right) \boldsymbol{\tau}_3$$

$$= \sum_{n,m=1}^{N} \left(\delta_{nm} \vec{\psi}_n^+ \vec{\psi}_m^{+\dagger} - \delta_{nm} \vec{\psi}_n^- \vec{\psi}_m^{-\dagger} \right) \boldsymbol{\tau}_3 = \mathbb{1}_{2N}.$$

Finally, (iii) can be shown by making use of the identity $\vec{\psi}_n^{\mp} = \boldsymbol{\tau}_1 (\vec{\psi}_n^{\pm})^*$. With this,

$$\begin{aligned}
\boldsymbol{\tau}_1 \mathbf{S} \boldsymbol{\tau}_1 &= \sum_{n=1}^{N} \boldsymbol{\tau}_1 \boldsymbol{\tau}_3 \left(\vec{\psi}_n^+ \vec{\psi}_n^{+\dagger} + \vec{\psi}_n^- \vec{\psi}_n^{-\dagger} \right) \boldsymbol{\tau}_3 \boldsymbol{\tau}_1 \\
&= (-1)^2 \sum_{n=1}^{N} \boldsymbol{\tau}_3 \boldsymbol{\tau}_1 \left(\vec{\psi}_n^+ \vec{\psi}_n^{+\dagger} + \vec{\psi}_n^- \vec{\psi}_n^{-\dagger} \right) \boldsymbol{\tau}_1 \boldsymbol{\tau}_3 \\
&= \sum_{n=1}^{N} \boldsymbol{\tau}_3 \left(\boldsymbol{\tau}_1 \vec{\psi}_n^+ \vec{\psi}_n^{+\dagger} \boldsymbol{\tau}_1 + \boldsymbol{\tau}_1 \vec{\psi}_n^- \vec{\psi}_n^{-\dagger} \boldsymbol{\tau}_1 \right) \boldsymbol{\tau}_3 \\
&= \sum_{n=1}^{N} \boldsymbol{\tau}_3 \left((\vec{\psi}_n^- \vec{\psi}_n^{-\dagger})^* + (\vec{\psi}_n^+ \vec{\psi}_n^{+\dagger})^* \right) \boldsymbol{\tau}_3 = \mathbf{S}^*.
\end{aligned}$$

□

In light of the third conclusion of Lemma 4.1.1, there is a sense in which the effective non-Hermiticity inherent to dynamically stable QBHs is artificial: **G** corresponds to a self-adjoint operator in a suitably modified Hilbert space. In particular, **G** possesses a **S**-orthonormal basis of eigenvectors (which, if **S** is the one we have explicitly constructed above, is the basis $\{\psi_n^{\pm}\}_{n=1}^N$, as one may verify). As a self-adjoint operator, any matrix representation of **G** in a **S**-orthonormal basis must be Hermitian. For example, the matrix representation of **G** in the basis $\{\psi_n^{\pm}\}_{n=1}^N$ is real and diagonal, and thus, Hermitian. Another choice of basis that need not require explicit diagonalization of **G** is $\{\vec{r}_j \equiv \mathbf{R}^{-1}\vec{e}_j\}_{j=1}^{2N}$, with **R** the unique positive-definite square root, i.e., $\mathbf{R}^2 = \mathbf{S}$ and $\mathbf{R} > 0$. By construction, $(\vec{r}_j, \mathbf{S}\vec{r}_k) = \vec{r}_j^\dagger \mathbf{S}\vec{r}_k = \vec{e}_j \mathbf{R}^{-1}\mathbf{R}^2\mathbf{R}^{-1}\vec{e}_k = \vec{e}_j^\dagger \vec{e}_k = \delta_{jk}$. The matrix representation of **G** in this basis is then given by

$$(\mathbf{G}_D)_{jk} = (\vec{r}_j, \mathbf{S}\mathbf{G}\mathbf{r}_k) = \vec{e}_j^\dagger \mathbf{R}\mathbf{G}\mathbf{R}^{-1}\vec{e}_k \Rightarrow \mathbf{G}_D \equiv \mathbf{R}\mathbf{G}\mathbf{R}^{-1}. \tag{4.2}$$

Hermiticity then follows straightforwardly from **S** pseudo-Hermiticity:

$$\mathbf{G}_D^\dagger = \mathbf{R}^{-1}\mathbf{G}^\dagger\mathbf{R} = \mathbf{R}^{-1}\mathbf{S}\mathbf{G}\mathbf{S}^{-1}\mathbf{R} = \mathbf{R}\mathbf{G}\mathbf{R}^{-1} = \mathbf{G}_D. \tag{4.3}$$

At this stage, however, it is not clear that $\mathbf{G}_D$ has any physical significance. For example, there is no guarantee that it can even be interpreted as a dynamical matrix of some QBH. If it can be, there need not be any relation between said QBH and the original one.

To proceed, we take $\mathbf{S}$ to be the specific choice of metric constructed in Eq. (4.1). With this, and by virtue of the uniqueness of the positive-definite square root, it follows that $\mathbf{R}$ (and $\mathbf{R}^{-1}$) satisfies properties (ii) and (iii) in Proposition 4.1.3. Remarkably, these properties are equivalent to the statement that the map $\Phi \mapsto \mathbf{R}^{-1}\Phi$ is a Gaussian canonical transformation (recall Eq. (2.18)). Thus, by the Stone-von Neumann theorem, there exists a (Gaussian) unitary operator U_D such that $U_D \Phi U_D^\dagger = \mathbf{R}^{-1}\Phi$. With this unitary in hand, we define the dual QBH:

$$H_D = U_D H U_D^\dagger = \frac{1}{2}\Phi^\dagger \boldsymbol{\tau}_3 \mathbf{G}_D \Phi. \tag{4.4}$$

It follows that $\mathbf{G}_D$ is a proper dynamical matrix that also happens to be Hermitian. Recalling that Hermiticity guarantees zero pairing, it follows that H_D has total number conservation: $[H, \sum_{j=1}^N a_j^\dagger a_j] = 0$. Several remarks are in order.

(i) Number conserving QBHs are, generically,[2] fixed points of the transformation constructed thus far. Since the dynamical matrix of a number-conserving QBH is Hermitian, we may take $\{\psi_n^\pm\}_{n=1}^N$ to be orthonormal (in addition to satisfying Eqs. (2.63)). Thus, from Eq. (4.1), $\mathbf{S} = \boldsymbol{\tau}_3 \mathbb{1}_{2N} \boldsymbol{\tau}_3 = \mathbb{1}_{2N}$. It follows that $\mathbf{R} = \mathbb{1}_{2N}$ and so $U_D = 1_{\mathcal{F}}$.

(ii) The fact that a dynamically stable QBH is unitarily equivalent to one with total number conservation may seem trivial, in hindsight. Taking U to be the unitary that takes the normal modes ψ_n (see Eq. (2.65)) to a_n, i.e., take U to be that of Eq. (2.67). In this case

$$U H U^\dagger = \frac{1}{2}\sum_{n=1}^{N} \omega_n \left(a_n^\dagger a_n + a_n a_n^\dagger\right), \tag{4.5}$$

which is manifestly number-conserving. However, to construct U, one must necessarily diagonalize the system. As we will see in examples, the transformation from H to H_D can often be identified *without* the need for full diagonalization. Moreover, in light of (i), number-conserving QBHs are not fixed points of this trivial 'duality'—diagonal Hamiltonians are. As such, our duality is optimal in the sense that it restores number conservation in this 'minimally invasive' way.

[2] The use of the word 'generically' here is to take into account the non-uniqueness of an eigenbasis, specifically in the presence of spectral degeneracies. For example, if there are degenerate eigenvalues, there may exist a basis of $\mathbf{G} = \mathbf{G}^\dagger$ that satisfies Eq. (2.63) but is not orthonormal. However, this is non-generic in the sense that spectral degeneracies are non-generic.

4.1.2 Physical Interpretation of the Duality

4.1.2.1 Symmetry Transmutation

Let us now explore the physical features of the number-symmetric dual Hamiltonian. The first notable feature is that the duality has brought forth a symmetry that did not previously exist in the system: number symmetry (equivalently, number conservation). Since our map is unitary, this symmetry must result from another symmetry intrinsic to dynamically stable QBHs. That is, if $U(\theta)$ is the number symmetry operator (Eq. (2.20)), then

$$[H_D, U(\theta)] = 0 \iff [H, U'(\theta)] = 0, \quad U'(\theta) \equiv U_D^\dagger U(\theta) U_D. \tag{4.6}$$

Thus, the continuous family of symmetries $U'(\theta)$ transmutes into number symmetry of the dual Hamiltonian. But what is $U'(\theta)$? To answer this, we write

$$U(\theta) = e^{i\theta \sum_{j=1}^N a_j^\dagger a_j} = e^{i\theta(\Phi^\dagger\Phi - N)/2}. \tag{4.7}$$

Using $U_D^\dagger \Phi U_D = \mathbf{R}\Phi$, we then obtain

$$U'(\theta) = U_D^\dagger e^{i\theta(\Phi^\dagger\Phi - N)/2} U_D = e^{i\theta(U_D^\dagger \Phi^\dagger \Phi U_D - N)/2} = e^{i\theta(\Phi^\dagger \mathbf{R}^2 \Phi - N)/2} \tag{4.8}$$

Now,

$$\begin{aligned}\Phi^\dagger \mathbf{R}^2 \Phi = \Phi^\dagger \mathbf{S} \Phi &= \sum_{n=1}^N \Phi^\dagger \boldsymbol{\tau}_3 \left(\vec{\psi}_n^+ \vec{\psi}_n^{+\dagger} + \vec{\psi}_n^- \vec{\psi}_n^{-\dagger}\right) \boldsymbol{\tau}_3 \Phi \\ &= \sum_{n=1}^N \left(\psi_n^\dagger \psi_n + \psi_n \psi_n^\dagger\right),\end{aligned} \tag{4.9}$$

where we have recalled the definition of the normal modes $\psi_n = \widehat{\vec{\psi}_n^+} = \vec{\psi}_n^{+\dagger} \boldsymbol{\tau}_3 \Phi = (-\vec{\psi}_n^{-\dagger} \boldsymbol{\tau}_3 \Phi)^\dagger$. Since the normal modes satisfy the CCRs, we have

$$U'(\theta) = e^{i\theta \sum_{n=1}^N \psi_n^\dagger \psi_n}, \tag{4.10}$$

which may be interpreted as the total quasiparticle number symmetry operator. Trivially, any dynamically stable QBH possesses this symmetry (as dynamical stability guarantees a normal mode basis in which H is diagonal). It so happens that this symmetry is mapped to the standard number symmetry under our duality.

Another example of symmetry transmutation is how translational symmetry transforms. For concreteness, let us consider periodic discrete translation symmetry,[3] i.e., $[H, V'_N] = 0$, with $V'_N \Phi V_N'^{\dagger} \equiv \mathbf{V}_N \Phi$ the many-body discrete translation operator. This symmetry is mapped to $V_N \equiv U_D V'_N U_D^{\dagger}$. This is best understood by its action on the Nambu array:

$$\begin{aligned} V_N \Phi V_N^{\dagger} &= U_D V'_N U_D^{\dagger} \Phi U_D V'_N U_D^{\dagger} = U_D \mathbf{R} V'_N \Phi V_N'^{\dagger} U_D^{\dagger} \\ &= \mathbf{R} \mathbf{V}_N U_D \Phi U_D^{\dagger} = \mathbf{R} \mathbf{V}_N \mathbf{R}^{-1} \Phi. \end{aligned} \tag{4.11}$$

Now, since $[H, V'_N] = 0$, we also have $[\mathbf{G}, \mathbf{V}_N] = 0$ (recall Eq. (2.77)). Therefore, the basis used to construct $\mathbf{S}$ (Eq. (4.1)), and thus $\mathbf{R}$, can be taken to be Bloch eigenstates. That is, $\vec{\psi}_n^{\pm} = \vec{\psi}_{n,k}^{\pm}$, with $\mathbf{V}_N \vec{\psi}_{n,k}^{\pm} = e^{ik} \vec{\psi}_{n,k}^{\pm}$, and k in the discrete Brillouin zone $\mathcal{K}_N$. Consequently, both $\mathbf{S}$ and $\mathbf{R}$ commute with $\mathbf{V}_N$, and so, $V_N = V'_N$. Thus, our duality preserves discrete translational symmetry.

4.1.2.2 Locality and Connections to the Vacuum Covariance Matrix

Suppose our original Hamiltonian is of range R, i.e., the hopping and pairing amplitudes $\mathbf{K}_{ij}$ and $\mathbf{\Delta}_{ij}$ (recall Eq. (2.38)) vanish for $|i - j| > R$. What then is the range of the dual Hamiltonian H_D? This is a rather subtle question, for which we can offer partial answers.

If one demands 'locality', i.e., that H_D be of range $R_D < R$, then an Ansatz for $\mathbf{S}$ may be developed. For example, consider a system with d_{int} degrees of freedom per lattice site and let us specify that $\mathbf{S}$ be site-local. That is, $\mathbf{S} = \sum_{j=1}^{N} \vec{e}_j \vec{e}_j^{\dagger} \otimes \mathbf{S}_j$, with $\mathbf{S}_j$ a $2d_{\text{int}} \times 2d_{\text{int}}$. These local matrices $\mathbf{S}_j$ are heavily constrained by the properties detailed in Proposition 4.1.3. For example, if $d_{\text{int}} = 1$, we have that[4]

$$\mathbf{S}_j = \cosh(\xi_j)\mathbb{1}_2 + \sinh(\xi_j)[\cos(\phi_j)\boldsymbol{\sigma}_1 + \sin(\phi_j)\boldsymbol{\sigma}_2], \tag{4.12}$$

with ξ_j and ϕ_j real constants to be determined from the constraint $\mathbf{S}^{-1}\mathbf{G}^{\dagger}\mathbf{S} = \mathbf{G}$. This procedure will find success in the BKC, as we will see. If less-strict locality requirements are implemented, this can be generalized to allow for $\mathbf{S}$ to be *quasi-local* (e.g., two-site local). Constraints on parameters can be found and used to solve for a valid $\mathbf{S}$.

For broader insight, a more concrete understanding of the matrices $\mathbf{S}$ is in order. First, let $\mathbf{T}^{-1}$ denote that matrix whose columns are eigenvectors of $\mathbf{G}$, i.e., Eqs. (2.62). Explicitly,

---

[3] We use the prime to be consistent with the number symmetry example, whereby the prime indicated the symmetry of the original Hamiltonian H.

[4] To see this, note that the most general single-mode Bogoliubov transformation is $a_j \mapsto \cosh(\xi_j)a_j + e^{i\phi_j}\sinh(\xi_j)a_j^{\dagger}$ modulo an overall phase. The matrix representation of this transformation is precisely $\mathbf{S}_j$. Validity of this as an Ansatz follows from positive-definitiness ($\mathbf{S}_j$ is clearly Hermitian with eigenvalues $e^{\pm\xi}$).

$$\mathbf{T}^{-1} = \sum_{j=1}^{N} \left(\vec{\psi}_n^{+} \vec{e}_{2j-1}^{\dagger} + \vec{\psi}_n^{-} \vec{e}_{2j}^{\dagger} \right). \tag{4.13}$$

It immediately follows that $\mathbf{S} = \boldsymbol{\tau}_3 \mathbf{T}^{-1} \mathbf{T}^{-1\dagger} \boldsymbol{\tau}_3 = \mathbf{T}^{\dagger}\mathbf{T}$. From here, we have a remarkable connection to the quasiparticle vacuum covariance matrix (Eq. (2.69)):

$$\mathbf{S}^{-1} = \boldsymbol{\tau}_3 \mathbf{S} \boldsymbol{\tau}_3 = \boldsymbol{\tau}_3 \mathbf{T}^{\dagger} \mathbf{T} \boldsymbol{\tau}_3 = 2\mathbf{C}_{|\widetilde{0}\rangle}, \quad \mathbf{C}_{|\widetilde{0}\rangle} = \frac{1}{2}\mathbf{S}^{-1}. \tag{4.14}$$

In words, the positive-definite metric in which $\mathbf{G}$ is self-adjoint with respect to is half the inverse of the quasiparticle covariance matrix of H. Thus, the locality properties of the duality are intrinsically tied to the locality properties of the quasiparticle vacuum correlations,[5] $(2\mathbf{C}_{|\widetilde{0}\rangle})_{ij} = \langle \widetilde{0} | \{\Phi_i, \Phi_j^{\dagger}\} | \widetilde{0} \rangle$. In the thermodynamically stable case, these are precisely the ground state correlations, which are particularly relevant to statistical mechanics-oriented investigations. In fact, the spatial decay of ground state correlations in harmonic lattices (thermodynamically stable QBHs of a particular form) has been a subject of much analysis [14, 15]. In particular, gapped systems are known to have ground states with exponentially decaying correlations.

4.1.2.3 Breakdown of the Duality at Stability Phase Transitions

Throughout this analysis, we have emphasized the necessity of dynamical stability. It is then of interest to understand how the duality transformation evolves as a dynamically stable QBH approaches a stability phase transition. For this, it is instructive to study tr[$\mathbf{S}$], which may be computed as

$$\text{tr}[\mathbf{S}] = \text{tr}[\boldsymbol{\tau}_3 \mathbf{S} \boldsymbol{\tau}_3] = \sum_{i=1}^{2N} \sum_{n=1}^{N} \left(|\vec{e}_i^{\dagger} \vec{\psi}_n^{+}|^2 + |\vec{e}_i^{\dagger} \vec{\psi}_n^{-}|^2 \right) = \sum_{n=1}^{N} \left((r_n^{+})^{-1} + (r_n^{-})^{-1} \right), \tag{4.15}$$

where we have inserted the identity $\mathbb{1}_N = \boldsymbol{\tau}_3^2$ and employed cyclicity of the trace in the first equality, and reintroduced the KPR $r_n^{\pm} = \|\vec{\psi}_n^{\pm}\|^{-2}$ (recall Eq. (3.6)) in the third equality.

Suppose we adjust the system parameters so that H smoothly approaches a stability phase transition. Leveraging Claim 3.2.5, we know that at least one $r_n^{\pm}$ must go to zero. Since $r_n^{\pm} \geq 0$, it follows that tr[$\mathbf{S}$] (and, accordingly, some eigenvalue of $\mathbf{S}$), diverges. In other words, $\mathbf{S}$ becomes unbounded, and hence ill-defined, as the transition is approached. Moreover, since tr[$\mathbf{S}$] = tr[$\boldsymbol{\tau}_3 \mathbf{S} \boldsymbol{\tau}_3$] = tr[$\mathbf{S}^{-1}$], we also have that $\mathbf{S}$ develops a zero eigenvalue. Thus, $\mathbf{S}$ loses invertibility at the transition point.

[5] Note that the mean vector of the quasiparticle vacuum is always zero.

Each conclusion we have made about $\mathbf{S}$ carries over to $\mathbf{R}$, and thus to the transformation $\Phi \mapsto U_D \Phi U_D^\dagger = \mathbf{R}^{-1}\Phi$. Physically, this means that the duality transformation requires more and more squeezing as the transition is approached. From the preceding section, we can further conclude that the quasiparticle correlations also become unbounded in the same limit. In this sense, our duality informs the behavior of the system in the vicinity of stability phase transitions.

4.1.2.4 Implications for Analog Quantum Simulation

The primary goal of analog (as opposed to gate-based) quantum simulation is to indirectly implement a given target Hamiltonian with an alternative (usually, more accessible) system [16]. Our duality is well-poised for such an application. Specifically, suppose we wish to realize a set of N, parametrically driven (i.e., paired) bosonic modes, described by a target QBH. While there exist techniques for realizing such terms [17–25], amplification necessarily introduces (undesired) complexities in the physical implementation. As long as the target Hamiltonian is dynamically stable, our duality transformation eliminates this issue by providing a unitarily equivalent Hamiltonian—*without* the need for driving. With sufficient knowledge of the duality transformation, one can map physical quantities of interest from one system to the other, allowing for indirect measurement of the target system. The implicit assumption here is that the transformation is sufficiently local. Explicitly, locality ensures that local quantities can be measured locally in the dual system. If the transformation is not local, we still expect that, in a large number of cases, the dual Hamiltonian will have couplings that drop off exponentially in distance. The reasoning comes from the connection between the duality transformation and the quasiparticle vacuum covariance matrix (Eq. (4.14)) and the fact that vacuum correlations are known to decay exponentially in a large class of QBHs [14, 15]. In these cases, the dual Hamiltonian should faithfully implement the original system when truncated to finite-range. This hypothesis will be explored in our first example below.

This proposal can be extended naturally to utilizing number-conserving QBHs for simulating *truly* non-Hermitian, PT-symmetric Hamiltonians. The implementations of such systems, which have been investigated across photonic, optomechanical, and cavity QED settings [26, 27], typically require precisely tuned (dissipative) loss and gain. Such mechanisms inevitably entail unwanted noise. Motivated by this challenge, a related question was addressed in Ref. [28]: under which conditions can a target PT-symmetric Hamiltonian be faithfully realized in a closed system of non-interacting bosons. In that work, it was found that there exists a class of PT-symmetric non-Hermitian Hamiltonians that can be unitarily mapped to the dynamical matrix of a QBH with pairing. Our duality can further simplify the implementation of said systems conditional upon dynamical stability. That is, if the given PT-symmetric Hamiltonian is in the PT-unbroken phase (recall Sect. 3.1.1), then its realization in a QBH will be dynamically stable. Applying our duality will then produce a unitarily equivalent QBH without the need for pairing. Not only does this dramatically reduce the complexity of implementation, but it also eliminates the extreme sensitivity to parameter variation characteristic of parametric amplification.

4.2 Examples

4.2.1 A Gapped Harmonic Chain and General Single-Band Pairing Chains

Our first example is a gapped harmonic chain (GHC) under PBCs:

$$H \equiv \sum_{j=1}^{N} \left(\frac{p_j^2}{2m} + \frac{C_o}{2} x_j^2 + \frac{C_{nn}}{2} (x_{j+1} - x_j)^2 \right), \tag{4.16}$$

where x_j (p_j) is the position (momentum) operator of the oscillator at site j, $m > 0$ is the uniform mass, and C_o, $C_{nn} \geq 0$ are onsite and nearest-neighbor (NN) stiffness constants, respectively. Physically, this Hamiltonian models independent oscillators of characteristic frequency $\sqrt{C_o/m}$ coupled harmonically to their NN with coupling constant C_{nn}. In particular, the limit $C_o = 0$ describes the well-known phonon chain which models vibrational excitations in a harmonically coupled lattice.[6] One such feature of this limit is the conservation of total momentum,

$$[H|_{C_o=0}, p_{tot}] = 0, \quad p_{tot} \equiv \sum_{j=1}^{N} p_j. \tag{4.17}$$

Thus, the center-of-mass, $x_{CM} \equiv N^{-1} \sum_{j=1}^{N} x_j$, is unstable in the sense that $\langle x_{CM} \rangle(t) = \langle p_{tot} \rangle t/M + \langle x_{CM} \rangle(0)$, with $M = Nm$ the total mass, diverges linearly in t in any state with non-zero total momentum. Recalling the dynamical behavior originating from Jordan chains in the normal-mode spectrum (Eq. (2.60)), this signals that $C_o = 0$ is an EP of the dynamical matrix. Clearly, being a sum of positive operators, H is thermodynamically stable.

By introducing the renormalized frequency $\Omega \equiv \sqrt{(2C_{nn} + C_o)/m}$, the coupling constant $J \equiv C_{nn}/m\Omega$, and the bosonic operators $a_j = \sqrt{m\Omega/2}(x_j + ip_j/m\Omega)$ describing phononic excitations, we can write the QBH as

$$H = \frac{1}{2} \sum_{j=1}^{N} \Omega \left(a_j^\dagger a_j + a_j a_j^\dagger \right) - J \left(a_{j+1}^\dagger a_j + a_{j+1}^\dagger a_j^\dagger + \text{H.c.} \right). \tag{4.18}$$

[6] We remark that the continuum limit of this model is the well-studied free real scalar field. The gap parameter C_o determines the mass parameter of the field theory.

In terms of the phonons, we have an on-site potential of frequency Ω with NN hopping and pairing, each of amplitude J. The pairing explicitly violates phonon number conservation. Moreover, the previously discussed EP corresponds to the limit $J = \Omega/2$. We further note that constraints on C_o and C_{nn} ensure that $J \le \Omega/2$.

Leveraging translation invariance, we move to the Fourier basis, $b_k = N^{-1/2}\sum_{j=1}^{N} e^{-ijk}a_j$, with $k \in \mathcal{K}_N$. It follows that $H = \sum_{k\in\mathcal{K}_N} H_k$, with

$$H_k = \frac{\Omega - J\cos(k)}{2}\left(b_k^\dagger b_k + b_{-k}b_{-k}^\dagger\right) - \frac{J\cos(k)}{2}\left(b_k^\dagger b_{-k}^\dagger + b_{-k}b_k\right), \quad k \in \mathcal{K}_N, \tag{4.19}$$

with the associated Bloch dynamical matrix given by $\mathbf{g}(k) = (\Omega - J\cos(k))\boldsymbol{\sigma}_3 - iJ\cos(k)\boldsymbol{\sigma}_2$. Away from the EP ($J < \Omega/2$), diagonalization reveals

$$\begin{aligned} H_k &\equiv \frac{\omega_k}{2}(\psi_k^\dagger\psi_k + \beta_{-k}\beta_{-k}^\dagger), \quad \omega_k \equiv \sqrt{\Omega^2 - 2J\Omega\cos(k)},\\ \psi_k &\equiv \mathcal{N}_k\left[(\Omega - J\cos(k) + \omega_k)b_k + J\cos(k)b_{-k}^\dagger\right], \end{aligned} \tag{4.20}$$

with $\mathcal{N}_k$ a normalization constant ensuring $[\psi_k, \beta_q^\dagger] = \delta_{kq}1_{\mathcal{F}}$. We observe that the system is gapped (in the thermodynamic limit $N \to \infty$), with energy $\Delta E = \omega_0\sqrt{C_o/m}$ separating the ground state $|\widetilde{0}\rangle$ from the first excited state $\beta_0^\dagger|\widetilde{0}\rangle$. In particular, the dynamically unstable limit $C_o \to 0$ represents a many-body gap closing.

To construct the dual of H, we will first construct the dual of the individual H_k's. Equivalently, we seek a (k-local) transformation $\mathbf{R}_k$ that removes the off-diagonal pairing term $-iJ\cos(k)\boldsymbol{\sigma}_2$ from $\mathbf{g}(k)$, i.e., $\mathbf{R}_k\mathbf{g}(k)\mathbf{R}_k^{-1}$ is diagonal. Since $\mathbf{g}(k)$ is 2×2, this boils down to diagonalizing $\mathbf{g}(k)$. Of course, we have already done this by computing ω_k and ψ_k. It follows that the duality is the map U_D defined by $U_D\psi_k U_D^\dagger = b_k$ so that

$$H_k^D = UH_kU_D^\dagger = \frac{\omega_k}{2}\left(b_k^\dagger b_k + b_{-k}b_{-k}^\dagger\right). \tag{4.21}$$

The full dual Hamiltonian in real space is then

$$\begin{aligned} H_D &= \sum_{k\in\mathcal{K}_N} H_k^D = \frac{1}{2}\sum_{r=0}^{N-1}\sum_{j=1}^{N-r}\left(K_r^D(a_{j+r}^\dagger a_j + a_j a_{j+r}^\dagger) + \text{H.c.}\right),\\ K_r^D &\equiv \frac{1}{N}\sum_{k\in\mathcal{K}_N}\omega_k e^{ikr}. \end{aligned} \tag{4.22}$$

The quantities K_r^D, which are the Fourier amplitudes of the normal mode frequencies ω_k, constitute the renormalized hopping amplitudes of the dual Hamiltonian.

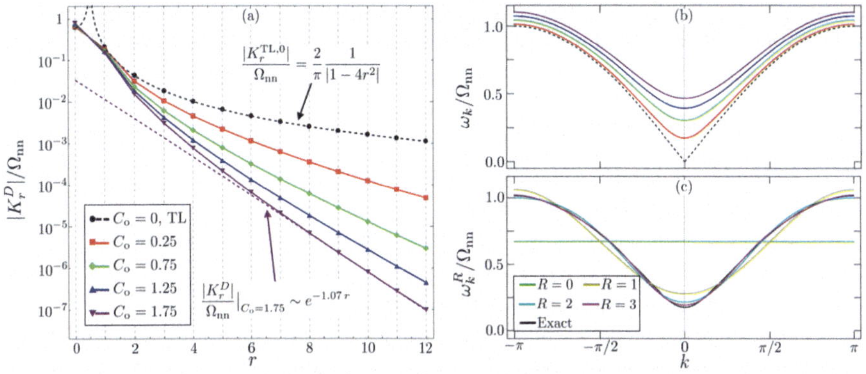

Fig. 4.1 (**a**) Log-plot of the rescaled hopping amplitude $|K_r^D|/\Omega_{\mathrm{nn}}$ for varying onsite stiffness C_o evaluated numerically with $N = 30$. The exact expression for the thermodynamical limit (TL, Eq. (4.23)) with $C_o = 0$ is plotted as a black dashed line. An exponential fit for the case $C_o = 1.75$ is plotted as a purple dashed line. (**b**) The band structure ω_k for the GHC for the same values of C_o in (**a**). The black dashed line, again, corresponds to the TL with $C_o = 0$. (**c**) The band structure of the Hamiltonian $H_D(R)$ with the real-space coupling truncated at ranges $R = 0, 1, 2, 3$. The exact band structure ω_k is shown in black. In (**a**)–(**c**), we take $m = 1$ and $C_{\mathrm{nn}} = 2$. Earlier versions of these figures first appeared in Ref. [1]

Remarkably, the quasiparticle gap is retained (by construction), despite vanishing pairing amplitudes.

While a closed-form expression for K_r^D is not obtained, we are able to evaluate it numerically for various values of system parameters, as shown in Fig. 4.1a. The most notable feature is the apparent exponential decay of the amplitudes as a function of r. Since it is known that the ground state of a gapped harmonic lattice always exhibits exponentially decaying correlations [14], this is consistent with previously identified connection between our duality transformation and the ground state covariance matrix (see Eq. (4.14)). Also plotted in Fig. 4.1a is an analytical expression for K_r^D taken in the thermodynamic limit $N \to \infty$, with $C_o = 0$. Explicitly,

$$K_r^{\mathrm{TL},0} \equiv \frac{1}{2\pi}\int_{-\pi}^{\pi} \left(\omega_k|_{C_o=0}\right) e^{ikr}\, dk = \frac{2}{\pi}\frac{\Omega_{\mathrm{nn}}}{1-4r^2}, \quad \Omega_{\mathrm{nn}} \equiv 2\sqrt{\frac{C_{\mathrm{nn}}}{m}}. \tag{4.23}$$

While it is true that the duality is not valid here (since $C_o = 0$ is manifestly dynamically unstable), the algebraic decay further reflects the connection between the duality and the ground state correlations. Specifically, the ground state of a gapless harmonic lattice is known to exhibit algebraically decaying correlations [14].

With the exponential decay of K_r^D observed numerically, and heuristically explained via the connection to ground state correlations, we expect that truncating the hopping of the dual Hamiltonian should produce a faithful finite-range approximation. Specifically, let $H_D(R)$ be H_D, with K_r^D taken to be zero for $r > R$. If ω_k^R is the associated band structure, how well does ω_k^R approximate the exact band structure ω_k (shown in Fig. 4.1b)? This question is answered in Fig. 4.1c, wherein we find excellent agreement between ω_k^R and ω_k for $R \gtrsim 1$. Theqrefore, even though the exact dual is not finite-range, we have an approximate finite-range dual in the sense that $U_D H U_D^\dagger = H_D \simeq H_D(R)$.

To conclude our analysis of the GHC, we remark that the procedure for computing H_D is a generically applicable one. Specifically, if H is a dynamically stable, translation invariant system with one degree of freedom per-site, the above procedure can be adapted by simply replacing ψ_k and ω_k with the appropriate quasiparticles and band structure, respectively, for the model of interest. Furthermore, based on the arguments we have given above, we conjecture that if H is gapped, then the dual Hamiltonian H_D will have hopping amplitudes K_r^D that decay (at worst) exponentially in r.

4.2.2 A Squeezing Duality for the Bosonic Kitaev Chain

In Sect. 3.3.3, it was found that the BKC under generalized BCs exhibits regions of dynamical stability as a function of boundary parameters (recall Fig. 3.6a–d). It follows that, within these regions, there exists a number-conserving dual. That is, to each value of s and φ where $H(s, \varphi)$ is dynamically stable, there is a unitarily equivalent Hamiltonian $H_D(s, \varphi)$ with vanishing pairing. There are two main approaches we can take to construct $H_D(s, \varphi)$. The first is to use the exact eigenvectors of $\mathbf{G}(s, \varphi)$ found in Appendix B.2 to construct $\mathbf{S}$ directly according to Eq. (4.1). The second is demand locality of the transformation and utilize the Ansatz in Eq. (4.12). This approach is mainly motivated by the observation in Ref. [24] that the pairing of the BKC under OBCs ($s = 0$), can be removed by a suitable squeezing transformation $a_j \mapsto \cosh(\xi_j)a_j + \sinh(\xi_j)a_j^\dagger$, with $\xi_j \in \mathbb{R}$. For this reason, it is natural to expect a local duality transformation at certain stable BCs. This second approach benefits from potentially revealing a duality beyond BCs that we have exactly solved. For concreteness, we will focus on $\pi/2$-twisted BCs and combine the analytical solutions found for $J > \Delta$ and Eq. (4.1). In this case, we find

$$\mathbf{S}(r) \equiv \sum_{j=1}^{N} \vec{e}_j \vec{e}_j^{\,\dagger} \otimes \mathbf{S}_j(r), \quad \mathbf{S}_j(r) \equiv \begin{bmatrix} \cosh[2(j - j_0)r] & -\sinh[2(j - j_0)r] \\ -\sinh[2(j - j_0)r] & \cosh[2(j - j_0)r] \end{bmatrix}, \tag{4.24}$$

with $j_0 = (N + 2)/2$ and $2r = \ln[(J + \Delta)/(J - \Delta)]$. *A priori*, this should only provide a valid positive-definite metric for $s = 1$. Remarkably, we find that its validity extends to arbitrary $s \in [0, 1]$. Specifically, the transformation $\Phi \mapsto U_D \Phi U_D^\dagger = \mathbf{R}^{-1}(r)\Phi$, with $\mathbf{R}^{-1}(r) \equiv \mathbf{S}^{-1/2}(r) = \mathbf{S}(-r/2)$, yields

$$H_D(s, \pi/2) = U_D H(s, \pi/2) U_D^\dagger = \frac{i\widetilde{J}}{2} \sum_{j=1}^{N-1} \left(a_{j+1}^\dagger a_j - a_j^\dagger a_{j+1} \right) - \frac{s\widetilde{J}}{2} \left(a_1^\dagger a_N - a_N^\dagger a_1 \right). \tag{4.25}$$

In the above, $\widetilde{J} \equiv \sqrt{J^2 - \Delta^2}$ is the renormalized hopping amplitude of the dual Hamiltonian. A number of remarks are in order.

(i) The fact that this duality removes pairing for arbitrary s at $\varphi = \pi/2$ constitutes analytical confirmation that the line $\varphi = \pi/2$ in boundary parameter space is always dynamically stable. That is, we have non-trivially extended the analytical stability analysis of Sect. 3.3.3 beyond BCs where exact solutions were obtained.

(ii) Locality of the duality ensures that the bulk (boundary) of $H(s, \pi/2)$ is mapped to the bulk (boundary) of $H_D(s, \pi/2)$. As previously mentioned, $H_D^{\text{OBC}} \equiv H_D(0, \pi/2)$ was obtained in Ref. [24] by leveraging insights from the physics of squeezing. In that case, however, the transformation coincides with ours only if one takes $j_0 = 0$. As it turns out, we find that our duality maps $H^{\text{OBC}} = H(0, \varphi)$ to H_D^{OBC} *regardless* of the choice of j_0, and, in fact, the choice $j_0 = (N+2)/2$ is only necessary to removing the boundary pairing for $s \neq 0$. This parametric freedom (which can even be generalized to allow for j_0 to vary spatially) in the OBC-only duality can be understood as a consequence of the chiral symmetry in the quasiparticle energy spectrum (recall Eq. (3.40)).

(iii) The duality exhibits rather dramatic behavior as the stability phase transition $J = \Delta$ is approached. Specifically, the degree of squeezing (quantified by r) needed to remove the pairing diverges. Such dramatic behavior is consistent with the discussion of Sect. 4.1.2.3, whereby it was noted that the duality diverges near such points. As J approaches Δ, the strength of the hopping $\widetilde{J}$ in the dual Hamiltonian decays algebraically, until ultimately $H^D(s, \pi/2) = 0$ at $J = \Delta$. Since $H^D(s, \pi/2)$ is not zero at $J = \Delta$, this further confirms the invalidity of the duality at this point.

We conclude our duality-focused analysis of the BKC with a remark regarding the potential use of the BKC for generating multipartite entanglement. It was noted in Ref. [24] that the identity $H^{\text{OBC}} = U_D^\dagger H_D^{\text{OBC}} U_D$ can be interpreted as follows: first, $U^\dagger$ implements a site-local squeezing transformation; second, H_D^{OBC} acts as a beam-splitter network (i.e., a photon-number-conserving QBH); and third, U_D undoes the local squeezing of the first step. Since U_D is perfectly local, it was observed that the entanglement properties generated via the beam-splitter step would be left untouched. This proposal can be greatly generalized through the lens of our more general duality framework. Firstly, such a proposal is valid whenever the associated duality is site-local (e.g., is of the form Eq. (4.12)). Second, whenever the associated duality is quasi-local, wherein the transformation couples (either exactly, or approximately) a finite number of subsystems, one can view the original Hamiltonian as generating multipartite entangled states with respect to a suitably generalized notion of entanglement. Specifically, by considering entanglement relative to a coarse-grained (e.g., two-site local) lattice partition to accommodate the locality structure of the duality transformation [29, 30], one can, in principle, generate multipartite generalized entangled states using dynamically stable QBHs with pairing.

References

1. V.P. Flynn, E. Cobanera, L. Viola, Europhys. Lett. **131**, 40006 (2020)
2. V.P. Flynn, E. Cobanera, L. Viola, Phys. Rev. B **108**, 214312 (2023)
3. R. Savit, Rev. Mod. Phys. **52**, 453 (1980)
4. E. Cobanera, G. Ortiz, Z. Nussinov, Phys. Rev. Lett. **104**, 020402 (2010)
5. E. Cobanera, G. Ortiz, Z. Nussinov, Adv. Phys. **60**, 679 (2011)
6. H. A. Kramers, G.H. Wannier, Phys. Rev. **60**, 252 (1941)
7. E. Cobanera, G. Ortiz, Z. Nussinov, Phys. Rev. B **87**, 041105 (2013)
8. E. Cobanera, G. Ortiz, Phys. Rev. B **92**, 155125 (2015)
9. C.M. Bender, S. Boettcher, Phys. Rev. Lett. **80**, 5243 (1998)
10. C.M. Bender, Rep. Prog. Phys. **70**, 947 (2007)
11. C. M. Bender, P.D., Mannheim, Phys. Lett. A **374**, 1616 (2010)
12. A. Mostafazadeh, J. Math. Phys. **43**, 205, ibid. 2814; ibid. 3944 (2002)
13. A. Mostafazadeh, J. Phys. A **36**, 7081 (2003)
14. N. Schuch, J.I. Cirac, M.M. Wolf, Commun. Math. Phys. **267**, 65 (2006)
15. M. Cramer, J. Eisert, New J. Phys. **8**, 71 (2006)
16. I.M. Georgescu, S. Ashhab, F. Nori, Rev. Mod. Phys. **86**, 153 (2014)
17. A. Kamal, J. Clarke, M.H. Devoret, Nat. Phys. **7**, 311 (2011)
18. B. Abdo, K. Sliwa, L. Frunzio, M. Devoret, Phys. Rev. X **3**, 031001 (2013)
19. K. M. Sliwa, M. Hatridge, A. Narla, S. Shankar, L. Frunzio, R. J. Schoelkopf, M.H. Devoret, Phys. Rev. X **5**, 041020 (2015)
20. A. Metelmann, A.A. Clerk, Phys. Rev. X **5**, 021025 (2015)
21. B. Abdo, A. Kamal, M. Devoret, Phys. Rev. B **87**, 014508 (2013)
22. N.E. Frattini, U. Vool, S. Shankar, A. Narla, K.M. Sliwa, M.H. Devoret, Appl. Phys. Lett. **110**, 222603 (2017)
23. D. Malz, L.D. Tóth, N. R. Bernier, A.K. Feofanov, T.J. Kippenberg, A. Nunnenkamp, Phys. Rev. Lett. **120**, 023601 (2018)
24. A. McDonald, T. Pereg-Barnea, A.A. Clerk, Phys. Rev. X **8**, 041031 (2018)
25. V.V. Sivak, N.E. Frattini, V.R. Joshi, A. Lingenfelter, S. Shankar, M.H. Devoret, Phys. Rev. Appl. **11**, 054060 (2019)
26. Y.K. Lu, P. Peng, Q.T. Cao, D. Xu, J. Wiersig, Q. Gong, Y.F. Xiao, Sci. Bull. **63**, 1096 (2018)

27. R. El-Ganainy, K.G. Makris, M. Khajavikhan, Z.H. Musslimani, S. Rotter, D.N. Christodoulides, Nat. Phys. **14**, 11 (2018)
28. Y.X. Wang, A.A. Clerk, Phys. Rev. A **99**, 063834 (2019)
29. H. Barnum, E. Knill, G. Ortiz, R. Somma, L. Viola, Phys. Rev. Lett. **92**, 107902 (2004)
30. S. Montangero, L. Viola, Phys. Rev. A **73**, 040302(R) (2006)

Chapter 5
Obstructions to SPT-Like Physics in QBHs

Abstract We apply the results of the previous two chapters to study certain topological aspects of QBHs. First, we leverage the duality of Chap. 4 to establish a connection between bosonic Berry phases and the more standard Berry phases encountered, for instance, in topological non-interacting fermionic systems. Following this, we recap the essential features of SPT phases in non-interacting fermions by means of a simple analysis of the fermionic Kitaev chain. We proceed to describe the practical and theoretical challenges towards realizing these features in QBHs. First, we describe the no-go theorems forbidding SPT phases in thermodynamically stable QBHs. Then, by forgoing the thermodynamic stability assumption, we explore the intrinsic instability of any zero energy edge modes that may arise. We illustrate these points by uncovering explicit bosonic "shadows" of Majorana fermions in a dynamically, and thermodynamically, unstable generalization of the BKC.

In this chapter, we dissect the role of topology in QBHs[1] SPT phases of quantum many-body systems present a paradigm shift from the celebrated Landau-Ginzburg approach to phase transitions. In the latter, inequivalent phases are characterized by *local* order parameters while the transitions between these phases are associated to spontaneous symmetry-breaking. In sharp contrast, SPT phases are defined via *non-local* order parameters, in addition to bulk *topological* invariants, and do not exhibit the breaking of symmetries across transitions. Free, or mean-field, fermionic matter provides one of the simplest, yet non-trivial, venues for exploring the nature of SPT phases whereby all such phases have been classified in the so-called "tenfold way"[7]. One of the key features, or *signatures*, of SPT physics in free fermions are the emergence of *robust* edge, or surface, states with various exotic properties.

[1] Section 5.1 is a result adapted from Ref. [1], which was co-authored with Emilio Cobanera & Lorenza Viola. The review of the FKC in Sect. 5.2.1 is a readaptation of Kitaev's original work in Ref. [2]. Section 5.2.2 covers three no-go theorems first proved in Ref. [3], which was jointly co-authored with Qiao-Ru Xu, Abhijeet Alase, Emilio Cobanera, Lorenza Viola, & Gerardo Ortiz. Finally, Sect. 5.2.3 contains the bosonic "shadows" first found in Ref. [4] in addition to perspectives on many-body (non-interacting) bosonic topology later developed in Refs. [5] and [6]—all three of which were jointly co-authored with Emilio Cobanera & Lorenza Viola.

V. P. Flynn, *Effective Non-Hermiticity and Topology in Markovian Quadratic Bosonic Dynamics*, Springer Theses, https://doi.org/10.1007/978-3-031-52045-7_5

With this comprehensive understanding of SPT phases of free fermions, one must naturally ask: what about free bosons? More precisely, to what extent can non-trivial topology (i) exist within, and (ii) influence the many-body physics of, QBHs.

In order to pursue bosonic analogues of non-interacting SPT phases, we must be careful to specify the Hamiltonians of interest. *A priori*, one should simply study the topological aspects of arbitrary QBHs. However, as we have seen, QBHs may exhibit instabilities (both thermodynamical and dynamical) that immediately invalidate the interpretation of them as describing "matter". After all, we expect bosonic matter to, at least, have a well-defined ground state. So let us then first focus on *thermodynamically stable QBHs*. Furthermore, we must enforce the many-body gap condition that is so central to non-interacting fermionic SPT physics. Remarkably, these simple restrictions, which allow for such a rich landscape of fermionic phases, completely obstruct the possibility for SPT phases in QBHs. This fundamental result manifests in the form of three no-go theorems which we will summarize in this chapter.

These no-go theorems may come as a surprise to those familiar with the fields of topological photonics, magnonics, and phononics. Within these fields, topological band structures analogous to those found in topological insulators and superconductors have been synthesized (theoretically, and experimentally) in non-interacting systems of these bosonic excitations [8–10]. However, all of these examples lack the many-body features of such phases in one or more phases. Either these systems are thermodynamically unstable (and thus lack any notion of a ground state or many-body gap) or feature topological bands away from zero energy (and thus, high energy surface states when subjected to hardwall boundaries) which, as it turns out, do not elicit the characteristic ground state degeneracy central to fermionic SPTs. These systems reveal, however, that topology certainly *can* play a role in non-interacting bosonic physics. Beyond the high energy surface states, it is well-known that, if one is willing to forgo thermodynamic stability and the many-body interpretation, topologically mandated surface states at *zero energy* can emerge. However, these modes are plagued by an intrinsic dynamical instability that cannot be avoided: they are either at the cusp of instability, or unstable to begin with. This fact, which we will formalize more clearly within this chapter, eliminates any potential robustness properties that one may expect from bosonic analogues to fermionic surface states. Ultimately, QBHs seem to be staunchly averse to anything resembling the free-fermionic SPT phases. For these reasons, we will argue that it is necessary to move beyond this, evidently, *overly restrictive*, closed-system paradigm, and instead advance one step further in dynamical complexity by instead considering genuinely open *Markovian* systems.

The outline of this chapter is as follows. In Sect. 5.1, we leverage the duality of Chap. 4 to establish a connection between the pseudo-Hermitian generalization of the Berry phase relevant to bosonic systems, to the standard Hermitian Berry phase encountered commonly in free fermionic systems. In Sect. 5.2 we assess the degree to which 'SPT physics', understood from the perspective of free fermions, can arise in free bosons modeled via QBHs. Specifically, after introducing the basic features of SPTs as exemplified by the fermionic Kitaev chain, we systematically address the failure of various classes of QBHs to exhibit any true analogous behavior. Along the way, we uncover bosonic "shadows" of Majorana edge modes present in a suitably generalized, unstable BKC. We conclude by emphasizing the need to forgo unitarity if we are to have any hope for realizing SPTs, or at a minimum, SPT-like physics in a quadratic bosonic setting.

5.1 Topological Invariants in the Presence of Effective Non-Hermiticity

The most commonly encountered topological invariants in translation invariant systems are derived from a Berry phase accumulated over closed paths in the Brillouin zone. These are well-known as the Zak phase in 1D, the Chern number in 2D, and so on [7]. A great deal of computational power is gained by focusing on non-interacting systems, whereby the many-body Berry phase can be computed in the single-particle, or more general, Nambu space. In fermionic systems, the adiabatic transport of a single-quasiparticle state around the Brillouin zone can be computed in terms of the Berry phase accumulated by a Bloch state. As it turns out, the effective non-Hermiticity, or more precisely, pseudo-Hermiticity, intrinsic to QBHs introduces a non-trivial modification of the standard fermionic paradigm.

Consider a QBH H that is translation invariant and defined on a D-dimensional infinite lattice. If H is dynamically stable, we may apply Bloch's theorem and diagonalize H, so that $H = \sum_{k \in \mathcal{K}} \omega_n(\mathbf{k}) \psi_{n,\mathbf{k}}^\dagger \psi_{n,\mathbf{k}} + E_0$, with $\mathcal{K}$ the D-dimensional Brillouin zone (torus), $\omega_n(\mathbf{k})$ the n'th band, $\psi_{n,\mathbf{k}}$ the (fermionic or bosonic) quasiparticle annihilation operator associated to band n and crystal momentum $\mathbf{k}$, and E_0 the vacuum energy. If $|\widetilde{0}\rangle$ denotes the quasiparticle vacuum, $\psi_{n,\mathbf{k}} |\widetilde{0}\rangle = 0$ for all n and $\mathbf{k}$, then we may define the single-quasiparticle state $|\widetilde{1}_{n,\mathbf{k}}\rangle = \psi_{n,\mathbf{k}}^\dagger |\widetilde{0}\rangle$. Thinking of the momentum $\mathbf{k}$ as a vector of parameters that we can vary, the Berry phase accumulated over a loop $\mathcal{C} \subset \mathcal{K}$ is given by

$$\gamma_n(\mathcal{C}) \equiv i \oint_{\mathcal{C}} \mathcal{A}_n(\mathbf{k}) \cdot d\mathbf{k}, \quad \mathcal{A}_n(\mathbf{k}) \equiv \langle \widetilde{1}_{n,\mathbf{k}} | \nabla_{\mathbf{k}} | \widetilde{1}_{n,\mathbf{k}} \rangle = \langle \widetilde{0} | \psi_{n,\mathbf{k}} \nabla_{\mathbf{k}} \psi_{n,\mathbf{k}}^\dagger | \widetilde{0} \rangle \,, \tag{5.1}$$

where we have introduced the many-body Berry connection $\mathcal{A}_n(\mathbf{k})$. First, note that

$$\psi_{n,\mathbf{k}} \nabla_{\mathbf{k}} \psi_{n,\mathbf{k}}^\dagger = [\psi_{n,\mathbf{k}}, \nabla_{\mathbf{k}} \psi_{n,\mathbf{k}}^\dagger] + (\nabla_{\mathbf{k}} \psi_{n,\mathbf{k}}^\dagger) \psi_{n,\mathbf{k}}. \tag{5.2}$$

Being the commutator of two linear forms, we know that the first term will be a constant multiple of the identity. Furthermore, since $\psi_{n,\mathbf{k}} |\widetilde{0}\rangle = 0$, it follows that

$$\mathcal{A}_n(\mathbf{k}) = \langle \widetilde{0} | [\psi_{n,\mathbf{k}}, \nabla_{\mathbf{k}} \psi_{n,\mathbf{k}}^\dagger] | \widetilde{0} \rangle \,. \tag{5.3}$$

We can simplify this greatly by making use of the hat notation. Recall that $\psi_{n,\mathbf{k}} = \widehat{\vec{\psi}}_{n,\mathbf{k}} \equiv \vec{\psi}^{\dagger}_{n,\mathbf{k}}\boldsymbol{\tau}_3\phi_{\mathbf{k}}$, where $\vec{\psi}_{n,\mathbf{k}}$ is the eigenvector of the Bloch dynamical matrix $\mathbf{g}(\mathbf{k})$ corresponding to the eigenvalue $\omega_n(\mathbf{k})$, and $\phi_{\mathbf{k}}$ is the $\mathbf{k}$'th Fourier mode of the bosonic Nambu array, i.e., $\phi_{\mathbf{k}} = (2\pi)^{-D}\int e^{i\mathbf{k}\cdot\mathbf{r}}\phi_{\mathbf{r}}\,d\mathbf{r}$ in terms of the real space-local Nambu array $\phi_{\mathbf{r}} \equiv [a_{1,\mathbf{r}}, a^{\dagger}_{1,\mathbf{r}}, \cdots]^T$. It follows that

$$
\begin{aligned}
\nabla_{\mathbf{k}}\psi^{\dagger}_{n,\mathbf{k}} = \nabla_{\mathbf{k}}\left(\phi_{\mathbf{k}}{}^{\dagger}\boldsymbol{\tau}_3\vec{\psi}_{n,\mathbf{k}}\right) &= \left(\nabla_{\mathbf{k}}\phi_{\mathbf{k}}{}^{\dagger}\right)\boldsymbol{\tau}_3\vec{\psi}_{n,\mathbf{k}} + \phi_{\mathbf{k}}{}^{\dagger}\boldsymbol{\tau}_3\left(\nabla_{\mathbf{k}}\vec{\psi}_{n,\mathbf{k}}\right)\\
&= (-i\mathbf{k})\phi_{\mathbf{k}}{}^{\dagger}\boldsymbol{\tau}_3\vec{\psi}_{n,\mathbf{k}} + \phi^{\dagger}_{\mathbf{k}}\nabla_{\mathbf{k}}\vec{\psi}_{n,\mathbf{k}}\\
&= (-i\mathbf{k})\widehat{\vec{\psi}_{n,\mathbf{k}}}^{\dagger} + \widehat{\nabla_{\mathbf{k}}\vec{\psi}_{n,\mathbf{k}}}^{\dagger}.
\end{aligned}
$$

Recalling that $[\widehat{\vec{\alpha}}, \widehat{\vec{\beta}}^{\dagger}] = (\vec{\alpha}^{\dagger}\boldsymbol{\tau}_3\vec{\beta})1_{\mathcal{F}}$, we may compute

$$
\begin{aligned}
[\psi_{n,\mathbf{k}}, \nabla_{\mathbf{k}}\psi^{\dagger}_{n,\mathbf{k}}] &= (-i\mathbf{k})[\widehat{\vec{\psi}_{n,\mathbf{k}}}, \widehat{\vec{\psi}_{n,\mathbf{k}}}^{\dagger}] + [\widehat{\vec{\psi}_{n,\mathbf{k}}}, \widehat{\nabla_{\mathbf{k}}\vec{\psi}_{n,\mathbf{k}}}^{\dagger}]\\
&= \left(-i\mathbf{k} + \vec{\psi}^{\dagger}_{n,\mathbf{k}}\boldsymbol{\tau}_3\nabla_{\mathbf{k}}\vec{\psi}_{n,\mathbf{k}}\right)1_{\mathcal{F}}.
\end{aligned}
\tag{5.4}
$$

Since $\oint \mathbf{k}\cdot d\mathbf{k} = 0$ over any closed contour, we obtain the final result in the form

$$
\gamma_n(C) = \oint_C \mathcal{A}^{\text{KB}}_n(\mathbf{k})\cdot d\mathbf{k}, \quad \mathcal{A}^{\text{KB}}(\mathbf{k}) \equiv \vec{\psi}^{\dagger}_{n,\mathbf{k}}\boldsymbol{\tau}_3\nabla_{\mathbf{k}}\vec{\psi}_{n,\mathbf{k}}. \tag{5.5}
$$

Here, we have introduced the *Krein-Berry connection* $\mathcal{A}^{\text{KB}}_n(\mathbf{k})$. This connection is the same one that would emerge if one were to consider parallel transport of a vector in an indefinite inner-product space with metric $\boldsymbol{\tau}_3$, i.e., $\mathcal{A}^{\text{KB}}_n(\mathbf{k}) = (\vec{\psi}_{n,\mathbf{k}}, \nabla_{\mathbf{k}}\vec{\psi}_{n,\mathbf{k}})_{\boldsymbol{\tau}_3}$ in terms of the $\boldsymbol{\tau}_3$ inner-product $(\vec{v}, \vec{w})_{\boldsymbol{\tau}_3} \equiv \vec{v}^{\dagger}\boldsymbol{\tau}_3\vec{w}$. Equivalently, $\gamma_n(C)$ is precisely the geometric phase accumulated by $\vec{\psi}_{n,\mathbf{k}}$, a solution to the "Schrödinger-like" equation $\mathbf{g}(\mathbf{k})\vec{\psi}_{n,\mathbf{k}} = \omega_n(\mathbf{k})\vec{\psi}_{n,\mathbf{k}}$, when adiabatically transported around the Brillouin zone.

A natural question immediately arises: to what extent does the presence of the indefinite inner-product in Eq. (5.5), or more broadly, *effective non-Hermiticity*, lead to nontrivial behavior of the many-body Berry phase. Dynamical stability affords us a method for answering this question. Building on the duality transformation discussed in Chap. 4, we can attempt to compute this Berry phase in terms of that of the unitarily equivalent number-conserving dual. Since the dynamical matrix of a number-conserving QBH is Hermitian, the Berry phase will be computed in a way analgous to fermionic systems. To be more specific, consider the number-conserving dual Hamiltonian $H^D = U_D H U^{\dagger}_D$ and its associated Bloch dynamical matrix $\mathbf{g}_D(\mathbf{k}) = \mathbf{R}(\mathbf{k})\mathbf{G}(\mathbf{k})\mathbf{R}(\mathbf{k})^{-1}$, with $\mathbf{R}(\mathbf{k})$ the unique positive-definite square root of the metric $\mathbf{S}(k)$ defining the duality transformation. It follows that $\vec{\psi}^{D}_{n,\mathbf{k}} \equiv \mathbf{R}(\mathbf{k})\vec{\psi}_{n,\mathbf{k}}$ is an eigenvector of $\mathbf{g}(\mathbf{k})$ corresponding to eigenvalue $\omega_n(\mathbf{k})$. Hermiticity of the full dynamical matrix ensures that $[\mathbf{g}(\mathbf{k}), \boldsymbol{\tau}_3] = 0$, so that we can simultaneously diagonalize $\mathbf{g}(\mathbf{k})$ and $\boldsymbol{\tau}_3$. We can then take, without loss of generality, $\vec{\psi}^D_{n,\mathbf{k}}$ to be an eigenstate of $\boldsymbol{\tau}_3$. Moreover, $\boldsymbol{\tau}_3\vec{\psi}_{n,\mathbf{k}} = +\vec{\psi}_{n,\mathbf{k}}$ since (i) the duality preserves Krein signatures,[2] and (ii) the association of the annihilation operator $\psi_{n,\mathbf{k}}$ with $\vec{\psi}_{n,\mathbf{k}}$ implies $\vec{\psi}_{n,\mathbf{k}}$ has positive Krein signature. Ultimately, the Berry phase in the dual system is

$$\gamma_{n,D}(\mathcal{C}) = i\oint_{\mathcal{C}} \vec{\psi}^{D\,\dagger}_{n,\mathbf{k}}\boldsymbol{\tau}_3\nabla_{\mathbf{k}}\vec{\psi}^D_{n,\mathbf{k}}\cdot d\mathbf{k} = i\oint_{\mathcal{C}} \vec{\psi}^{D\,\dagger}_{n,\mathbf{k}}\nabla_{\mathbf{k}}\vec{\psi}^D_{n,\mathbf{k}}\cdot d\mathbf{k} = i\oint_{\mathcal{C}} \mathcal{A}^B_{n,D}(\mathbf{k}), \tag{5.6}$$

where we have introduced the standard Berry connection of the dual eigenvector $\mathcal{A}^B_{n,D}(\mathbf{k}) \equiv \vec{\psi}^{D\,\dagger}_{n,\mathbf{k}}\nabla_{\mathbf{k}}\vec{\psi}^D_{n,\mathbf{k}}$. This connection is precisely what we would encounter when considering parallel transport in a standard Hilbert space (e.g., in the fermionic case). So then, how do $\gamma_{n,D}(\mathcal{C})$ and $\gamma_n(\mathcal{C})$ differ? To answer this, we analyze the connections. First,

$$\begin{aligned}\mathcal{A}^B_{n,D}(\mathbf{k}) &= \vec{\psi}^{D\,\dagger}_{n,\mathbf{k}}\nabla_{\mathbf{k}}\vec{\psi}^D_{n,\mathbf{k}} = \vec{\psi}_{n,\mathbf{k}}^{\dagger}\mathbf{R}(\mathbf{k})\nabla_{\mathbf{k}}[\mathbf{R}(\mathbf{k})\vec{\psi}_{n,\mathbf{k}}] \\ &= \vec{\psi}_{n,\mathbf{k}}^{\dagger}\mathbf{S}(\mathbf{k})[\nabla_{\mathbf{k}}\vec{\psi}_{n,\mathbf{k}}] + \vec{\psi}_{n,\mathbf{k}}^{\dagger}\mathbf{R}(\mathbf{k})[\nabla_{\mathbf{k}}\mathbf{R}(\mathbf{k})]\vec{\psi}_{n,\mathbf{k}}.\end{aligned}$$

Referring to the construction of $\mathbf{S}(\mathbf{k})$ (see Eq. (4.1)), it follows that $\boldsymbol{\tau}_3\mathbf{S}(\mathbf{k})\vec{\psi}_{n,\mathbf{k}} = \vec{\psi}_{n,\mathbf{k}}$. Rearranging this identity and applying it to the first term in the above, we obtain

$$\begin{aligned}\mathcal{A}^B_{n,D}(\mathbf{k}) &= \vec{\psi}_{n,\mathbf{k}}^{\dagger}\boldsymbol{\tau}_3[\nabla_{\mathbf{k}}\vec{\psi}_{n,\mathbf{k}}] + \vec{\psi}_{n,\mathbf{k}}^{\dagger}\mathbf{R}(\mathbf{k})[\nabla_{\mathbf{k}}\mathbf{R}(\mathbf{k})]\vec{\psi}_{n,\mathbf{k}} \\ &= \mathcal{A}^{KB}_n(\mathbf{k}) + \vec{\psi}_{n,\mathbf{k}}^{\dagger}\mathbf{R}(\mathbf{k})[\nabla_{\mathbf{k}}\mathbf{R}(\mathbf{k})]\vec{\psi}_{n,\mathbf{k}},\end{aligned}$$

or more succinctly,

$$\mathcal{A}^B_{n,D}(\mathbf{k}) - \mathcal{A}^{KB}_n(\mathbf{k}) = \vec{\psi}_{n,\mathbf{k}}^{\dagger}\mathbf{R}(\mathbf{k})[\nabla_{\mathbf{k}}\mathbf{R}(\mathbf{k})]\vec{\psi}_{n,\mathbf{k}}. \tag{5.7}$$

In a sense, we have isolated precisely the influence of the indefinite metric, and thus the effective non-Hermiticity, on the connection. The duality transformation $\mathbf{R}(\mathbf{k})$ defines the degree to which the Krein-Berry connection of the original system differs from the usual Berry connection of the dual system. With an eye towards topological physics, this can have concrete physical consequences. For example, suppose one has a QBH whereby the right hand-side of Eq. (5.7) not only does not vanish, but cannot be expressed as a total derivative of some function of $\mathbf{k}$. Thus the Berry phase, which may correspond to a topological invariant like the Zak phase or Chern number, between the original system and its dual will differ. Employing the bulk-boundary correspondence (which is valid in this effective non-Hermitian

[2] Explicitly, if $\vec{v}^{\dagger}\boldsymbol{\tau}_3\vec{v} = \kappa$, then $\vec{v}'^{\dagger}\boldsymbol{\tau}_3\vec{v}' = \kappa$, with $\vec{v}' \equiv \mathbf{R}\vec{v}$ follows directly from $\mathbf{R}^{\dagger}\boldsymbol{\tau}_3\mathbf{R} = \mathbb{1}$.

context [11]) means that, generically, the number of boundary modes that emerge on the edge of the two systems will differ. Necessarily, imposing the BCs must violate the duality since unitary equivalence would require that the same number of boundary modes. This could suggest, for instance, that these additional boundary modes may be associated to insabilities [12, 13], ultimately preventing the existence of a number conservation-restoring duality.

As a final remark, the two connections can be shown to coincide whenever we have $[\mathbf{R}(\mathbf{k}), \nabla_{\mathbf{k}}\mathbf{R}(\mathbf{k})] = 0$. Under this (rather opaque) assumption, it follows that $\mathbf{R}(\mathbf{k})\nabla_{\mathbf{k}}\mathbf{R}(\mathbf{k}) = \nabla_{\mathbf{k}}\mathbf{S}(\mathbf{k})/2$. Furthermore,

$$\vec{\psi}_{n,\mathbf{k}}^{\dagger}[\nabla_{\mathbf{k}}\mathbf{S}(\mathbf{k})]\vec{\psi}_{n,\mathbf{k}} = \nabla_{\mathbf{k}}\left[\vec{\psi}_{n,\mathbf{k}}^{\dagger}\mathbf{S}(\mathbf{k})\vec{\psi}_{n,\mathbf{k}}\right] - 2\mathrm{Re}\left[\vec{\psi}_{n,\mathbf{k}}^{\dagger}\mathbf{S}(\mathbf{k})\nabla_{\mathbf{k}}\vec{\psi}_{n,\mathbf{k}}\right] \tag{5.8}$$

Recalling that the eigenbasis of the dynamical matrix is orthonormal in the $\mathbf{S}(\mathbf{k})$ inner-product, the first term vanishes. Again using $\boldsymbol{\tau}_3\mathbf{S}(\mathbf{k})\vec{\psi}_{n,\mathbf{k}} = \vec{\psi}_{n,\mathbf{k}}$, the second term can simplified to the real part of $\vec{\psi}_{n,\mathbf{k}}^{\dagger}\boldsymbol{\tau}_3\nabla_{\mathbf{k}}\vec{\psi}_{n,\mathbf{k}}$. However, this is manifestly purely imaginary. To see this, note that since $\vec{\psi}_{n,\mathbf{k}}^{\dagger}\boldsymbol{\tau}_3\vec{\psi}_{n,\mathbf{k}} = 1$, we have

$$\begin{aligned} 0 = \nabla_{\mathbf{k}}\left[\vec{\psi}_{n,\mathbf{k}}^{\dagger}\boldsymbol{\tau}_3\vec{\psi}_{n,\mathbf{k}}\right] &= \vec{\psi}_{n,\mathbf{k}}^{\dagger}\boldsymbol{\tau}_3\nabla_{\mathbf{k}}\vec{\psi}_{n,\mathbf{k}} + [\nabla_{\mathbf{k}}\vec{\psi}_{n,\mathbf{k}}]^{\dagger}\boldsymbol{\tau}_3\vec{\psi}_{n,\mathbf{k}} \\ &= \mathrm{Re}\left[\vec{\psi}_{n,\mathbf{k}}^{\dagger}\mathbf{S}(\mathbf{k})\nabla_{\mathbf{k}}\vec{\psi}_{n,\mathbf{k}}\right]. \end{aligned} \tag{5.9}$$

5.2 The Search for SPT Physics in QBHs

5.2.1 *Primer: The Fermionic Kitaev Chain*

The results of the previous section indicate that effective non-Hermiticity can manifest in concrete differences between topological features of (quadratic) bosonic and fermionic systems at the level of edge physics. However, the distinctions are much more significant. To appreciate this further, let us identify some of the key features of non-interacting fermionic SPT phases present in a quintessential topological superconductor: the fermionic Kitaev chain [2]. In terms of Dirac fermionic creation and annihilation operators $c_j^{\dagger}$ and c_j, the Hamiltonian is given by

$$H_{\mathrm{FKC}} \equiv -\sum_{j=1}^{N}\mu c_j^{\dagger}c_j - \sum_{j=1}^{N}\left(Jc_j^{\dagger}c_{j+1} - \Delta c_j^{\dagger}c_{j+1}^{\dagger} + \mathrm{H.c.}\right), \tag{5.10}$$

with $\mu, J, \Delta \in \mathbb{R}$ the chemical potential, hopping amplitude, and pairing amplitudes respectively, and OBCs (PBCs) imposed by taking $c_{N+1} = 0$ (c_1). The associated Bloch Hamiltonian is

$$\mathbf{h}_{\mathrm{FKC}}(k) \equiv -2\Delta\sin(k)\boldsymbol{\sigma}_2 - (\mu + 2J\cos(k))\boldsymbol{\sigma}_3 = \vec{d}(k)\cdot\vec{\boldsymbol{\sigma}}, \tag{5.11}$$

where we have defined the row vectors $\vec{d}(k) = [0, -2\Delta \sin(k), -(\mu + 2J\cos(k))]$ and $\vec{\sigma} = [\sigma_1, \sigma_2, \sigma_3]$. In terms of the real vector $\vec{d}(k)$, the topological phase of the FKC can be most concisely characterized as the values of μ, J, and Δ, such that $\vec{d}(k)$ winds around the x-axis. This condition is equivalent to $|\mu/2J| < 1$ when $\Delta \neq 0$. Consequentially, the quasiparticle energy gap (which can be computed as the minimum difference between the two eigenvalues of $h_{\text{FKC}}(k)$) closes at $|\mu| = 2|J|$, signaling a quantum phase transition.

The implications of this phase transition can be seen most dramatically in the OBC system. Let us restrict to the special case $\Delta = J$ and change to a Majorana basis. That is, we define the Majorana fermions γ_ℓ via $c_j = (\gamma_{2j-1} + i\gamma_{2j})/2$. The OBC Hamiltonian follows as

$$H_{\text{FKC}} = -\frac{i\mu}{2}\sum_{j=1}^{i} \gamma_{2j-1}\gamma_{2j} + \frac{iJ}{2}\sum_{j=1}^{N-1} i\gamma_{2j}\gamma_{2j+1} + \text{const.} \tag{5.12}$$

Right away, the signifance of the point $|\mu| = 2|J|$ is made clear—it signals the point in which the two terms in Eq. (5.12) are of “equal magnitude”. When $|\mu| < 2|J|$ (in the topological phase), one may verify that the operators

$$\gamma_L \equiv \mathcal{M}(N)\sum_{j=1}^{N} \delta^{j-1}\gamma_{2j-1}, \quad \gamma_R \equiv \mathcal{M}(N)\sum_{j=1}^{N} \delta^{N-j}\gamma_{2j},$$
$$\mathcal{M}(N) \equiv \sqrt{\frac{1-\delta^2}{1-\delta^{2(N-1)}}}, \qquad \delta \equiv \frac{\mu}{2J}, \quad |\delta| < 1, \tag{5.13}$$

satisfy

$$i[H_{\text{FKC}}, \gamma_L] = 2\mathcal{M}(N)\delta^N J\gamma_{2N}, \quad i[H_{\text{FKC}}, \gamma_R] = 2\mathcal{M}(N)\delta^N J\gamma_1, \tag{5.14}$$

along with

$$\{\gamma_L, \gamma_R\} = 0, \quad \gamma_L^2 = \gamma_R^2 = 1_{\mathcal{F}}, \tag{5.15}$$

with $1_{\mathcal{F}}$ now denoting the fermionic Fock space identity. Let us interpret each equation separately. Specifying that the system is in the topological phase means that Eq. (5.13) define two Hermitian operators that are localized on opposite sides of the chain, each with localization length $[\ln(|\delta|)]^{-1}$. Their commutator with the Hamiltonian in Eq. (5.14) imply, in particular, that they are exact ZMs of the system in the infinite-size limit, $N \to \infty$. For this reason, we call γ_L and γ_R *Majorana zero modes*. Finally, Eq. (5.15) states that these operators satisfy the Majorana algebra or, equivalently, $c \equiv \gamma_L + i\gamma_R$ is a proper Dirac fermion operator. To adopt bosonic language, we say that these are then *canonically conjugate* modes. Remarkably then, the operator $c^\dagger$ creates an (approximately, for finite N) zero-energy fermionic (quasi)particle that is split between the two macroscopically separated ends of the chain. Finally, it may be shown that no such operators arise in the trivial phase $|\delta| > 1$. Let us summarize the key properties of these Majorana ZMs.

(i) They are a pair of Hermitian operators that (approximately, for finite N) commute with the Hamiltonian.
(ii) They are normalized to satisfy the Majorana algebra.
(iii) One member of the pair is exponentially localized on the left half of the chain while the other is exponentially localized on the right half. Both have a characteristic localization length that diverges near the topological phase boundary.
(iv) Combining (i)–(iii) allows us to construct a spatially split (Dirac) fermionic degree of freedom $\gamma_L + i\gamma_R$ whose real and imaginary components are these Majorana fermions. This "split" fermion corresponds to a (approximate, for finite N) zero energy quasiparticle.

Thus far, the Majorana ZMs can be understood as the boundary manifestation of bulk topology, i.e., a consequence of the bulk-boundary correspondence (BBC). What, then, are the many-body implications of their existence? Focusing entirely on the thermodynamic limit $N \to \infty$ (which corresponds physically to SIBCs, recall Sect. 2.2), let us consider the ground state of H_{FKC} at two extremes: (i) $\mu < 0$, $J = \Delta = 0$ (trivial), and (ii) $\mu = 0$, $J = \Delta \neq 0$ (non-trivial). In case (i) $H_{\text{FKC}} = -\mu \sum_j c_j^\dagger c_j$, the ground state is $|0\rangle$, the zero-particle fermionic Fock state. In case (ii), it is useful to define new fermionic operators $d_j \equiv \gamma_{2j-1} + i\gamma_{2(j+1)}$, for $j = 1, \ldots, N-1$. It follows that

$$H_{\text{FKC}} = J \sum_{j=1}^{N-1} \left(2d_j^\dagger d_j - 1\right) \tag{5.16}$$

and, more importantly, $\gamma_L = \gamma_1$ and $\gamma_R = \gamma_{2N}$ now *commute* with the Hamiltonian (see Eq. (5.14) with $\mu = 0$). This commutation implies we have two orthogonal ground states: the state with zero d-fermions, which we denote by $|\widetilde{0}\rangle$, and

$$|\widetilde{1}_N\rangle = d_N^\dagger |\widetilde{0}\rangle, \quad d_N \equiv \gamma_L + i\gamma_R = \gamma_1 + i\gamma_{2N}. \tag{5.17}$$

A key physical difference between these two states is that they have opposite fermionic parity. Specifically, $|\widetilde{0}\rangle$ has even parity and $|\widetilde{1}_N\rangle$ has odd parity. This ground state distinction between the trivial and non-trivial phase is surprising because it is not reflected in the bulk physics. The bi-infinite (or periodic) system has a unique ground state with even parity on each side of the quantum phase transition. The only distinction is that of the topological winding number. In this sense, we say that bulk ground state has a non-trivial topological invariant in the non-trivial phase. This analysis extends to general values of μ, J, and Δ.

Let us highlight the broad many-body features of the FKC:

(1) The Hamiltonian is bulk-translation invariant and gapped in the bi-infinite limit on each side of a phase transition.
(2) There is a topological invariant that distinguishes the bulk ground states on each side of the phase transition.
(3) When boundaries are imposed, the system has a unique even parity ground state on one side of the transition, and two orthogonal ground states with opposite parity on the other side.

In addition to these, we add two points that we have not yet mentioned:

(4) The system enjoys certain many-body symmetries (in this case, particle-hole $c_j \mapsto c_j^\dagger$).
(5) The bulk topological invariant cannot be changed without either (i) breaking some of the above many-body symmetries or (ii) undergoing a quantum phase transition. Consequently, the Majorana edge modes are robust against symmetry-preserving perturbations.

These final two points capture the concept of a *symmetry-protected topological* (SPT) *phase*. Features (1)–(5) are generic in non-interacting fermionic SPT phases, and in fact, all such phases have been classified in the so-called *tenfold way* [7]. Retrospectively, we identify the emergence of the edge modes with the properties (i)–(iv) above as a *signature* of SPT physics in the model. With the lessons of the FKC under our belt, let us begin our search for SPT physics in QBHs.

5.2.2 *Thermodynamically Stable Systems*

Central to the preceding discussion was the emphasis on ground-state physics. Unlike fermions, bosonic Hamiltonians need not afford us a ground state to begin with. If we first demand a comprehensive bosonic analogue to free-fermionic SPT phases, we naturally must consider only those systems that are thermodynamically stable. The next two ingredients to add are the conditions laid out in (1) of the preceding section: bulk-translation invariance and a many-body gap condition. Specifically, we consider those QBHs whose quasiparticle energy bands $\omega_n(\mathbf{k})$ are *strictly positive* (or strictly negative).

With all of these constraints laid out, we are faced with three no-go theorems (see Ref. [3] for precise technical statements and proofs). Let H be one such QBH. Then,

- H can be adiabatically deformed into *any* other QBH belonging to the same class without breaking any many-body symmetries or closing the many-body energy gap.

- ZMs and their higher-dimensional analogues (e.g, surface bands) cannot emerge upon terminating the system along any hyperplane. Equivalently, the lowest energy band is always topologically trivial.
- The ground state of any such QBH is always of even bosonic parity, even upon termination.[3]

In short: *there are no SPT phases in thermodynamically stable, gapped, bulk-translation invariant QBHs*. Does this mean topology plays no role in such systems? Absolutely not—in fact, higher energy bands can be topologically non-trivial (hence our foray into the Krein-Berry phase in Sect. 5.1) and, as such, they interesting edge physics [8–11]. These non-trivial bands, and their associated surface modes, necessarily cannot impact the ground state physics, however. Furthermore, one may consider the possibility of zero energy modes emerging as a result of other, non-trivial BCs (beyond OBCs and its higher-dimensional analagoues). Such modes have been found to emerge as a result of a nontrivial kernel-preserving map between fermionic BdG Hamiltonians and bosonic dynamical matrices. However, as we will see in a more general context below, bosonic ZMs are *intrinsically unstable*. So even if non-trivial BCs may host them (e.g., atop boundary impurities), they lack dynamical robustness to perturbations. Such instability necessarily carries over to the degenerate ground states formed by exciting these modes on top of the quasiparticle vacuum. Thus, it is inappropriate to compare them to their fermionic counterparts which, via symmetry-protection, are intrinsically robust to (symmetry-preserving) perturbations.

5.2.3 Beyond Thermodynamic Stability: Bosonic Shadows of Majorana Fermions

The ground state features of SPT phases are impossible to realize in thermodynamically stable QBHs. If we liberate ourselves from this constraint, can we at least find bosonic signatures of SPT physics? For example, can we find bosonic analogues of the Majorana fermions satisfying properties similar to (i)–(iv) from the FKC? In terms of the other ingredients, bulk-translation invariance can straightforwardly be retained, whereas the many-body gap condition clearly departs alongside thermodynamic stability. In fact, things are more dire. The combination of thermodynamic stability and the many-body gap condition implicitly lead to a dynamically stable system (recall that only zero energy can support dynamical instabilities in the form of free particles while demanding thermodynamic stability). Our search thus bifurcates based on whether we allow for dynamical instabilities or not.

[3] More generally, any thermodynamically stable QBH with *strictly* positive quasiparticle energies will always have an even parity ground state (see also Eq. (2.68)).

In order to not wander too far from the fermionic paradigm, let us begin by demanding dynamical stability. The first step is to generalize the many-body gap condition to a quasiparticle energy gap condition. That is, we will demand that the (now positive and negative) energy bands $\omega_n(\mathbf{k})$ do not cross zero. We justify this generalization by noting that the existence of a quasiparticle energy gap is equivalent to the existence of a many-body gap in thermodynamically stable systems. With this generalization, it turns out that the energy bands around zero energy can become topologically non-trivial. Consequently, ZMs emerge upon termination of the system and are well-known, e.g., in photonic systems [12, 14]. Do these modes escape the intrinsic instabilities of those found in thermodynamically stable systems under generalized BCs? No. More specifically, we have the following corollary of Theorem 3.2.2.

Corollary 5.2.1 (Adapted Partially from Ref. [11]) *Let H be a dynamically stable QBH that hosts ZMs. Then there are arbitrarily small perturbations that split the frequencies of these modes into the complex plane, rendering the system dynamically unstable.*

Proof Let $\mathbf{G}$ be the associated dynamical matrix. The assumptions imply that $\mathbf{G}$ is diagonalizable and it has a non-trivial kernel $\ker\mathbf{G} \equiv \{\vec{\psi} : \mathbf{G}\vec{\psi} = 0\}$. In fact, diagonalizability, in addition to the four-fold symmetry of the spectrum, together imply that the kernel is even-dimensional. Let $\vec{\psi} \in \ker\mathbf{G}$. If $\vec{\psi}$ has Krein signature ± 1, then $\boldsymbol{\tau}_1\vec{\psi}^*$ has Krein signature -1 and is also in the kernel. Thus, the zero eigenvalue hosts a Krein collision and we are done. If $\vec{\psi}$ has zero Krein signature, then $\ker\mathbf{G}$ is a $\boldsymbol{\tau}_3$-indefinite eigenspace, and we are done.[4] □

We conclude that, even if such modes are found, they do not constitute valid bosonic analogues of the fermionic ZMs. They are not robust in any analogous way.

As a last stand, we may liberate ourselves from the constraints of both thermodynamical and dynamical stability. This freedom allows us to welcome the intrinsically unstable modes as *loose* analogues to their fermionic cousins. In fact, we can find "shadows" of the FKC's Majorana fermions in a modified version of the generalized BKC of Eq. (3.25). To accomplish this, it helps to recall origin of the FKC's topological invariant: it is the winding number of the vector $\vec{d}(k)$, which we will denote by $\vec{d}_f(k)$ henceforth. What is the analogous vector for the BKC? Well, the Bloch dynamical matrix Eq. (3.44) can be written as

$$\mathbf{g}(k) = \vec{d}_b(k) \cdot \vec{\boldsymbol{\sigma}}', \tag{5.18}$$

with $\vec{d}_b(k) \equiv [J\sin(k), \Delta\cos(k), 0]$ and $\vec{\boldsymbol{\sigma}} \equiv i[\mathbb{1}_2, \boldsymbol{\sigma}_1, \boldsymbol{\sigma}_2]^T$. While there are clearly differences, the vectors $\vec{d}_b(k)$ and $\vec{d}_f(k)$ bear resemblance. In fact, if we add a term to the Hamiltonian, such that $\vec{d}_b(k) = [J\sin(k), \mu + \Delta\cos(k), 0]^T$, we

[4] One may additionally prove that if an indefinite eigenspace does not host Jordan chains, then it hosts a Krein collision. This is unnecessary for our applications, however.

can strengthen the analogy. The term that implements this is a uniform degenerate (onsite) parametric amplification (DPA) term:

$$H_{\text{DPA}} = \frac{i\mu}{2}\sum_{j=1}^{N}\left(a_j^{\dagger 2} - a_j^2\right), \tag{5.19}$$

where $\mu \in \mathbb{R}$. The full Hamiltonian becomes (under OBCs, for concreteness)

$$\begin{aligned} H^{\text{OBC}} \equiv H_{\text{BKC}}^{\text{OBC}} + H_{\text{DPA}} &= \frac{1}{2}\sum_{j=1}^{N-1}\left(iJa_{j+1}^{\dagger}a_j + i\Delta a_{j+1}^{\dagger}a_j^{\dagger} + \text{H.c.}\right) \\ &+ \frac{i\mu}{2}\sum_{j=1}^{N}\left(a_j^{\dagger 2} - a_j^2\right). \end{aligned} \tag{5.20}$$

It follows that the dynamical matrix is shifted by $i\mu\boldsymbol{\tau}_1$. Recalling that this matrix commutes with $\boldsymbol{\tau}_1$, this implies the normal mode frequencies are split into the complex plane $\omega_n \mapsto \omega_n \pm i\mu$. Ultimately, the chain is unstable for $\mu \neq 0$. Taking $J = \Delta$ for simplicity, we find that $\vec{d}_b(k)$ winds around the z-axis whenever $|\mu| < |J|$. Modulo factors of 2, we see that the DPA term plays *exactly the same role* as the chemical potential in the FKC. Consequentially, we uncover two Hermitian modes

$$\begin{aligned} \gamma_L &\equiv \mathcal{M}(N)\sum_{j=1}^{N}\delta^{j-1}x_j, \quad \gamma_R \equiv \mathcal{M}(N)\sum_{j=1}^{N}\delta^{N-j}p_j, \\ \mathcal{M}(N) &\equiv \sqrt{\frac{1}{N\delta^{N-1}}}, \qquad \delta \equiv -\frac{\mu}{2J}, \quad |\delta| < 1 \end{aligned} \tag{5.21}$$

which satisfy

$$i[H^{\text{OBC}}, \gamma_L] = -\mathcal{M}(N)\delta^N Jx_N, \quad i[H^{\text{OBC}}, \gamma_R] = -\mathcal{M}(N)\delta^N Jp_1, \tag{5.22}$$

along with

$$[\gamma_L, \gamma_R] = i1_{\mathcal{F}}. \tag{5.23}$$

These equations constitute direct bosonic analogues to Eqs. (5.13)–(5.15), with two accommodations made. First, the odd and even Majoranas fermions γ_{2j-1} and γ_{2j} are replaced with the real and imaginary quadratures x_j and p_j. Second, the normalization constant is modified in order to accommodate Heisenberg-Weyl commutation relations, i.e., the bosonic analogues of the Majorana fermion algebra. With these, we obtain bosonic "shadows" of Majorana fermions in a generalized bosonic Kitaev chain. Retrospectively, however, we have abandoned almost all other connections to free-fermionic SPT physics.

To what extent can we restore these connections? Is there any sense in which a quadratic bosonic system can display signatures of SPT physics and at least *some* of the properties (1)–(5) detailed before? The answer is Yes, but requires we forgoe a more subtle assumption made so far: unitarity. In the next part of this thesis, we will consider *open* bosonic quadratic systems. By establishing a conceptual correspondence between the ground state and the steady state, and the many-body gap with a suitably defined Lindblad gap, we will ultimately uncover *tight* bosonic analogues of fermionic edge modes while maintaining a compelling many-body picture.

References

1. V.P. Flynn, E. Cobanera, L. Viola, Europhys. Lett. **131**, 40006 (2020)
2. A. Kitaev, Phys. Usp. **44**, 131 (2001)
3. Q.R. Xu, V.P. Flynn, A. Alase, E. Cobanera, L. Viola, G. Ortiz, Phys. Rev. B **102**, 125127 (2020)
4. V.P. Flynn, E. Cobanera, L. Viola, New J. Phys. **22**, 083004 (2020)
5. V.P. Flynn, E. Cobanera, and L. Viola, Phys. Rev. Lett. **127**, 245701 (2021).
6. V.P. Flynn, E. Cobanera, L. Viola, Phys. Rev. B **108**, 214312 (2023)
7. C.K. Chiu, J.C.Y. Teo, A.P. Schnyder, S. Ryu, Rev. Mod. Phys. **88**, 035005 (2016)
8. T. Ozawa, H.M. Price, A. Amo, N. Goldman, M. Hafezi, L. Lu, M. C. Rechtsman, D. Schuster, J. Simon, O. Zilberberg, I. Carusotto, Rev. Mod. Phys. **91**, 015006 (2019)
9. R. Shindou, R. Matsumoto, S. Murakami, J. Ohe, Phys. Rev. B **87**, 174427 (2013)
10. V. Peano, C. Brendel, M. Schmidt, F. Marquardt, Phys. Rev. X **5**, 031011 (2015)
11. V. Peano, H. Schulz-Baldes, J. Math. Phys. **59**, 031901 (2018)
12. R. Barnett, Phys. Rev. A **88**, 063631 (2013)
13. B. Galilo, D.K.K. Lee, R. Barnett, Phys. Rev. Lett. **115**, 245302 (2015)
14. V. Peano, M. Houde, C. Brendel, F. Marquardt, A.A. Clerk, Nat. Commun. **7**, 1 (2016)

Part II
Signs of Genuine SPT Physics in Open Bosonic Systems

Chapter 6
Background: Quadratic Bosonic Lindbladians

Abstract This chapter provides an extension of Chap. 2 to the Markovian setting. We first introduce the basic properties of quantum Markovian dynamical semigroups and, in particular, cover the basic notions of conservation laws, symmetries, and the breakdown of Noether's theorem. Then, we introduce the specific class of Markovian systems we are interested in, i.e., those systems whose Lindblad generator is a QBL. Such generators are defined by a QBH, paired with Lindblad dissipators that are linear in creation and annihilation operators. In particular, we highlight the key differences with the Hamiltonian formalism used up until this point. To this end, we emphasize the changes to the relevant equations of motion for linear and quadratic observables, and discuss the appropriate extensions of the stability notions in QBHs and their criteria. Of particular interest are the steady states, which we characterize in detail. The chapter concludes with a simple single-mode example that exemplifies the key parts of the formalism, in addition to an account of the implications of bulk-translation invariance for these systems.

This chapter constitutes the open (Markovian) counterpart to Chap. 2.[1] Leading up to this point, we have ultimately concluded that *tight* bosonic analogues of free-fermionic SPT physics cannot exist in any satisfactory way in a strictly closed-system, or unitary setting. While one-parameter unitary families can describe continuous-time quantum dynamics, they are not the most general model. Instead, one may consider a one-parameter of *completely positive trace-preserving maps*, which in turn, are the most general maps that preserve the interpretation of a density operator as a quantum state. While characterizations of such families exist (e.g., via a Kraus representation), they are generically difficult to describe

[1] This chapter contains an elementary introduction to quantum Markovian systems that can be found in standard texts on open quantum systems, e.g., Ref. [1]. The discussion of conserved quantities and symmetries summarizes key results and observations from Refs. [2–6]. While our approach to quadratic Lindbladians more closely follows works like Ref. [7], there exists an alternative approach, known as third quantization, first developed in Refs. [8, 9] for fermions, and Ref. [10] for bosons. Notably, our use of the word *rapidities* is specifically derived from these works.

V. P. Flynn, *Effective Non-Hermiticity and Topology in Markovian Quadratic Bosonic Dynamics*, Springer Theses, https://doi.org/10.1007/978-3-031-52045-7_6

without additional restrictions. A mathematically natural requirement, which is also physically motivated for a large class of dynamical systems, is that our family obey a forward composition, or Markov, law. Such a law engenders a "memoryless" nature to the dynamics, i.e., the future configuration of a quantum state is determined exclusively by its present configuration. These families, also called quantum dynamical semigroups (as they obey the mathematical axioms of a semigroup), are in turn characterized by an infinitesimal generator known as the Lindbladian. Specifically, the quantum state obeys a master equation, called the *Lindblad master equation*, defined by this generator. Notably, such master equations arise naturally when evaluating the reduced system dynamics of a system coupled to a bath, specifically when various physically motivated approximations (e.g., Born, Markov, secular) are made.

One immediate consequence of forgoing unitarity is the loss of a simple characterization of the relationship between conserved quantities and symmetries. For unitary systems, a conserved quantity always generates a one-parameter (unitary) symmetry group, while to each continuous symmetry group, there exists a family of conserved generators. This latter statement is essentially a non-relativistic quantum incarnation of Noether's theorem. In Markovian systems, conserved quantities are described as those operators whose expectation values are time-independent in arbitrary states. Symmetries, on the other hand, bifurcate into "weak" and "strong" symmetries [2–4]. Both incarnations are represented by unitary or antiunitary operators (via Wigner's theorem) that leave the dynamics invariant. However, unlike strong symmetries, weak symmetries do not separately leave the "coherent" and "incoherent" components of the dynamics invariant. As a consequence, continuous families of weak symmetries need not be generated by conserved quantities, while families of strong symmetries are. Surprisingly, however, this does not mean that every conserved quantity generates a family of strong symmetries, or any family of symmetries, for that matter. The lacking of any clear correspondence between conserved quantities and SGs, in this sense, represent a sort-of "breakdown" of Noether's theorem in Markovian systems. To this end, various attempts have been made to characterize this breakdown and explore to what extent it may be restored [5].

Refocusing on bosonic systems, Markovian dynamics on our multimode bosonic Fock spaces should be modeled by a Lindblad generator that is defined in terms of the relevant creation and annihilation operators. If we are to maintain the "non-interacting" essence of QBHs, it is natural to consider those Lindbladians that generate *Gaussian* transformations on the Fock space. Such Lindladians generate *Gaussian quantum Markov semigroups* [11, 12] which are especially relevant for modeling dissipation in quantum optical systems [13], noise processes in continuous variable quantum information [14], and the dynamical behavior bosonic matter coupled to thermal baths [15], among other things. Techniques for studying such systems vary greatly. One of the more popular approaches is known as *third quantization* which begins by defining superoperator analogues of the creation and annihilation operators, and then diagonalizing the Lindbladian in terms of these third quantized "quasiparticles" [10]. The notion of quasiparticle energies generalizes to what are known as *rapidities*, which constitute the fundamental building blocks of the Lindbladian spectrum. While this formalism is extremely useful, we will instead focus explicitly on the Heisenberg equations of motion for the creation and annihilation operators (and their quadratic counterparts), and thus, remain in a "second quantized framework". Such a perspective will ultimately prove sufficient for uncovering non-trivial topological physics in these systems. Thankfully, these equations of motion are well-studied [7] and are known to contain all of the relevant information (for our purpose) about the dynamics.

The outline of this chapter is as follows. In Sect. 6.1 we cover the basic features of Markovian open quantum systems, including spectral and convergence properties in addition to the essential aspects of their conserved quantities and symmetries. In Sect. 6.2, we identify the Markovian open-system dynamics we are interested in, i.e., those whose dynamical maps are defined by Lindbladians that are quadratic in creation and annihilation operators. Working primarily in the Heisenberg picture, we write down the equations of motion for linear and quadratic forms (with applications of the "hat map" of Sect. 2.3.3 specifically emphasized), discuss stability and relaxation criteria, and present a simple single-mode example to demonstrate the key aspects of the formalism. Finally, we briefly describe the roles of translational symmetry and BCs in these Lindbladians and thus generalize Sect. 2.5 to the genuinely open Markovian setting.

6.1 The Lindblad Formalism

The dynamics of Markovian ("memoryless") open quantum systems are described by *quantum dynamical semigroups* (QDSs). Such semigroups are built from a one-parameter family completely positive trace-preserving (CPTP) maps $\{\mathcal{E}_t\}_{t\geq t_0}$ that constitute a semigroup, namely, in the simplest time-homogeneous case, they obey a forward (Markov) composition law, $\mathcal{E}_t \circ \mathcal{E}_s = \mathcal{E}_{t+s}$, with $\mathcal{E}_{t_0} = \mathcal{I}$. Underlying this notion are technical assumptions on continuity and boundedness that we will not state, nor dwell on. Notably, unitary dynamics provide QDSs, but a given QDS need not, and typically does not, take the form $\mathcal{E}_t(\rho) = U(t)\rho U^\dagger(t)$ with $U(t)$ unitary. Moreover, one may arrive at this class of dynamics in two distinct ways: (i) as a description for the reduced system dynamics starting from a microscopic system-bath Hamiltonian and making a series of approximations (Born, Markov, secular), or (ii) as a phenomenological description of a given open quantum system. The class of bosonic models we will consider arise most commonly following approach (ii).

Henceforth, we will take $t_0 = 0$ for convenience, so that the initial state $\rho(0) = \rho$. Given a QDS, the time-dependence of ρ is given by $\rho(t) = \mathcal{E}_t(\rho) \equiv e^{t\mathcal{L}}(\rho)$, with the Markovian generator $\mathcal{L}$ (or "Lindbladian") obeying the *Lindblad master equation* (LME)

$$\frac{d}{dt}\rho(t)=\mathcal{L}(\rho(t)), \quad t \geq 0, \quad \mathcal{L}(\rho)=-i[H,\rho]+\sum_{n=1}^{d}\left(L_n\rho L_n^\dagger - \frac{1}{2}\{L_n^\dagger L_n, \rho\}\right). \tag{6.1}$$

Here, H is self-adjoint and the $\{L_n\}_{n=1}^d$ is a set of d (typically not self-adjoint) operators called Lindblad, or noise operators. Physically, the operators L_n encode the dissipative effect of the coupling to the environment. It is often convenient to separate the "unitary" contribution, i.e., the Hamiltonian commutator, and the "non-unitary", or dissipative, portion, which we will denote by $\mathcal{D}(\rho)$.

While specifying H and $\{L_n\}_{n=1}^d$ yields a particular Lindbladian, such a representation need not be unique. Two well-known invariance properties are that of unitary rotation of the Lindblad operators $L_n \mapsto L_n' = \sum_{m=1}^d \mathbf{U}_{nm}L_m$, with $\mathbf{U}$ a $d \times d$ unitary matrix, and the more complicated inhomogeneous transformation

$$H \mapsto H' = H + \frac{1}{2i}\sum_{n=1}^{d}\left(z_n L_n - z_n^* L_n^\dagger\right) + a 1_S, \quad L_n \mapsto L_n' = L_n + z_n^* 1_S, \tag{6.2}$$

with $z_n \in \mathbb{C}$, $a \in \mathbb{R}$, and 1_S the system identity operator. However, it is often convenient to work within a fixed representation.

Just as for closed-system dynamics, we can define an equivalent Heisenberg picture of the dynamics. We define the time-dependence of an arbitrary observable $B = B^\dagger \mapsto B(t)$ implicitly by stipulating that the expectation value $\text{tr}[B(t)\rho]$ must be equal to $\text{tr}[B\rho(t)]$ for any initial state $\rho(0) = \rho$. Defining the Hilbert-Schmidt inner-product $\langle X, Y\rangle_{\text{HS}} \equiv \text{tr}[X^\dagger Y]$, we see[2] that $B(t) = \mathcal{E}_t^\star(B)$, with $\star$ denoting the Hilbert-Schmidt adjoint, satisfies the stated requirement of a Heisenberg picture. One may verify that $\{\mathcal{E}_t^\star\}_{t\geq 0}$ is a unital ($\mathcal{E}_t^\star(1_S) = 1_S$) semigroup that may be written as $\mathcal{E}_t^\star = e^{t\mathcal{L}^\star}$. The Heisenberg EOM follows naturally as

$$\frac{d}{dt}B(t) = \mathcal{L}^\star(B(t)), \quad t \geq 0, \tag{6.3}$$

with the Hilbert-Schmidt adjoint of the Lindbladian taking the form

$$\mathcal{L}^\star(B) = i[H, B] + \sum_{n=1}^{d}\left(L_n^\dagger B L_n - \frac{1}{2}\{L_n^\dagger L_n, B\}\right), \tag{6.4}$$

[2] To be more specific, expectation values of observables $B = B^\dagger$ may be written as $\langle B\rangle(t) = \langle B, \rho(t)\rangle_{\text{HS}} = \langle B, \mathcal{E}_t(\rho)\rangle_{\text{HS}} = \left\langle \mathcal{E}_t^\star(B), \rho\right\rangle_{\text{HS}}$, by definition of the adjoint of an operator in an inner-product space.

which consequently satisfies $\mathcal{L}^\star(1_S) = 0$ and $\mathcal{L}^\star(B^\dagger) = [\mathcal{L}^\star(B)]^\dagger$. These two properties are Heisenberg-picture restatements of trace preservation and adjoint preservation, respectively.

Before moving to specific features of Lindbladian dynamics, let us consider an alternative form for Lindbladians. Namely, suppose we have a set of system operators $\{A_j\}_{j=1}^{d'}$, that have particular physical relevance to the problem at hand (for example, the system Hamiltonian[3] may be a simple function of these operators). Suppose further than we can express the Lindblad operators L_n in terms of these operators,[4] i.e., $L_n = \sum_{j=1}^{d'} \ell_{jn} A_j$, with $\ell_{jn} \in \mathbb{C}$. With this, we obtain the Gorini-Kossakowski-Sudarshan (GKS) representation of the Lindbladian

$$\mathcal{L}(\rho) = -i[H, \rho] + \sum_{j,k=1}^{d'} \mathbf{M}_{jk} \left(A_k \rho A_j^\dagger - \frac{1}{2}\{A_j^\dagger A_k, \rho\} \right), \quad \mathbf{M}_{jk} = \sum_{n=1}^{d} \ell_{jn}^* \ell_{kn}. \tag{6.5}$$

Here, the GKS matrix $\mathbf{M}$ is a $d' \times d'$ positive-semidefinite matrix. Equivalently, $\mathbf{M}$ is defined implicitly via the equation

$$\sum_{n=1}^{d} L_n^\dagger L_n = \sum_{j,k=1}^{d'} \mathbf{M}_{jk} A_j^\dagger A_k, \tag{6.6}$$

where the object on the left hand-side is sometimes known as the *parent Hamiltonian* of the dissipator [16]. In the Heisenberg picture,

$$\mathcal{L}^\star(B) = i[H, B] + \sum_{j,k=1}^{d'} \mathbf{M}_{jk} \left(A_j^\dagger B A_k - \frac{1}{2}\{A_j^\dagger A_k, B\} \right). \tag{6.7}$$

Here it is convenient to introduce more notation, namely, $\mathcal{D}[A, B](\rho) = A\rho B - \{BA, \rho\}/2$ and $\mathcal{D}[A] = \mathcal{D}[A, A^\dagger]$. Now, unlike the unitary case, Markovian dynamics generically do not factor. That is, if B and C are fixed operators, then, generically, $(BC)(t) \neq B(t)C(t)$. An important consequence is the following: if $\{B_j\}$ is a set of operators that generate the observable algebra of interest (i.e., every observable of interest is a linear combination of arbitrary degree products of the B_j

[3] We will often refer to H as the "system Hamiltonian". This should be understood simply as shorthand, and not as a statement that H is the system contribution to the hypothetical full system-bath Hamiltonian. Typically, due to the so-called "Lamb shift" correction, microscopic descriptions of LME's produce representations where H differs from the bare system Hamiltonian.

[4] In a finite-dimensional setting (say, with dimension D), it is common practice to choose the A_j to be an orthonormal (with respect to the Hilbert-Schmidt inner-product) basis for the $(D^2 - 1)$-dimensional operator space. It follows that the Hamiltonian and the Lindblad operators are expressible as linear combinations of these operators.

operators), then the dynamics of all observables need not follow immediately from the dynamics of the generators $B_j(t)$.

6.1.1 Spectral and Convergence Properties

If the system Hilbert space is finite dimensional, say of dimension D, then the LME describes an LTI equation of motion whose generator $\mathcal{L}$ may be thought of as non-Hermitian linear operator acting on a D^2-dimensional space. As such, we always can find a discrete set of distinct eigenvalues $\lambda_1, \ldots, \lambda_M$, with $M \leq D^2$, of $\mathcal{L}$. It is well-known that (i) zero is always one of these eigenvalues, (ii) $\text{Re}[\lambda_m] \leq 0$ for all $m = 1, \ldots, N$, and (iii) for each eigenvalue λ_m, there is another eigenvalue λ_m^*. The first property implies that there exists at least one steady state ρ_{ss} satisfying $\mathcal{L}(\rho_{\text{ss}}) = 0$. The second property ensures that any initial condition $\rho(0) = \rho$ will asymptotically relax to a particular manifold built from those (possibly generalized) eigenvectors with purely imaginary eigenvalues. If the only eigenvalue of $\mathcal{L}$, with $\text{Re}[\lambda_m] = 0$ is zero, then property (ii) guarantees all initial states will relax to the steady state manifold, i.e., the kernel of $\mathcal{L}$. To this end, it is useful to define the *Lindblad, spectral*, or *dissipative gap*,

$$\Delta_{\mathcal{L}} = |\sup \text{Re}[\sigma(\mathcal{L}) \setminus \{0\}]| \geq 0, \tag{6.8}$$

i.e., the closest distance between the imaginary axis and the set of nonzero eigenvalues of $\mathcal{L}$. Physically, $\Delta_{\mathcal{L}}$ sets the asymptotic decay rate through the inequality

$$d_{\text{mix}}(t) = \sup_{\rho(0)} \inf_{\sigma} \{\|\mathcal{E}_t(\rho(0)) - \sigma\|_{\text{tr}} : \mathcal{L}(\sigma) = 0\} \leq K e^{-\Delta_{\mathcal{L}} t}. \tag{6.9}$$

That is, the worst case distance (here, the trace distance $\|A\|_{\text{tr}} = \text{tr}[\sqrt{A^\dagger A}]$) from the steady state manifold is bounded above by an unspecified constant K times a factor decaying exponentially when $\Delta_{\mathcal{L}} > 0$. In short, the asymptotic convergence to the steady state manifold is exponential when the gap is nonzero. From here it is standard to define the mixing time[5] $t_{\text{mix}}(\epsilon)$ which is defined as the time it takes for the worst-case distance $d_{\text{mix}}(t)$ to fall below a prescribed accuracy ϵ. Such a quantity is a crude, but useful tool for characterizing the transient (pre-asymptotic) dynamics.

A crucial assumption made so far is that of finite dimensionality. At each stage of the QDS discussion, major results breakdown once the possibility for infinite dimensionality is opened up. At the fundamental level, it is not even necessarily

[5] The mixing time is a rather general concept in dynamical (both continuous and discrete) dynamical system theory. Here, we only discuss its specific definition for (quantum) Markovian semigroups.

the case that there exists a generator of the form in Eq. (6.1) if certain boundedness assumptions on the QDS are not obeyed [17]. Even when one stipulates a generator of that form, infinite-dimensionality allows for major differences in the spectral properties. Firstly, the spectrum may develop a continuum component (something that is not-so surprising from the Hamiltonian perspective). Secondly, of the three stated properties (i–iii) of the Lindblad spectrum, only (iii) carries over, in general. As we will see with bosonic systems, an unbounded Lindbladian acting on an infinite-dimensional space can lack any steady state and its adjoint can possess spectra in the right half plane. These possibilities, while seemingly "bugs" in the approach, have actually turned out to be extremely useful features for modeling common systems of interest. For example, these more "exotic" LMEs are often employed to model a wide class of amplification phenomena arising in quantum optical systems due to both coherent and incoherent gain mechanisms (e.g., two- and one- photon driving, respectively). Since our primary focus is bosonic physics, we will be forced to grapple with these exotic features and explore to what extent infinite-dimensionality may break commonly held intuitions about Markovian dynamics.

6.1.2 Symmetries and Conserved Quantities in Markovian Systems

The definitions of symmetries and conserved quantities in closed systems is straightforward textbook material. By Wigner's theorem, a symmetry of a physical system is represented by a unitary or antiunitary operator that commutes with the Hamiltonian. Similarly, a conserved quantity is an observable (a self-adjoint operator) that commutes with the Hamiltonian. There are even correspondences between the two concepts. The first is that, to each continuous family of (necessarily) unitary symmetries, there is an associated conserved quantity. That is to say, such a family can be written as $U(\theta) = e^{i\theta Q}$, with $\theta \in \mathbb{R}$ and Q a conserved quantity. This constitutes a realization of Noether's theorem in non-relativistic quantum mechanics. More or less surprising is the converse: to each conserved quantity Q, there is a continuous family of symmetries $U(\theta) = e^{i\theta Q}$. Proofs of both directions are trivial in standard Hamiltonian quantum mechanics.

The landscape of symmetries and conserved quantities becomes much richer for open quantum systems, even in the simplest case of Markovian dissipation [4, 6]. Immediately, one is met with two distinct notions of symmetry: weak and strong. A weak symmetry is a unitary or antiunitary operator S that leaves the dynamics invariant, i.e.,

$$\mathcal{E}_t(S\rho S^{-1}) = S\mathcal{E}_t(\rho)S^{-1}. \tag{6.10}$$

Equivalently, the superoperator $\mathcal{S}(\rho) \equiv S\rho S^{-1}$ commutes with $\mathcal{E}_t$ for all t. Continuity of $\mathcal{E}_t$ allows us to restate this in terms of the generator $\mathcal{L}$, namely $[\mathcal{S}, \mathcal{L}] = 0$. When further specification is not particularly relevant, we simply refer to weak symmetries as symmetries. Suppose we have a continuous family of (necessarily) unitary operators $U(\theta) = e^{i\theta G}$, with $G = G^\dagger$. Each $U(\theta)$ is a weak symmetry if and only if

$$\mathcal{L}([G, \rho]) = [G, \mathcal{L}(\rho)], \quad \forall \rho. \tag{6.11}$$

That is, G generates a family of weak symmetries if and only if the adjoint action of G, i.e., $[G, \cdot]$, commutes with $\mathcal{L}$. The equivalent condition in the Heisenberg picture is

$$\mathcal{L}^\star([G, A]) = [G, \mathcal{L}^\star(A)], \quad \forall A, \tag{6.12}$$

where A is an arbitrary system operator.

Given a Lindbladian with a representation in terms of H and $\{L_n\}_{n=1}^d$, we further define a strong symmetry as a unitary or antiuntary operator S that satisfies

$$[H, S] = 0, \quad [L_n, S] = 0, \; n = 1, \ldots, d. \tag{6.13}$$

One can immediately verify that every strong symmetry is a weak symmetry. Furthermore, if H' and $\{L'_n\}_{n=1}^{d'}$ is a representation obtained via the invariance transformation mentioned in the previous section, then $[S, H'] = [S, L'_n] = 0$. A continuous family of (necessarily) unitary operators $U(\theta) = e^{i\theta G}$ provide a family of strong symmetries if and only $[G, H] = [G, L_n] = 0$ for all n.

Finally, a conserved quantity is a self-adjoint operator Q that satisfies

$$\mathcal{L}^\star(Q) = 0 \tag{6.14}$$

The immediate implication is that $\langle Q \rangle (t) = \langle Q \rangle (0)$ in any state. In stark contrast to Hamiltonian systems, this does not imply all moments $\langle Q^n \rangle (t)$ are time-independent. Conservation of all moments instead requires that both Q and Q^2 are conserved [5].

The connection between conserved quantities and symmetries is not nearly as straightforward as it is in the Hamiltonian case. Firstly, there is no direct correspondence between generators of weak symmetries and conserved quantities, as evidenced by the differences in conditions Eq. (6.11) and Eq. (6.14). We refer to this lack of correspondence as a breakdown of Noether's theorem (and its converse) in Markovian systems. However, there is a one-to-one correspondence between generators of strong symmetries and conserved quantities Q with all moments conserved.

6.2 Quadratic Bosonic Lindbladians

The open system dynamics we concern ourselves with are generated by *purely quadratic* bosonic Lindbladians. Such Lindbladians may be represented by a purely QBH H and Lindblad operators L_n that are linear in creation and annihilation operators. The restriction that the L_n are linear ensures that the Lindbladian itself, which is bilinear in the L_n's, will be quadratic. We further comment that specifying that the Lindbladian be purely quadratic, rather than at-most quadratic (e.g., by including linear terms in the Hamiltonian or constant shifts in the Lindblad operators), follows from the same motivation as in the Hamiltonian case: (i) the inhomogeneous transformation Eq. (6.2), combined with the methods detailed in Sect. 2.3 for removing linear terms in quadratic Hamiltonians, can typically be combined to remove linear terms and (ii) restricting to purely quadratic systems ensures parity is a (weak) symmetry, thus ensuring we are on level ground for direct comparison to fermionic systems. Such purely quadratic Lindbladians will be henceforth referred to as quadratic bosonic Lindbladians (QBLs).

6.2.1 *Equations of Motion*

To be more concrete, our Lindbladians are defined by a QBH $H = \frac{1}{2}\Phi^\dagger \boldsymbol{\tau}_3 \mathbf{G}_0 \Phi$, with $\mathbf{G}_0$ the Hamiltonian dynamical matrix and Lindblad operators $L_n = \sum_{j=1}^{N} \ell_{jn} \Phi_j$, with $\ell_{jn} \in \mathbb{C}$ and d the number of Lindblad operators. The $2N \times 2N$ GKS representation of $\mathcal{L}$ is then

$$\mathcal{L}(\rho) = -i[H, \rho] + \mathcal{D}(\rho) = -i[H, \rho] + \sum_{j,k=1}^{2N} \mathbf{M}_{jk} \left(\Phi_k \rho \Phi_j^\dagger - \frac{1}{2}\{\Phi_j^\dagger \Phi_k, \rho\} \right), \tag{6.15}$$

with the $2N \times 2N$ GKS matrix defined according to the previous section, i.e., $\mathbf{M}_{jk} = \sum_n \ell_{jn}^* \ell_{kn}$. The Heisenberg picture Lindbladian follows as in Eq. (6.7).

The formulation of Markovian bosonic dynamics is more efficiently captured in the Heisenberg picture. In particular, the dynamics of the Nambu array Φ are given by

$$\frac{d}{dt}\Phi(t) = \mathcal{L}^\star(\Phi(t)) = -i\mathbf{G}\Phi(t), \quad \mathbf{G} = \mathbf{G}_0 - i\boldsymbol{\tau}_3 \mathcal{F}(\mathbf{M}). \tag{6.16}$$

In terms of similarities with the Hamiltonian case (Eq. (2.43)), we have an LTI system with a non-Hermitian generator $\mathbf{G}$, for which we will retain the name dynamical matrix. Furthermore we retain the property $\boldsymbol{\tau}_1 \mathbf{G}^* \boldsymbol{\tau}_1 = -\mathbf{G}$ and hence the associated spectral symmetry $\sigma(\mathbf{G}) = -\sigma(\mathbf{G})^*$. One important difference reveals itself in the generic lack of pseudo-Hermiticity (and hence the lack of conjugate spectral symmetry). Specifically, the dissipator contributes an anti-pseudo-Hermitian term $-i\boldsymbol{\tau}_3\mathcal{F}(\mathbf{M})$, i.e.,

$$\boldsymbol{\tau}_3\mathbf{G}^\dagger\boldsymbol{\tau}_3 = \mathbf{G}_0 + i\boldsymbol{\tau}_3\mathcal{F}(\mathbf{M}). \tag{6.17}$$

In the case where this term vanishes (i.e., the fermionic projection of $\mathbf{M}$ vanishes), pseudo-Hermiticity is retained. In this case one may verify that $\mathcal{L}(1_\mathcal{F}) = 0$, i.e., the dynamics is unital. However, infinite-dimensionality prevents the interpretation of $1_\mathcal{F}$ as a quantum state, as it is not trace-class. At the opposite end of the spectrum, if $\mathbf{G}_0 = 0$ (or equivalently $H = 0$), the dynamical matrix is purely anti-pseudo-Hermitian. The mathematical development of pseudo-Hermitian matrices can then be brought over without issue.

The bosonic and fermionic projections of $\mathbf{M}$ play distinct roles in the dissipator. Specifically, if we write $\mathbf{M} = \mathcal{B}(\mathbf{M}) + \mathcal{F}(\mathbf{M})$, and leverage the CCRs, one may verify that

$$\mathcal{D}^\star(B) = \frac{1}{2}\sum_{i,j=1}^{2N} \mathcal{B}(\mathbf{M})_{ij}\left(\left[[\Phi_i^\dagger, B], \Phi_j\right] + \left[\Phi_i^\dagger, [A, \Phi_j]\right]\right) \tag{6.18}$$

$$+\mathcal{F}(\mathbf{M})_{ij}\left(\left\{[\Phi_i^\dagger, B], \Phi_j\right\} + \left\{\Phi_i^\dagger, [A, \Phi_j]\right\}\right). \tag{6.19}$$

If B is a monomial in Φ_j of degree d, i.e., B is a product of d creation and annihilation operators, then the CCRs imply that the first term reduces the degree by 2 while the second preserves the degree. This follows from the algebraic fact that (anti)commuting a degree d operator with a creation or annihilation operator will output a degree $d - 1$ ($d + 1$) operator. Thus, the bosonic projection of $\mathbf{M}$ is responsible for connecting the degree d and degree $d - 2$ sectors of the operator algebra while the fermionic projections leave the degree d sector invariant. While this understanding is consistent with the in-built (weak) parity symmetry, there is a notable exclusion of coupling between degree d and $d + 2$ operators.[6] Consequentially, the dynamics of linear forms can only contain contributions from $\mathcal{F}(\mathbf{M})$, as we have seen in Eq. (6.16). The equations of motion for quadratic operators have no such restriction. Let $Q_{ij} = \Phi_i\Phi_j^\dagger$ be the array of all products of creation and annihilation operators. Observe that the CCRs yield $\mathcal{F}(Q) = (\boldsymbol{\tau}_3/2)1_\mathcal{F}$. Thus expectation values of Q are uniquely determined by those of $\mathcal{B}(Q)_{ij} = \{\Phi_i, \Phi_j^\dagger\}/2$. The equations of motion for these operators are

$$\frac{d}{dt}\mathcal{B}(Q(t)) = \mathcal{L}^\star(\mathcal{B}(Q(t))) = -i\left(\mathbf{G}\mathcal{B}(Q(t)) - \mathcal{B}(Q(t))\mathbf{G}^\dagger\right) + \boldsymbol{\tau}_3\mathcal{B}(\mathbf{M})\boldsymbol{\tau}_3\, 1_\mathcal{F}. \tag{6.20}$$

[6] Such an exclusion is not present in quadratic fermionic Lindbladians.

As expected from Eq. (6.18), $\mathcal{F}(\mathbf{M})$ (present in $\mathbf{G}$) connects the degree 2 operators $\mathcal{B}(Q_{ij})$ to other degree 2 operators and $\mathcal{B}(\mathbf{M})$ connects $\mathcal{B}(Q)$ to the degree $2-2=0$ operator $1_{\mathcal{F}}$.

Equations (6.16) and (6.20) can be leveraged to find the time-dependence of the mean vector and covariance matrix of a given state ρ. Following from Eq. (6.16), the mean vector at time t is given by $\vec{m}_\rho(t) = e^{-i\mathbf{G}t}\vec{m}_\rho(0)$. The covariance matrix is a bit less straightforward. Firstly, note that $\mathbf{C}_\rho(t) = \text{tr}[\rho(t)\mathcal{B}(Q)] - \vec{m}_\rho(t)\vec{m}_\rho(t)^\dagger$ from which it follows that

$$\frac{d}{dt}\mathbf{C}_\rho(t) = -i\left(\mathbf{G}\mathbf{C}_\rho(t) - \mathbf{C}_\rho(t)\mathbf{G}^\dagger\right) + \boldsymbol{\tau}_3\mathcal{B}(\mathbf{M})\boldsymbol{\tau}_3. \tag{6.21}$$

The solution to this equation is

$$\mathbf{C}_\rho(t) = e^{-i\mathbf{G}t}\mathbf{C}_\rho(0)e^{i\mathbf{G}^\dagger t} + \int_0^t e^{-i\mathbf{G}s}\boldsymbol{\tau}_3\mathcal{B}(\mathbf{M})\boldsymbol{\tau}_3 e^{i\mathbf{G}^\dagger s}\,ds. \tag{6.22}$$

In particular, if $\rho(0) = \rho$ is a Gaussian state with mean vector $\vec{m}_\rho(0)$ and covariance matrix $\mathbf{C}_\rho(0)$, then $\rho(t)$ will be a Gaussian state with mean vector $\vec{m}_\rho(t)$ and covariance matrix $\mathbf{C}_\rho(t)$. Gaussianity preservation follows from the quadratic nature of the generator. That is to say, the semigroups generated by QBLs are examples of *Gaussian quantum Markov semigroups* [11, 12].

We conclude by exploring the interplay between the mappings of Sect. 2.3.3 and the Lindbladian. Given a vector $\vec{\alpha}$ and the associated linear form $\widehat{\vec{\alpha}}$, we have

$$\mathcal{L}^\star(\widehat{\vec{\alpha}}) = i\widehat{\widetilde{\mathbf{G}}\vec{\alpha}}, \quad \widetilde{\mathbf{G}} = \boldsymbol{\tau}_3\mathbf{G}^\dagger\boldsymbol{\tau}_3, \tag{6.23}$$

in terms of the *associated dynamical matrix* $\widetilde{\mathbf{G}}$. In other words, the (contravariant) dynamics of $\widehat{\vec{\alpha}}$ are determined by the equation of motion

$$\frac{d}{dt}\vec{\alpha}(t) = i\widetilde{\mathbf{G}}\vec{\alpha}(t). \tag{6.24}$$

Here we see that the pseudo-Hermitian conjugate of $\mathbf{G}$, i.e., $\widetilde{\mathbf{G}} = \boldsymbol{\tau}_3\mathbf{G}^\dagger\boldsymbol{\tau}_3$ plays an explicit role in determining the dynamics of generic linear forms. Only when $\mathcal{F}(\mathbf{M}) = 0$ (e.g., in the dissipation-free case) is the EOM of $\vec{\alpha}$ generated by the same matrix as that of Φ. The dynamical matrix itself instead appears in a more subtle identity:

$$\mathcal{L}([\widehat{\vec{\alpha}}, \rho]) - [\widehat{\vec{\alpha}}, \mathcal{L}(\rho)] = [\widehat{i\mathbf{G}\vec{\alpha}}, \rho] \text{ or } \mathcal{L}^\star([\widehat{\vec{\alpha}}, A]) - [\widehat{\vec{\alpha}}, \mathcal{L}^\star(A)] = -[\widehat{i\mathbf{G}\vec{\alpha}}, A], \tag{6.25}$$

where ρ (A) is an arbitrary state (system operator). This identity serves as the starting point for assessing the invariance of the dynamics under certain Weyl displacements (see Sect. 7.1).

6.2.2 Stability Criteria and Asymptotic Relaxation

The rich stability landscape of QBHs is modified in a non-trivial way due to the presence of Markovian dissipation. While, there is no clear generalization of thermodynamic stability for QBLs, the definition of dynamical stability formally carries over without issue. We say a QBL is dynamically stable if the expectation values of arbitrary observables in an arbitrary state remain bounded for all time. What changes, in comparison to the Hamiltonian case, is the approach to diagnosing dynamical stability.

At this stage it is convenient to introduce the notion of a *rapidity*. The rapidities, or rapidity spectrum, of a QBL are defined to be the eigenvalues of [7] $-i\mathbf{G}$. The rapidities serve as a useful generalization of the normal mode frequencies in QBHs. In fact, they occupy a role similar to that of quasiparticle energies in quadratic Hamiltonians. Techniques such as "third quantization" [10, 18] reveal that, if a QBL has a steady state, then the spectrum of the Lindblad operator itself is given by

$$\sigma(\mathcal{L}) = \Big\{ \sum_j n_j \lambda_j \ : \ n_j \in \mathbb{Z}_{\geq 0}, \lambda_j \in \sigma(-i\mathbf{G}) \Big\}. \tag{6.26}$$

Just as quasiparticle energies are the building blocks of the many-body energies, i.e., the Hamiltonian spectrum, the rapidities are the building blocks of the Lindblad spectrum. In the Hamiltonian case, the mapping between quasiparticle energies and many-body energies relied heavily on dynamical stability. Here, it appears that the condition "there exists a steady state" serves an analogous purpose. As it turns out, the existence of a steady state implies dynamical stability for QBLs [18]. Equivalently, dynamically unstable QBLs always lack a steady state. This is actually rather surprising. One may expect that a system can support both states that amplify and states that relax, but this is not the case.

In an effort to be more quantitative in our assessment, we define the *stability gap*,

$$\Delta_S \equiv \max \mathrm{Re}(\sigma(-i\mathbf{G})). \tag{6.27}$$

From Eq. (6.16), we see that if $\Delta_S > 0$, then the system is dynamically unstable and, consequentially, lacks a steady state. Beyond this case, we must assess what is required of a hypothetical steady state ρ_{ss}. Such a state must have a stationary mean vector $\vec{m}_{\mathrm{ss}} = \vec{m}_{\rho_{\mathrm{ss}}}$ and a stationary covariance matrix $\mathbf{C}_{\mathrm{ss}} = \mathbf{C}_{\rho_{\mathrm{ss}}}$. If such a pair exists, then the Gaussian state uniquely defined by $\vec{m}_{\mathrm{ss}}$ and $\mathbf{C}_{\mathrm{ss}}$ is automatically a steady state. The explicit stationarity conditions are

[7] The factor of $-i$ distinguishing rapidities from the normal mode frequencies of QBHs follows from the natural notion that the Lindbladian, in the closed-system limit, involves $-iH$ rather than H itself. This factor also implies that the symmetry $\sigma(\mathbf{G}) = -\sigma(\mathbf{G})^*$ leads to a conjugate symmetry in the rapidity spectrum.

Table 6.1 The relationships between the stability gap Δ_S of a QBL, its dynamical stability, and the nature of its steady state(s)

Stability gap	Dynamical stability	Steady state(s)
$\Delta_S < 0$	Stable	One
$\Delta_S = 0$	Stable or unstable	Infinitely many or none
$\Delta_S > 0$	Unstable	None

$$-i\mathbf{G}\vec{m}_{ss} = 0, \quad -i(\mathbf{G}\mathbf{C}_{ss} - \mathbf{C}_{ss}\mathbf{G}^\dagger) + \boldsymbol{\tau}_3\mathcal{B}(\mathbf{M})\boldsymbol{\tau}_3 = 0. \tag{6.28}$$

The first condition just says that $\vec{m}_{ss}$ is either a zero eigenvector of $\mathbf{G}$, or simply $\vec{m}_{ss} = 0$. The second condition is an example of a Lyapunov equation, $\mathbf{AX} + \mathbf{XA}^\dagger + \mathbf{Q} = 0$, with $\mathbf{A} = -i\mathbf{G}$ and $\mathbf{Q} = \boldsymbol{\tau}_3\mathcal{B}(\mathbf{M})\boldsymbol{\tau}_3 \geq 0$. A given Lyapunov equation, which may be understood as a linear system of equations for the matrix elements of $\mathbf{X}$, has either (i) a unique positive-semidefinite solution, (ii) infinitely many solutions, or (iii) zero solutions. Case (i) explicitly requires that the coefficient matrix $\mathbf{A}$ is Hurwitz, i.e., its spectrum lies strictly in the left half-complex plane (each eigenvalue has a strictly negative real part). For QBLs, this means that if $\Delta_S < 0$ then there is a unique positive-semidefinite solution $\mathbf{C}_{ss}$. Furthermore, if $\Delta_S < 0$, then $\vec{m}_{ss} = 0$. Consequently the unique steady state is the Gaussian state with vanishing mean vector and covariance matrix given by the unique solution to the Lyapunov equation. If $\Delta_S = 0$, then $-i\mathbf{G}$ is not Hurwitz and we have either case (ii) or (iii). In any case where a steady state exists, Eq. (6.26) reveals that $\Delta_{\mathcal{L}} = |\Delta_S|$, so that the stability gap sets the asymptotic relaxation rate. Thus, there are either infinitely many steady states (and hence the system is stable) or no steady states. The results of this analysis are summarized in Table 6.1.

6.2.3 An Elementary Example

To ground the discussion of QBLs, let us take a moment to emphasize the key concepts in a simple single-mode setting. First, consider the QBL defined by a harmonic oscillator Hamiltonian $H = \omega(a^\dagger a + aa^\dagger)/2$, with $\omega \geq 0$, and two Lindblad operators $L_- \equiv \sqrt{2\kappa_-}a$ and $L_+ \equiv \sqrt{2\kappa_+}a^\dagger$, with $\kappa_\pm \geq 0$. Physically, L_- (L_+) encodes single-photon loss (gain) at a rate κ_- (κ_+). The dynamical and GKS matrices follow as

$$\mathbf{G} = \begin{bmatrix} \omega - i\kappa & 0 \\ 0 & -\omega - i\kappa \end{bmatrix}, \quad \mathbf{M} = \begin{bmatrix} 2\kappa_- & 0 \\ 0 & 2\kappa_+ \end{bmatrix}, \tag{6.29}$$

where we have defined the *net loss rate* $\kappa \equiv \kappa_- - \kappa_+$. Notably, $\mathbf{G}$ only depends on the net loss rate and not the individual rates themselves. The rapidities follow as $\lambda_\pm = -\kappa \pm i\omega$ and so the stability gap is $\Delta_S = -\kappa$. The system is then dynamically stable for $\kappa_- > \kappa_+$ and unstable for $\kappa_- < \kappa_+$. The case $\kappa_- = \kappa_+$ is more subtle and requires we assess the existence of a steady state.

Consider first the case where $\kappa_- > \kappa_+$ (so that $\Delta_S < 0$). It follows that the Lindblad spectrum is given by $\sigma(\mathcal{L}) = \{n_+\lambda_+ + n_-\lambda_-\}_{n_\pm \in \mathbb{Z}_{\geq 0}}$. Furthermore, the steady state is unique and is Gaussian with mean vector $\vec{m}_{\text{ss}} = [0, 0]^T$ and covariance matrix given by,

$$\mathbf{C}_{\text{ss}} = \frac{\kappa_- + \kappa_+}{2(\kappa_- - \kappa_+)}\mathbb{1}_2, \tag{6.30}$$

which one may verify satisfies Eq. (6.28). In particular, the total boson number converges to $\langle a^\dagger a \rangle_{\text{ss}} = (\mathbf{C}_{\text{ss}})_{11} - 1/2 = \kappa_+/(\kappa_- - \kappa_+)$.

The case where $\kappa_- = \kappa_+$ (i.e., $\kappa = 0$) is more subtle. If it happens that $\kappa_- = \kappa_+ = 0$, then we are left with only coherent dynamics of the oscillator. Formally speaking, there are then infinitely many steady (or perhaps more properly in this context, stationary) states. For example, $\mathcal{L}(|n\rangle \langle n|) = 0$ for all Fock states $|n\rangle$. However, if $\kappa_\pm \neq 0$, then the system is dynamically unstable, and thus, lacks a steady state. Instability may be verified explicitly by solving Eq. (6.21) when $\kappa_+ = \kappa_- \neq 0$. Doing so, we find that, for example, the total number is given by

$$\langle a^\dagger a\rangle(t) = 2\kappa_+ t + \langle a^\dagger a\rangle(0), \quad \kappa_+ = \kappa_-, \tag{6.31}$$

which diverges as long as $\kappa_- \neq 0$, *independently* of the initial condition. One may loosely say that the “steady state” is one of infinitely many excitiations.

We note that, in this special case $\kappa = 0$, we have that $\langle a(t)\rangle = e^{-i\omega t} \langle a\rangle (0)$. Thus, any linear form will remain bounded, even if the overall system is dynamically unstable. This demonstrates how, if $\Delta_S = 0$, dynamical instabilities can occur without any instability in the LTI system Eq. (6.16). This should be contrasted with QBHs, whereby dynamical instability is completely determined by stability of Eq. (2.43).

6.2.4 Translationally Invariant QBLs and Arbitrary BCs

We now consider QBLs possessing discrete translational symmetry. The nature of open dynamics immediately forces us to answer the question: should we impose weak or strong translational symmetry? A quintessential example of a QBL with weak translational symmetry, but not strong, is one in which we have zero Hamiltonian and one Lindblad operator per site $L_j = \sqrt{2\kappa}a_j$, $j = 1, \ldots, N$. The isotropic nature of the dissipation rate κ ensure weak translational invariance. However, the L_j's clearly do not individually commute with the discrete translation operators. Instead, consider the QBL with zero Hamiltonian and just one Lindblad operator $L = \sqrt{2\kappa}\sum_j a_j$. In this case, L commutes with discrete translations and thus we have a strong translational symmetry. With the distinction in mind, we will require only weak translation-symmetry so that we may consider the largest possible class of QBLs.

As in the Hamiltonian case, we first consider a system on a finite ring under PBCs. In the end, weak translational symmetry demands that the Hamiltonian H be translation invariant (i.e., it must be of the form Eq. (2.76)) and that the dissipator takes the form

$$\mathcal{D}^{\star}(A) = \sum_{j=1}^{N} \sum_{r=-R}^{R} \left(\phi_j^{\dagger} A \mathbf{m}_r \phi_{j+r} - \frac{1}{2} \{ \phi_j^{\dagger} \mathbf{m}_r \phi_{j+r}, A \} \right), \tag{6.32}$$

with $\mathbf{m}_r$ the $2d_{\text{int}} \times 2d_{\text{int}}$ incoherent coupling matrix encoding dissipation, pumping, in incoherent pairing between sites j and $j+r$. These stipulations ensure that the dynamical matrix and the GKS matrix are both block-circulant matrices

$$\mathbf{G}_N^{\text{PBC}} = \mathbb{1}_N \otimes \mathbf{g}_0 + \sum_{r=1}^{R} \left(\mathbf{V}_N^r \otimes \mathbf{g}_r + \mathbf{V}_N^{\dagger\, r} \otimes \mathbf{g}_{-r} \right), \tag{6.33}$$

$$\mathbf{M}_N^{\text{PBC}} = \mathbb{1}_N \otimes \mathbf{m}_0 + \sum_{r=1}^{R} \left(\mathbf{V}_N^r \otimes \mathbf{m}_r + \mathbf{V}_N^{\dagger\, r} \otimes \mathbf{m}_{-r} \right), \tag{6.34}$$

where now $\mathbf{g}_r = \boldsymbol{\tau}_3 \mathbf{h}_r - i\boldsymbol{\tau}_3 \left(\mathbf{m}_r - \boldsymbol{\tau}_1 \mathbf{m}_r^* \boldsymbol{\tau}_1 \right)/2$. Block diagonalization of the dynamical matrix via Fourier transform proceeds exactly as in the Hamiltonian case. Ultimately, we obtain the Bloch dynamical matrix $\mathbf{g}(k)$ (defined exactly as in Eq. (2.80)) which determines the dynamics of the Fourier modes $\widetilde{\phi}_k$. Since we are in an open setting, we consider the eigenvalues of $-i\mathbf{g}(k)$ rather than $\mathbf{g}(k)$ itself. We call these eigenvalues, which we label $\lambda_n(k)$, $n = 1, \dots, 2d_{\text{int}}$, rapidity bands. The full rapidity spectrum is then $\{\lambda_n(k) : n = 1, \dots, 2d_{\text{int}},\ k \in \mathcal{K}_N\}$. The bi-infinite case then follows by once again adjusting the Fourier transform appropriately and taking $k \in [-\pi, \pi]$ to obtain a continuum of rapidities.

Finally, BCs can be imposed in a way analogous to the Hamiltonian case in Sect. 2.5.2. The only modification is that, in addition to the Hamiltonian, the dissipator must also respect the BCs. Focusing on OBCs for concreteness, this is done by removing dissipative couplings, encoded by $\mathbf{m}_r$, $r = 1, \dots, R$, between the internal degrees of freedom at site $N - R + 1, \dots, N$ and those at site $1, 2, \dots, R$. Thus, the resulting GKS matrix will be block-Toeplitz. In fact, the correspondence between BCs and the QBH dynamical matrix structure shown in Table 2.1 exactly extend to the QBL case. Under the BC in the first column, *both* the QBL dynamical matrix[8] and the GKS matrix fall into the corresponding class in the second column.

[8] While it is clear that the Hamiltonian contribution to the QBL dynamical matrix will be of the appropriate type to encode BCs, it may not be obvious that the addition of dissipation (via the term $-i\boldsymbol{\tau}_3\mathcal{F}(\mathbf{M})$) will leave this property invariant. However, since we have argued that the GKS matrix will be of the appropriate type, it follows that $\boldsymbol{\tau}_3\mathcal{F}(\mathbf{M})$ will be as well. This is because $\boldsymbol{\tau}_3$ and $\boldsymbol{\tau}_1$

The generalized Bloch theorem discussed in Sect. 2.5.2, and covered in detail in Appendix A.3, can then be applied to one, or both, of the dynamical matrix and GKS matrix.

References

1. R. Alicki, K. Lendi, *Quantum Dynamical Semigroups and Applications* (Springer, Berlin, 1987)
2. B. Berislav, T. Prosen, New J. Phys. **14**, 073007 (2012)
3. B. Baumgartner, H. Narnhofer, J. Phys. A **41**, 395303 (2008)
4. V.V. Albert, L. Jiang, Phys. Rev. A **89**, 022118 (2014)
5. J. Gough, J. Math. Phys. **56**, 022108 (2015)
6. P.D. Johnson, F. Ticozzi, L. Viola, Quant. Info. Comput. **16**, 657 (2016)
7. A.E. Teretenkov, Infin. Dimens. Anal. Quantum Probab. Relat. Top. **22**, 1930001 (2019)
8. T. Prosen, New J. Phys. **10**, 043026 (2008)
9. T. Prosen, J. Stat. Mech. **2010**, P07020 (2010)
10. T. Prosen, T.H. Seligman, J. Phys. A **43**, 392004 (2010)
11. P. Blanchard, M. Hellmich, P. Lugiewicz, R. Olkiewicz, J. Func. Anal. **256**, 1453 (2009)
12. D. Poletti, *Infinite Dimensional Analysis, Quantum Probability and Applications: QP41 Conference, Al Ain, UAE, March 28–April 1, 2021* (Springer, Berlin, 2022), p. 197
13. H. Carmichael, *An Open Systems Approach to Quantum Optics: Lectures Presented at the Université Libre de Bruxelles, October 28 to November 4, 1991*, vol. 18 (Springer, Berlin, 2009)
14. V.V. Albert, K. Noh, K. Duivenoorden, D.J. Young, R.T. Brierley, P. Reinhold, C. Vuillot, L. Li, C. Shen, S.M. Girvin, B.M. Terhal, L. Jiang, Phys. Rev. A **97**, 032346 (2018)
15. A. D'Abbruzzo, D. Rossini, Phys. Rev. A **103**, 052209 (2021)
16. C.E. Bardyn, M.A. Baranov, C.V. Kraus, E. Rico, A. İmamoglu, P. Zoller, S. Diehl, New J. Phys. **15**, 085001 (2013)
17. I. Siemon, A.S. Holevo, R.F. Werner, Open Syst. Inf. Dyn. **24**, 1740015 (2017)
18. T. Barthel, Y. Zhang, J. Stat. Mech. **2022**, 113101 (2022)

(which arises in the fermionic projection) act trivially on the lattice, i.e., $\boldsymbol{\tau}_j = \mathbb{1}_{\text{lattce}} \otimes \mathbb{1}_{2d_{\text{int}}} \otimes \boldsymbol{\sigma}_j$. Thus, the matrix and operator classes (Toeplitz, circulant, Laurent, etc), which are determined by lattice symmetries, are left invariant.

Chapter 7
Zero Modes, Weyl Symmetries, and QBL Design

Abstract We present a set of new results related to general QBLs with an eye towards the search for non-interacting bosonic SPT physics. First, we describe the breakdown of Noether's theorem in QBLs within the space of linear forms. We define *ZMs* of QBLs and, additionally, *Weyl symmetry generators* (SGs), which are generators of displacements in phase space that leave the overall dynamics invariant. Despite no obvious connection between these two operators in an open context, we prove a *one-to-one canonical correspondence* between them: To each zero mode, there must exists a canonically conjugate Weyl SG, and vice versa. We further generalize this result to accommodate *approximate* zero modes and generators of *approximate* symmetries. We proceed to discuss two novel design protocols for QBLs with certain desirable properties. The first takes, as an input, a quadratic fermionic Hamiltonian (QFH) with edge-localized, Hermitian, (possibly approximate) zero modes and provides, as an output, *purely dissipative QBL* (i.e., a QBL whose system Hamiltonian vanishes) possessing bosonic analogues of these (possibly approximate) zero modes. The second takes, as an input, a dynamically stable QBH and utilizes the duality of Chap. 4 to engineer a QBL whose unique steady state is given by the corresponding quasiparticle vacuum.

In this chapter, we prove a general correspondence between ZMs and generators of Weyl displacement symmetries in QBLs and, in addition, provide two QBL design protocols for engineering (i) Hermitian edge modes descendent from topological free-fermionic Hamiltonians and (ii) convergence to pure steady states of interest.[1] If we are to uncover SPT phases in QBLs, the first natural thing to investigate is the potential for topologically-mandated zero "energy", or more properly, *rapidity* edge modes. To this end, we must define precisely what we are looking for. While the precise definitions will be given later, we will think of *ZMs* as conserved quantities that are linear in creation and annihilation operators. This, of course,

[1] All results from this section can be found in Ref. [1]. Additionally, early incarnations of Theorems 7.1.3 and 7.1.4 can be found in the supplementary material of Ref. [2]. Both of these works were co-authored with Emilio Cobanera & Lorenza Viola.

V. P. Flynn, *Effective Non-Hermiticity and Topology in Markovian Quadratic Bosonic Dynamics*, Springer Theses, https://doi.org/10.1007/978-3-031-52045-7_7

naturally generalizes the standard fermionic notion of ZMs. We will see, however, that these modes capture only "half of the picture". After all, Hamiltonian ZMs may alternatively interpreted as the generators of one-parameter symmetry groups. In light of Sect. 6.1.2, we should not expect that our QBL ZMs possess any such property. Thus, we are lead to define *Weyl SGs* as operators that are (i) linear in creation and annihilation operators and (ii) generate a one-parameter family of weak symmetries. The nomenclature here refers to the fact that the corresponding symmetries take precisely the form of a displacement operator in phase space. These two families of operators together make up what we call *Noether operators* and, a priori, have no interdependence. While there are situations (e.g., in the purely dissipative limit) where one operator can play both roles (in which case, we say the are "non-split" Noether modes), these are highly non-generic. Remarkably, however, we will prove that, not only are these objects in one-to-one correspondence, they are actually in canonical correspondence. That is, to each ZM, there exists a *canonically conjugate* Weyl SG, and vice-versa. In this sense, there is a partial "restoration" of Noether's theorem for QBLs within the Nambu space. Furthermore, we are able to generalize this result to *approximate* ZMs and SGs, which are defined in a fairly natural way.

To further aid in our search for signatures of SPT physics in QBLs, we develop two design protocols. The first consists of a recipe for "embedding" arbitrary quadratic fermionic Hamiltonians into the dissipator of a QBLs. This procedure builds off two key observations: (i) If $H = 0$, the dynamical matrix of the QBL is defined uniquely by a fermionic matrix (specifically, the fermionic projection of the GKS matrix); and (ii) to each (possibly approximate) ZM in the original fermionic Hamiltonian, there is a corresponding (possibly, approximate) ZM-SG pair in the associated QBL. This allows us to import the Majorana edge states of topologically non-trivial fermionic Hamiltonians directly into a QBL setting. The second design protocol takes as an input a dynamically stable QBH and provides, as an output, a QBL that asymptotically relaxes any initial condition to the quasiparticle vacuum of said QBH. The QBL is defined by the QBH, in addition to a dissipator whose GKS matrix is defined by the duality transformation of Chap. 4 associated to the QBH.

The outline of this chapter is as follows. In Sect. 7.1, we state the precise definitions of ZMs and Weyl SGs and introduce the unifying concept of a *Noether mode*. We then prove that there is a one-to-one *canonical* correspondence between these objects. Following this, we generalize these concepts to allow for *approximate* ZMs and SGs and, again, prove a canonical correspondence. In Sect. 7.2, we present the two QBL design protocols. The first details how one may embed a fermionic Hamiltonian into the dissipator of QBL, while the second leverages the duality of Chap. 4 to design Lindbladians that converge to the quasiparticle vacuum of a given dynamically stable Hamilton of interest.

7.1 A Partial Restoration of Noether's Theorem

Signatures of SPT phases in QFHs come in many forms. However, as a consequence of the BBC, the most ubiquitous signature is the emergence of robust, symmetry protected, edge-localized ZMs. For example, consider the Majorana fermions γ_L and γ_R of the FKC (recall Sect. 5.2.1). The "*ZM*" property is precisely that (i) these are normal *modes* of the Hamiltonian and (ii) they correspond to *zero* energy (equivalently, the normal mode frequency is zero). This is summarized concisely in the equations $[H, \gamma_R] = [H, \gamma_L] = 0$. Beyond the ZM property, these equations, plus Hermiticity of $\gamma_{L,R}$, engender a secondary meaning: the Majorana fermions generate *symmetries* of the Hamiltonian. Specifically, $U_{L,R}(\theta) = e^{i\gamma_{L,R}\theta}$, with $\theta \in \mathbb{R}$ provide two continuous families of unitaries that commute with the Hamiltonian. Such is a property of Hamiltonian systems: if $Q = Q^\dagger$ is a conserved quantity, then $e^{iQ\theta}$ commutes with the Hamiltonian for all $\theta \in \mathbb{R}$. After all, the conservation condition and the SG condition are identical: $[H, Q] = 0$. Interpreting Majorana ZMs as SGs is essentially non-existent in literature, however. The reason is simple: the corresponding symmetriehs $U_{L,R}$ violate fermionic parity superselection (recall the discussion of Sect. 2.3).[2] Again, no such restriction applies to closed bosonic systems. If one had a Hermitian linear form that commuted with the Hamiltonian, it may be safely interpreted as a ZM *and* as the generator of a one-parameter symmetry. The latter interpretation would mean that the Hamiltonian is invariant under a certain displacement in phase space (to be explained momentarily).

7.1.1 The Exact Case

We have learned in Sect. 6.1.2 that there is no direct correspondence between conserved quantities and (weak) SGs in genuinely open Markovian systems. We then ask ourselves, if we are able to find signatures of SPT physics in QBLs, which role will, or should, they occupy: the role of a ZM, or the role of SG? To answer this, let us explore in more detail the ZM and SG conditions. A linear form γ^z is a ZM of a QBL if $\mathcal{L}^\star(\gamma^z) = 0$. Using the Nambu representation $\gamma = \widehat{\vec{\gamma}^z}$ and Eq. (6.23), we see that

$$\mathcal{L}^\star(\gamma^z) = 0 \iff \widetilde{\mathbf{G}}\vec{\gamma}^z = 0, \quad (\gamma^z = \widehat{\vec{\gamma}^z}), \tag{7.1}$$

in terms of the associated dynamical matrix $\widetilde{\mathbf{G}}$. Following from the general Lindbladian property $\mathcal{L}^\star(A^\dagger) = [\mathcal{L}^\star(A)]^\dagger$, which is manifests at the level of linear forms as $\boldsymbol{\tau}_1\widetilde{\mathbf{G}}^*\boldsymbol{\tau}_1 = -\widetilde{\mathbf{G}}$, we see that we are free to take $\gamma = \gamma^\dagger$ or, equivalently

[2] Despite the non-physical nature of these symmetries, they are useful for describing the ground state degeneracy. Specifically, if $|\widetilde{0}\rangle$ is the even-parity ground state, then $|\widetilde{1}_N\rangle = [U_R(\pi/2) - iU_L(\pi/2)]|\widetilde{0}\rangle$ is the degenerate, odd-parity ground state.

$\vec{\gamma}^z = -\boldsymbol{\tau}_1\vec{\gamma}^{z*}$. In words, the *ZMs of a QBL* are defined by kernel vectors of $\widetilde{\mathbf{G}}$ satisfying the Hermiticity condition $\vec{\gamma}^z = -\boldsymbol{\tau}_1\vec{\gamma}^{z*}$. We may formalize this by introducing the real vector space of ZMs:

$$\mathcal{Z} \equiv \{\text{Hermitian linear forms } \gamma^z : \mathcal{L}^\star(\gamma^z) = 0\}. \tag{7.2}$$

Now, consider $\ker\widetilde{\mathbf{G}}$. The property $\boldsymbol{\tau}_1\widetilde{\mathbf{G}}^*\boldsymbol{\tau}_1 = -\widetilde{\mathbf{G}}$ implies that $\ker\widetilde{\mathbf{G}}$ is invariant under the antilinear operation $\vec{v} \mapsto \boldsymbol{\tau}_1\vec{v}^*$. It follows that the real vector space[3]

$$\ker_-\widetilde{\mathbf{G}} \equiv \{\vec{\gamma}^z \in \ker\widetilde{\mathbf{G}} : \vec{\gamma}^z = -\boldsymbol{\tau}_1\vec{\gamma}^{z*}\}, \tag{7.3}$$

as the same dimension (as a real vector space) as $\ker\widetilde{\mathbf{G}}$ (as a complex vector space). We summarize these arguments in a Lemma.

Lemma 7.1.1 *Given a QBL with associated dynamical matrix* $\widetilde{\mathbf{G}}$*, we have that* $\mathcal{Z}$ *and* $\ker_-\widetilde{\mathbf{G}}$ *are isomorphic as real vector spaces.*

Moving on to SGs, a linear form $\gamma^s = \gamma^{s\dagger}$ generates a (weak) symmetry of a QBL if $\mathcal{L}([\gamma^s, \rho]) - [\gamma^s, \mathcal{L}(\rho)] = 0$ for all ρ. We will call such operators *Weyl SGs* and the corresponding symmetries $e^{i\theta\gamma^s}$ *Weyl symmetries.* This nomenclature is motivated by the fact that these operators generate displacements in phase space. Specifically, if $\gamma^s = \widehat{\vec{\gamma}^s}$ is Hermitian, then we can write $\vec{\gamma}^s = i[\alpha_1, \alpha_1^*, \ldots, \alpha_N, \alpha_N^*]^T$ in terms of some vector $\vec{\alpha} \in \mathbb{C}^N$. With this, the one-parameter family of symmetries can be written as $e^{i\theta\gamma} = D(\theta\vec{\alpha})$ in terms of the displacement operator Eq. (2.12). The statement that these displacements provide (weak) symmetries of the QBL is then

$$\mathcal{E}_t(D(\theta\vec{\alpha})\rho D(\theta\vec{\alpha})^\dagger) = D(\theta\vec{\alpha})\mathcal{E}_t(\rho)D(\theta\vec{\alpha})^\dagger, \quad (\mathcal{E}_t(\rho) = e^{t\mathcal{L}}(\rho)), \tag{7.4}$$

for all θ and ρ. Leveraging Eq. (6.25), we can express the Weyl SG condition in terms of the Nambu representation $\gamma^s = \widehat{\vec{\gamma}^s}$, i.e.,

$$\mathcal{L}([\gamma^s, \rho]) - [\gamma^s, \mathcal{L}(\rho)] = 0, \quad \forall\rho \iff \mathbf{G}\vec{\gamma}^s = 0, \quad (\gamma^s = \widehat{\vec{\gamma}^s}). \tag{7.5}$$

We note that any operator satisfying the condition on the left hand-side may be taken to be Hermitian. As in the ZM case, this follows from the dagger-preservation of the Lindbladian. Of course, we automatically assume a SG is Hermitian in order to ensure unitarity of the symmetry. Following the ZM analysis, we can formalize this by introducing the real vector space of Weyl SGs,

[3] More specifically, the invariance under this antilinear involution imposes what is sometimes known as a *real structure* on the kernel. Given a complex vector space $\mathcal{V}$ with a real structure defined by an antilinear involution Θ with $\Theta\mathcal{V} = \mathcal{V}$, it follows that $\mathcal{V} \simeq \mathcal{V}_+ \oplus \mathcal{V}_-$, with $\mathcal{V}_\pm \equiv \{\vec{v} \pm \Theta\vec{v} : \vec{v} \in \mathcal{V}\}$ One may further verify that the complex dimension of $\mathcal{V}$ is equal to the real dimension of each of the $\mathcal{V}_\pm$.

$$\mathcal{W} \equiv \{\text{Hermitian linear forms } \gamma^s : \mathcal{L}([\gamma^s, \rho]) - [\gamma^s, \mathcal{L}(\rho)] = 0, \quad \forall \rho\}, \tag{7.6}$$

as well as the modified kernel

$$\ker_- \mathbf{G} \equiv \{\vec{\gamma}^s \in \ker \mathbf{G} : \vec{\gamma}^z = -\boldsymbol{\tau}_1 \vec{\gamma}^{s*}\}. \tag{7.7}$$

The following lemma follows immediately

Lemma 7.1.2 *Given a QBL with associated dynamical matrix* $\mathbf{G}$, *we have that* $\mathcal{W}$ *and* $\ker_- \mathbf{G}$ *are isomorphic as real vector spaces.*

In this way, we have characterized the sets of ZMs and Weyl SGs, which we will jointly refer to as *Noether modes*, of a given QBL in terms of the kernels of two, *generically distinct*, matrices $\widetilde{\mathbf{G}}$ and $\mathbf{G}$, respectively. Thus, any correspondence between ZMs and Weyl SGs must be encoded in a correspondence between these two kernels. Let's explore the simplest case where $\widetilde{\mathbf{G}} = \mathbf{G}$. From the definition $\widetilde{\mathbf{G}} = \boldsymbol{\tau}_3 \mathbf{G}^\dagger \boldsymbol{\tau}_3$, it follows that $\mathbf{G}$ is pseudo-Hermitian in this case. In terms of the QBL, this is equivalent to $\mathcal{F}(\mathbf{M}) = 0$. One notable example where this is the case is when $\mathbf{M} = 0$, i.e., the closed-system limit. The recovers the well-known correspondence in Hamiltonian systems (one direction of which is a non-relativistic, quantum mechanical statement of Noether's theorem). However, $\mathcal{F}(\mathbf{M}) = 0$ does *not* imply $\mathbf{M}$, and hence the dissipator $\mathcal{D} = 0$. Thus it is possible to find instances where these matrices coincide while retaining non-trivial dissipation. See, for example, the $\kappa_+ = \kappa_- \neq 0$ case discussed in Sect. 6.2.3. Beyond equality of the two matrices, we may ensure that $\mathcal{Z}$ and $\mathcal{W}$ coincide by ensuring the two kernels are equal. Another simple example where this is the case is when $\widetilde{\mathbf{G}} = -\mathbf{G}$, which corresponds to the case where the Hamiltonian contribution $\mathbf{G}_0 = 0$, and hence $H = 0$. This Hamiltonian-free (purely dissipative) case will be particularly relevant in upcoming sections. Generally speaking, if it turns out that a particular Noether mode is *both* a ZM *and* a SG, we say that it is *non-split*.

Beyond these special cases, we may establish a more general correspondence between $\mathcal{Z}$ and $\mathcal{W}$. The following theorem establishes not only a one-to-one correspondence but, in fact, a *canonical* one.

Theorem 7.1.3 *For a given QBL, the space of ZMs* $\mathcal{Z}$ *and the space of Weyl SGs* $\mathcal{W}$ *are isomorphic as real vector spaces. Moreover, if the zero rapidity hosts only length on Jordan chains, then for each ZM, there is a canonically conjugate Weyl SG.*

Proof To establish an isomorphism, we first note that, in terms of complex dimension, we have that $\dim \ker \widetilde{\mathbf{G}} = \dim \ker \mathbf{G}^\dagger$ follows from $\widetilde{\mathbf{G}} = \boldsymbol{\tau}_3 \mathbf{G}^\dagger \boldsymbol{\tau}_3$ and the invertibility of $\boldsymbol{\tau}_3$. From basic linear algebra, the kernel of a matrix and its Hermitian conjugate are of equal dimension. Thus we may conclude $\dim \ker \widetilde{\mathbf{G}} = \dim \ker \mathbf{G}^\dagger = \dim \ker \mathbf{G}$. Since the real dimensions of $\ker_- \widetilde{\mathbf{G}}$ and $\ker_- \mathbf{G}$ are equal to the complex dimensions of $\ker \widetilde{\mathbf{G}}$ and $\ker \mathbf{G}$, respectively, we ultimately conclude

$\dim \mathcal{Z} = \dim \mathcal{W}$ and so $\mathcal{Z} \simeq \mathcal{W}$. This proof can be summarized in the chain of equalities

$$\dim \mathcal{Z} = \dim \ker_{-} \widetilde{\mathbf{G}} = \dim \ker \widetilde{\mathbf{G}} = \dim \ker \mathbf{G}^{\dagger} = \dim \ker \mathbf{G}$$
$$= \dim \ker_{-} \mathbf{G} = \dim \mathcal{W}.$$

For the canonical correspondence, we use the tool of a biorthogonal basis. Let $\{\vec{\gamma}_j^z, \vec{\eta}_j\}$ be one such basis for $\ker \widetilde{\mathbf{G}}$. Specifically, $\vec{\gamma}_j^z$ span $\ker \widetilde{\mathbf{G}}$ and $\vec{\eta}_j$ span $\ker \widetilde{\mathbf{G}}^{\dagger}$, while $\vec{\eta}_j^{\dagger}\vec{\gamma}_k^z = \delta_{jk}$. The existence of such a basis follows from the Jordan chain assumption[4] . The invariance of each of these spaces under the antilinear involution $\vec{v} \mapsto \boldsymbol{\tau}_1 \vec{v}^*$ allows us to further impose the restrictions $\vec{\gamma}_j^z = -\boldsymbol{\tau}_1 \vec{\gamma}_j^{z*}$ and $\vec{\eta}_j = -\boldsymbol{\tau}_1 \vec{\eta}_j^*$. Now let $\vec{\gamma}_j^s = -i\boldsymbol{\tau}_3 \vec{\eta}_j$. It follows that $\mathbf{G}\vec{\gamma}_j^s = 0$. Thus, we have constructed ZMs $\gamma_j^z \equiv \widehat{\vec{\gamma}_j^z}$ and SGs $\gamma_j^s \equiv \widehat{\vec{\gamma}_j^s}$. Further, they satisfy the HWRs: $[\widehat{\vec{\gamma}_j^s}, \widehat{\vec{\gamma}_k^z}] = \vec{\gamma}_j^{s\dagger}\boldsymbol{\tau}_3 \vec{\gamma}_k^z 1_{\mathcal{F}} = i\vec{\eta}_j^{\dagger}\vec{\gamma}_k^z 1_{\mathcal{F}} = i\delta_{jk}$. □

In the sense of this theorem, we say that we have established a partial restoration of Noether's theorem—a general correspondence between ZMs and SGs—within the space of linear forms. The algebraic relationship between the two spaces is an additional (rather remarkable!) "cherry on top".

7.1.2 *The Approximate Case*

An extremely useful extension of the above result is to instead consider *approximate* ZMs and Weyl SGs. At the operator level, these may be loosely defined by first prescribing an accuracy $\epsilon > 0$. Then we say, for example, γ^z provides an *approximate ZM* if $\mathcal{L}(\gamma^z) = K\alpha$, for some constant K satisfying $|K| < \epsilon$ and α a linear form with appropriately bounded (in norm) Nambu representation $\vec{\alpha}$. Similarly, an *approximate Weyl SG* γ^s is defined by the identity $\mathcal{L}([\gamma^s, \rho]) - [\gamma^s, \mathcal{L}(\rho)] = [K\alpha, \rho]$ for all ρ, with the same restrictions placed on K and α. The kernel conditions of the preceding discussion are replaced with *approximate* kernel conditions. To be more concrete, we will generally define an approximate ZM (Weyl SG) as having a Nambu representation $\vec{\gamma}^z$ ($\vec{\gamma}^s$) satisfying the approximate kernel condition $\|\widetilde{\mathbf{G}}\vec{\gamma}^z\| < \epsilon$ ($\|\mathbf{G}\vec{\gamma}^s\| < \epsilon$) for some to-be-specified norm $\|\cdot\|$ and the Hermiticity condition being $\vec{\gamma}^z = -\boldsymbol{\tau}_1\vec{\gamma}^{z*}$ ($\vec{\gamma}^z = -\boldsymbol{\tau}_1\vec{\gamma}^{z*}$). We will call an upper bound on K the accuracy of the approximation. It turns out that, in this

[4] Proof: Extend $\{\vec{\gamma}_j^z\}$ to a basis for $\mathbb{C}^{2N}$ and consider the associated dual basis. This dual basis contains the functionals ϕ_j which satisfy $\phi_j(\vec{\gamma}_k^z) = \delta_{jk}$. From the Riesz representation theorem, we associate to each ϕ_j a vector $\vec{\eta}_j$ such that $\phi_j(\vec{v}) = \vec{\eta}_j^{\dagger}\vec{v}$. We claim that $\phi_j(\widetilde{\mathbf{G}}\vec{v}) = \vec{\eta}_j^{\dagger}\widetilde{\mathbf{G}}\vec{v} = 0$ for all $\vec{v} \in \mathbb{C}^{2N}$, or equivalently, $\widetilde{\mathbf{G}}^{\dagger}\vec{\eta}_j = 0$. Note that $\phi_j(\widetilde{\mathbf{G}}\vec{v}) \neq 0$ if and only if $\widetilde{\mathbf{G}}\vec{v} \propto \vec{\gamma}_j^z$. But then $\vec{\gamma}_j^z$ and $\vec{v}$ constitute a length two Jordan chain at zero, contradicting our assumption.

approximate setting, Theorem 7.1.3 has a natural generalization. Henceforth, we will only consider the 2-norm.

Theorem 7.1.4 *Let* $\mathbf{G}$ *be a dynamical matrix of a QBL and* γ^s *be an approximate SG of accuracy* $\epsilon > 0$ *(in the sense that its Nambu representation* $\vec{\gamma}^s$ *satisfies* $\mathbf{G}\vec{\gamma}^s = \vec{\alpha}$ *for some vector* $\vec{\alpha}$*, with* $\|\vec{\alpha}\| < \epsilon$*), Then, if the matrix* $\mathbf{G}' \equiv \mathbf{G} - \vec{\alpha}\vec{\gamma}^{s\dagger}/\|\vec{\gamma}^z\|$ *hosts only Jordan chains of length one at* 0*, there exists a canonically conjugate approximate ZM* γ^z *of accuracy* $\epsilon' = \epsilon\|\vec{\gamma}^z\|^2$*. The converse holds as well.*

Proof Let $\vec{\alpha}$ be as stated in the theorem. Note that Hermiticity of γ^s implies that the Nambu representation satisfies $\vec{\gamma}^s = -\boldsymbol{\tau}_1\vec{\gamma}^{s*}$. Using this, and the fact that $\boldsymbol{\tau}_1\mathbf{G}^*\boldsymbol{\tau}_1 = -\mathbf{G}$, we have

$$\boldsymbol{\tau}_1\vec{\alpha}^* = \boldsymbol{\tau}_1\mathbf{G}^*\vec{\gamma}^{s*} - \mathbf{G}\boldsymbol{\tau}_1\vec{\gamma}^{s*} = \mathbf{G}\vec{\gamma}^s = \vec{\alpha}. \tag{7.8}$$

Now, consider the matrix $\mathbf{G}'$ defined in the theorem statement. We claim that $\mathbf{G}'$ may be interpreted as a dynamical matrix of some other QBL and that $\mathbf{G}'\vec{\gamma}^s = 0$. The first claim follows because $\mathbf{G}'$ obeys the only constraint set on dynamical matrices:

$$\begin{aligned}\boldsymbol{\tau}_1\mathbf{G}'^*\boldsymbol{\tau}_1 &= \boldsymbol{\tau}_1\mathbf{G}^*\boldsymbol{\tau}_1 - \boldsymbol{\tau}_1\vec{\alpha}^*\vec{\gamma}^{sT}\boldsymbol{\tau}_1 = -\mathbf{G} - (\boldsymbol{\tau}_1\vec{\alpha})^*(\boldsymbol{\tau}_1\vec{\gamma}^{s*})^\dagger \\ &= -\mathbf{G} + \vec{\alpha}\vec{\gamma}^{s\dagger} = -\mathbf{G}',\end{aligned} \tag{7.9}$$

where we have used Eq. (7.8) in the third equality. The second claim is verified directly:

$$\mathbf{G}'\vec{\gamma}^s = \mathbf{G}\vec{\gamma}^z - \vec{\alpha}\frac{\vec{\gamma}^{s\dagger}\vec{\gamma}^s}{\|\vec{\gamma}^s\|^2} = \vec{\alpha} - \vec{\alpha} = 0. \tag{7.10}$$

Now, any QBL with dynamical matrix $\mathbf{G}'$ will have γ^s as an exact SG. Since we assume $\mathbf{G}'$ is diagonalizable, we may then directly apply Theorem 7.1.3 to find a canonically conjugate *exact* ZM γ^z. In Nambu space, this means there is a vector $\vec{\gamma}^z = -\boldsymbol{\tau}_1\vec{\gamma}^{z*}$, such that $\widetilde{\mathbf{G}}'\vec{\gamma}^z = 0$, with

$$\widetilde{\mathbf{G}}' = \boldsymbol{\tau}_3\mathbf{G}'^\dagger\boldsymbol{\tau}_3 = \widetilde{\mathbf{G}} - \boldsymbol{\tau}_3\vec{\gamma}^s\vec{\alpha}^\dagger\boldsymbol{\tau}_3, \tag{7.11}$$

and $\vec{\gamma}^{s\dagger}\boldsymbol{\tau}_3\vec{\gamma}^z = i$ so that $[\gamma^s, \gamma^z] = i$. It follows that

$$\|\widetilde{\mathbf{G}}\vec{\gamma}^z\| = \|\widetilde{\mathbf{G}}'\vec{\gamma}^z + \frac{\boldsymbol{\tau}_3\vec{\gamma}^s\vec{\alpha}^\dagger\boldsymbol{\tau}_3}{\|\vec{\gamma}^s\|^2}\vec{\gamma}^z\| = \frac{1}{\|\vec{\gamma}^s\|^2}\|\boldsymbol{\tau}_3\vec{\gamma}^s\vec{\alpha}^\dagger\boldsymbol{\tau}_3\vec{\gamma}^z\| = \frac{|\vec{\alpha}^\dagger\boldsymbol{\tau}_3\vec{\gamma}^z|}{\|\vec{\gamma}^s\|}. \tag{7.12}$$

Using the Cauchy-Schwarz inequality, we may bound the numerator of the right hand-side from above by $\|\vec{\alpha}\|\|\vec{\gamma}^z\| < \epsilon\|\vec{\gamma}^z\|$. Thus,

$$\|\widetilde{\mathbf{G}}\vec{\gamma}^z\| < \epsilon \frac{\|\vec{\gamma}^z\|}{\|\vec{\gamma}^s\|}. \tag{7.13}$$

An additional application of the Cauchy-Schwarz inequality to the identity $\vec{\gamma}^{s\dagger}\boldsymbol{\tau}_3\vec{\gamma}^z = i$ yields $1/\|\vec{\gamma}^s\| < \|\vec{\gamma}^z\|$ so that

$$\|\widetilde{\mathbf{G}}\vec{\gamma}^z\| < \epsilon\|\vec{\gamma}^z\|^2, \tag{7.14}$$

as claimed. The converse holds by simply replacing $\mathbf{G}$ with $\widetilde{\mathbf{G}}$ and $\vec{\gamma}^s$ with $\vec{\gamma}^z$. □

We remark that canonically conjugate modes can always be rescaled in a natural way to make the accuracies ϵ and ϵ' equal. Generally, if $[\gamma^s, \gamma^z] = i$, then $[\gamma^{s\prime}, \gamma^{z\prime}] = i$, with $\gamma^{s\prime} = \mathcal{M}\gamma^s$ and $\gamma^{z\prime} = \gamma^z/\mathcal{M}$, for any $\mathcal{M} > 0$. Taking $M = \|\vec{\gamma}^z\|$ provides approximate SGs and ZMs of accuracy $\mathcal{M}\epsilon$. The most interesting examples will be those in which $\epsilon \ll 1/\mathcal{M}$.

Let us give a heuristic recipe for constructing approximate ZMs and SGs Approximate ZMs can be built from eigenvectors of the (positive-semidefinite) matrix $\widetilde{\mathbf{A}} \equiv \widetilde{\mathbf{G}}^\dagger\widetilde{\mathbf{G}}$. That is, if $\widetilde{\mathbf{A}}\vec{\gamma}^z = s^2\vec{\gamma}^z$ for some $s > 0$, then $\|\mathbf{G}\vec{\gamma}^z\| = s\,\|\vec{\gamma}^z\|$, which can be chosen to be less than a desired ϵ if there exists a sufficiently small s. We may equivalently understand s as a singular value of $\widetilde{\mathbf{G}}$. Hermiticity can be guaranteed by noting that $\boldsymbol{\tau}_1\widetilde{\mathbf{A}}\boldsymbol{\tau}_1 = \widetilde{\mathbf{A}}^*$ implies that the eigenspaces of $\widetilde{\mathbf{A}}$ are invariant under the antilinear involution $\vec{v} \mapsto \boldsymbol{\tau}_1\vec{v}^*$. Ultimately, we construct a linearly independent set of approximate ZMs by taking a linear independent set of eigenvectors $\{\vec{\gamma}_j^z\}$ satisfying $\widetilde{\mathbf{A}}\vec{\gamma}_j^z = s_j^2\vec{\gamma}_j^z$, with $s_j^2 < \epsilon$ and $\vec{\gamma}_j^z = -\boldsymbol{\tau}_1\vec{\gamma}_j^{z*}$.

Approximate SGs are similarly constructed out of eigenvectors of $\mathbf{A} = \mathbf{G}^\dagger\mathbf{G}$. Now, note that $\widetilde{\mathbf{A}} = \widetilde{\mathbf{G}}^\dagger\widetilde{\mathbf{G}} = \boldsymbol{\tau}_3\mathbf{G}\mathbf{G}^\dagger\boldsymbol{\tau}_3$. Furthermore, it can be shown (for instance, by performing a singular value decomposition on $\mathbf{G}$) that $\mathbf{G}^\dagger\mathbf{G} = \mathbf{V}\mathbf{G}\mathbf{G}^\dagger\mathbf{V}^\dagger$, for some unitary $\mathbf{V}$. Given the approximate ZM vector $\{\vec{\gamma}_j^z\}$, we propose a corresponding SG vectors $\vec{\gamma}_j^s \equiv \boldsymbol{\tau}_3\mathbf{V}\vec{\gamma}_j^z$. It follows that $\|\vec{\gamma}^s\| = \|\vec{\gamma}^z\|$ and

$$\mathbf{A}\vec{\gamma}_j^s = \mathbf{V}\mathbf{G}\mathbf{G}^\dagger\mathbf{V}^\dagger\vec{\gamma}_j^s = \mathbf{V}\boldsymbol{\tau}_3\widetilde{\mathbf{A}}\boldsymbol{\tau}_3\mathbf{V}^\dagger\vec{\gamma}_j^s = \mathbf{V}\boldsymbol{\tau}_3\widetilde{\mathbf{A}}\vec{\gamma}_j^z = s_j^2\mathbf{V}\boldsymbol{\tau}_3\vec{\gamma}_j^z = s_j^2\vec{\gamma}_j^s. \tag{7.15}$$

In other words, $\vec{\gamma}_j^s$ are, by construction, approximate kernel vectors for $\mathbf{G}$. It follows that $\gamma^s \equiv \widehat{\vec{\gamma}^s}$ is an approximate SG.

Ultimately, we have established that there is a *one-to-one correspondence* between approximate ZMs and SGs. To ensure canonical commutations, we consider the matrix

$$\mathbf{F}_{jk} \equiv \vec{\gamma}_j^s\boldsymbol{\tau}_3\vec{\gamma}_k^z = \vec{\gamma}_j^z\mathbf{V}^\dagger\vec{\gamma}_k^z. \tag{7.16}$$

If $\mathbf{F}_{jk} = i\delta_{jk}$ we are done. If $\mathbf{F}$ is diagonalizable, we can rearrange the approximate kernel vectors so that canonical commutations are met. While we do not have a generic condition for diagonalizability of $\mathbf{F}$, we conjecture it is rather generic. Ultimately, this procedure succeeds in every example considered later in this thesis.

Importantly, it turns out that remark that the approximate kernel conditions are effectively captured by the mathematical concept of *pseudospectra*. While we defer the more expansive discussion on the utility of pseudospectra for describing dynamical features of QBLs to Sect. 8.1.2, we simply acknowledge here that the Nambu vectors associated to approximate ZMs and Weyl SGs correspond to *pseudoeigenvectors* at zero *pseudoeigenvalue* for the matrices $\widetilde{\mathbf{G}}$ and $\mathbf{G}$, respectively.

7.2 Two Protocols for QBL Design

To aid in our search for signatures of SPT phases, it will be convenient to investigate QBLs with certain desirable properties. One such property is that the QBL possesses either exact, or approximate, Noether modes. We will accomplish this by essentially "importing" the edge modes of QFHs into the bosonic arena. A second desirable property will be that the QBL possesses a unique, pure steady state ($\text{tr}[\rho_{\text{ss}}^2] = 1$). Then, if we can (and we will) find a QBL with a pure steady state that also possesses Noether modes, we can think of the steady state as a closer analogue to the fermionic ground state than the non-pure case. Engineering such QBLs will be possible using the duality transformation of Chap. 4.

7.2.1 Embedding Fermionic Hamiltonians in Bosonic Dissipators

Our procedure for embedding QFHs into the dissipator of a QBL hinges upon a single observation: the dynamical matrix of a *purely dissipative* ($H = 0$) QBL is *uniquely* defined by a fermionic matrix $\mathbf{H}_F = \mathcal{F}(\mathbf{H}_F)$. Let's dissect this. Firstly, a QBL may be uniquely associated to a dynamical matrix $\mathbf{G}$ and a GKS matrix $\mathbf{M}$. Referring to Eq. (6.16), the dynamical matrix of a purely dissipative QBL is given by $\mathbf{G} = -i\tau_3\mathcal{F}(\mathbf{M})$. The aforementioned fermionic matrix is then $\mathbf{H}_F = \mathcal{F}(\mathbf{M})$.

Now, consider a QFH $H = \Psi^\dagger \mathbf{H}_F \Psi/2$ defined in terms of the fermionic Nambu array $\Psi = [c_1, c_1^\dagger, \ldots, c_N, c_N^\dagger]^T$ and a Hermitian matrix $\mathbf{H}_F$. From the canonical anticommutation relations (CARs) $\{c_i, c_j^\dagger\} = \delta_{ij} 1_{\mathcal{F}}$ (with $1_{\mathcal{F}}$ the fermionic Fock space identity, in this context) and $\{c_i, c_j\} = 0$, it follows that

$$H = \frac{1}{2}\Psi^\dagger \mathbf{H}_F \Psi = \frac{1}{2}\Psi^\dagger \mathcal{F}(\mathbf{H}_F)\Psi + \frac{1}{2}\text{tr}[\mathcal{B}(\mathbf{H}_F)]\, 1_{\mathcal{F}}. \tag{7.17}$$

Since we are generally uninterested in constant shifts to the Hamiltonian, we can take, without loss of generality, $\mathcal{B}(\mathbf{H}_F) = 0$. Thus, the complementary nature of the bosonic and fermionic projectors allow us to conclude that a QFH is defined uniquely (modulo constant shifts) by a fermionic matrix $\mathbf{H}_F = \mathcal{F}(\mathbf{H}_F)$.

Combining these two observations leads to the following constructive procedure. Given a QFH of interest, defined by a fermionic matrix $\mathbf{H}_F$, let us consider the QBL defined by the zero Hamiltonian and the GKS matrix $\mathbf{M} \equiv \mathbf{B}+\mathbf{H}_F$, with $\mathbf{B} = \mathcal{B}(\mathbf{B})$ a Hermitian bosonic matrix chosen so that $\mathbf{M} \geq 0$. It follows that $\mathcal{F}(\mathbf{M}) = \mathbf{H}_F$ and that the dynamical matrix is $\mathbf{G} = -i\boldsymbol{\tau}_3\mathbf{H}_F$. Which QFH input would lead to the desired Noether modes in the QBL output? For concreteness, let us consider an N-mode QFH under OBCs with (perhaps topological in origin) ZMs. The generic feature of such a Hamiltonian is the existence of a uni-norm eigenvector $\vec{\psi}$ satisfying $\mathbf{H}_F\vec{\psi}_+ = \epsilon_N\vec{\psi}_+$, with $\epsilon_N \sim \mathcal{O}(e^{-N})$ positive, but decreasing exponentially in system size. If ϵ_N lies in-between the bulk bands, then $\vec{\psi}$ will be localized on the edges. The fermionic nature of $\mathbf{H}_F$ implies that $\vec{\psi}_- \equiv \boldsymbol{\tau}_1\vec{\psi}_+^*$ satisfies $\mathbf{H}_F\vec{\psi}_- = -\epsilon_N\vec{\psi}_-$.

Let us now move to the QBL. Consider the *bosonic* linear form $\gamma = \widehat{\vec{\gamma}} = \vec{\gamma}^\dagger\boldsymbol{\tau}_3\Phi$, in terms of the vector $\vec{\gamma} \equiv (\vec{\psi}_+ - \vec{\psi}_-)/\sqrt{2}$. We claim that this operator is both an approximate SG *and* an approximate ZM of the QBL. The first follows from

$$\begin{aligned}\|\mathbf{G}\vec{\gamma}\| &= \|-i\boldsymbol{\tau}_3\mathbf{H}_F\vec{\gamma}\| = \|\mathbf{H}_F\vec{\gamma}\| = \frac{1}{\sqrt{2}}\|\mathbf{H}_F(\vec{\psi}_+ - \vec{\psi}_-)\| \\ &= \frac{\epsilon_N}{\sqrt{2}}\|\vec{\psi}_+ + \vec{\psi}_-\| = \epsilon_N \sim \mathcal{O}(e^{-N}).\end{aligned}$$

In addition, $\gamma = \gamma^\dagger$ follows from $\vec{\gamma} = -\boldsymbol{\tau}_3\vec{\gamma}^*$ which, in turn, immediately follows from the relationship between $\vec{\psi}_+$ and $\vec{\psi}_-$. The ZM claim follows immediately from the fact that Noether modes in purely dissipative QBLs are non-split, as explained in Sect. 7.1. To summarize, this procedure yields a QBL with a non-split Noether mode descendent from a ZM of the initial QFH. Furthermore, the localization properties of $\vec{\psi}_\pm$ descend to those of γ in a clear way. We remark that this procedure need not yield a dynamically stable QBL.

7.2.2 Reservoir Engineering Pure Steady States via Dualities

Consider a dynamically stable QBH H with dynamical matrix $\mathbf{G}_0$. Dynamically stability ensures that there exists a set of bosonic quasiparticles that diagonalize the Hamiltonian. In particular, this set of quasiparticles defines the quasiparticle vacuum $|\widetilde{0}\rangle$ (see Eq. (2.68)). Moreover, when H breaks total number symmetry, we have seen that there exists a duality transformation mapping H to a number-conserving dual H^D (recall Chap. 4). This transformation is given by $\Phi \mapsto \mathbf{R}^{-1}\Phi$, with $\mathbf{R} = \mathbf{S}^{1/2}$ defined in terms of the metric $\mathbf{S}$ in Eq. (4.1). This metric is closely tied to the quasiparticle vacuum. As we have seen, the covariance matrix of the vacuum is given by $\mathbf{S}^{-1/2}$ (recall Eq. (4.14)).

Is it possible to extend H to a QBL, by introducing dissipation, such that $|\widetilde{0}\rangle$ is the steady state? Indeed, this is possible, and may be accomplished by utilizing the aforementioned duality. We claim that the QBL defined by the Hamiltonian H and the GKS matrix $\mathbf{M} \equiv \kappa(\mathbf{S} + \boldsymbol{\tau}_3)$, with $\kappa > 0$, does the job. Specifically, we must check that (i) the QBL is well-defined; (ii) it is dynamically stable; and (iii) it possesses a unique steady state given by $\rho_{\text{ss}} = |\widetilde{0}\rangle\langle\widetilde{0}|$. Let us prove these claims one by one.

First, the QBL is well-defined as long as $\mathbf{M}$ is positive-semidefinite. To see this, recall the resolution of the identity Eq. (2.63) in terms of the eigenvectors $\vec{\psi}_n^{\pm}$ of $\mathbf{G}_0$. Then,

$$\boldsymbol{\tau}_3 = \boldsymbol{\tau}_3 \mathbb{1}_{2N} = \sum_{n=1}^{N} \boldsymbol{\tau}_3 \left(\vec{\psi}_n^{+} \vec{\psi}_n^{+\dagger} - \vec{\psi}_n^{-} \vec{\psi}_n^{-\dagger} \right) \boldsymbol{\tau}_3 \tag{7.18}$$

Given this, and Eq. (4.1), we have

$$\mathbf{M} = \kappa(\mathbf{S} + \boldsymbol{\tau}_3) = 2\kappa \sum_{n=1}^{N} \boldsymbol{\tau}_3 \vec{\psi}_n^{+} \vec{\psi}_n^{+\dagger} \boldsymbol{\tau}_3, \tag{7.19}$$

which is evidently positive-semidefinite.

Secondly, we may diagnose dynamical stability from the dynamical matrix $\mathbf{G} = \mathbf{G}_0 - i\boldsymbol{\tau}_3 \mathcal{F}(\mathbf{M})$. Now, $\mathbf{S}$ is Hermitian and satisfies $\mathbf{S}^* = \boldsymbol{\tau}_1 \mathbf{S} \boldsymbol{\tau}_1$ (recall Prop. 4.1.3). Combining these yields $\mathbf{S} = \boldsymbol{\tau}_1 \mathbf{S}^T \boldsymbol{\tau}_1$ or, equivalently, $\mathbf{S} = \mathcal{B}(\mathbf{S})$. It immediately follows that $\mathcal{F}(\mathbf{M}) = \kappa\boldsymbol{\tau}_3$, from which we conclude

$$\mathbf{G} = \mathbf{G}_0 - i\boldsymbol{\tau}_3 \mathcal{F}(\mathbf{M}) = \mathbf{G}_0 - i\kappa \mathbb{1}_{2N}. \tag{7.20}$$

From here we can compute the rapidities. Let ω_n be the (necessarily real, by the dynamical stability assumption on H) eigenvalues of $\mathbf{G}_0$. Then the rapidities are $\lambda_n = -i(\omega_n - i\kappa) = -\kappa - i\omega_n$. Thus the stability gap is $\Delta_S = -\kappa < 0$, and so our QBL is dynamically stable.

Finally, we note that the strict negativity of the stability gap implies there is a steady state and that it is unique (recall Table 6.1). From the general properties of QBLs, we know it is Gaussian with mean vector $\vec{m}_{\text{ss}} = 0$ and covariance matrix $\mathbf{C}_{\text{ss}}$ satisfying the Lyapunov equation (recall Eq. (6.28))

$$-i(\mathbf{G}\mathbf{C}_{\text{ss}} - \mathbf{C}_{\text{ss}}\mathbf{G}^\dagger) + \kappa\mathbf{S}^{-1} = 0, \tag{7.21}$$

where we have used $\boldsymbol{\tau}_3 \mathcal{B}(\mathbf{M}) \boldsymbol{\tau}_3 = \kappa \boldsymbol{\tau}_3 \mathbf{S} \boldsymbol{\tau}_3 = \kappa \mathbf{S}^{-1}$. Since $-i\mathbf{G}$ is Hurwitz, we obtain the solution by taking $t \to \infty$ in Eq. (6.22), yielding

$$\mathbf{C}_{\text{ss}} = \kappa \int_0^\infty e^{-i\mathbf{G}t} \mathbf{S}^{-1} e^{i\mathbf{G}^\dagger t}\, dt. \tag{7.22}$$

Since $\mathbf{G} = \mathbf{G}_0 - i\kappa\mathbb{1}_{2N}$, we may simplify this as

$$\mathbf{C}_{\text{ss}} = \kappa \int_0^\infty e^{-2\kappa t} e^{-i\mathbf{G}_0 t} \mathbf{S}^{-1} e^{i\mathbf{G}_0^\dagger t}\, dt. \tag{7.23}$$

Recalling that $\mathbf{G}_0 = \mathbf{S}^{-1}\mathbf{G}^\dagger\mathbf{S}$ (Prop. 4.1.3), it follows that

$$e^{-i\mathbf{G}_0 t}\mathbf{S}^{-1}e^{i\mathbf{G}_0^\dagger t} = e^{-i\mathbf{G}_0 t}e^{i\mathbf{S}^{-1}\mathbf{G}_0^\dagger \mathbf{S} t}\mathbf{S}^{-1} = e^{-i\mathbf{G}_0 t}e^{i\mathbf{G}_0 t}\mathbf{S}^{-1} = \mathbf{S}^{-1}. \tag{7.24}$$

Altogether,

$$\mathbf{C}_{\text{ss}} = \kappa \int_0^\infty e^{-2\kappa t}\mathbf{S}^{-1}\, dt = \frac{1}{2}\mathbf{S}^{-1} = \mathbf{C}_{|\widetilde{0}\rangle}. \tag{7.25}$$

Since both states are Gaussian, and since both the mean vectors and covariance matrices coincide, we conclude that $\rho_{\text{ss}} = |\widetilde{0}\rangle\langle\widetilde{0}|$, as claimed.

This fact can actually be seen in a more straightforward manner. From Eq. (7.19),

$$\begin{aligned} \mathcal{D}(\rho) &= \sum_{j=1}^{N} \mathbf{M}_{jk}\left(\Phi_k \rho \Phi_j^\dagger - \frac{1}{2}\{\Phi_j^\dagger \Phi_k, \rho\}\right) \\ &= 2\kappa \sum_{n=1}^{N}\left(\psi_n \rho \psi_n^\dagger - \frac{1}{2}\{\psi_n^\dagger \psi_n, \rho\}\right) = 2\kappa \sum_{n=1}^{N} \mathcal{D}[\psi_n](\rho), \end{aligned} \tag{7.26}$$

where, per usual, $\psi_n = \widehat{\vec{\psi}_n^{+}}$ are the quasiparticle annihilation operators of the Hamiltonian. The final equality says that the dissipator is diagonal in the quasiparticle basis, and more over, induces uniform loss of rate κ on each quasiparticle mode. From this perspective, it is easy to see that the vacuum state $|\widehat{0}\rangle$ is the steady state. Moreover, the dynamics of the quasiparticle modes follow in a straightforward fashion: $\psi_n(t) = e^{-i\omega t}e^{-\kappa t}\psi_n(0)$.

We conclude this section with a comment regarding symmetries. Suppose H has a Gaussian unitary symmetry. It follows that $\mathbf{G}_0$ commutes with the pseudo-unitary representation $\mathbf{U}$ of this symmetry operator. Since the dynamical matrix $\mathbf{G}$ of the QBL is equal to $\mathbf{G}_0$ plus a constant shift, this commutation property is retained. However, we *cannot* then conclude that the QBL has the same symmetry. After all, the GKS matrix need not commute with $\mathbf{U}$. In this sense, the QBL has a sort-of "partial symmetry" at the level of the dynamical matrix. We will see an explicit example of this in Sect. 9.3, whereby the dynamical matrix possesses a translational symmetry that the overall QBL lacks.

References

1. V.P. Flynn, E. Cobanera, L. Viola, Phys. Rev. B **108**, 214312 (2023)
2. V.P. Flynn, E. Cobanera, L. Viola, Phys. Rev. Lett. **127**, 245701 (2021)

Chapter 8
Signatures of SPT Physics in 1D Bulk-Translationally Invariant QBLs

Abstract This chapter contains the main result of Part II, namely, the discovery of *SPT-like bosonic edge zero modes and symmetries* that manifest, and persist, in a *transient dynamical regime whose duration increases with system size*. We begin by spelling out the necessary features any system supporting such entities must possess: *Non-trivial spectral topology*, *bulk-instabilities*, and *highly non-normal dynamical matrices*. We naturally arrive at the mathematical theory of pseudospectra. Following a brief mathematical digression, we proceed to uncover conditions for *anomalous transient dynamics in QBLs*. Two relevant dynamical phases emerge: an *anomalously relaxing* phase, and a *dynamically metastable* phase. The anomalously relaxing phase is characterized by exponential relaxation of generic observables in a two-step manner. The first step is characterized by a decay rate set by the bulk (infinite-size) Lindblad gap, while the second rate is set by the finite-size Lindblad gap which, remarkably, differs from its infinite-size counterpart by a non-zero amount, independent of system-size. A dynamically metastable system is one for which all finite open boundary truncations are dynamically stable, despite possessing an unstable infinite-size limit. Such systems possess bulk instabilities that are suppressed by imposing hard-wall boundaries. The evolution of dynamically metastable systems is characterized by a transient regime whereby generic observables are amplified, followed by an asymptotic exponential relaxation at a rate set by the finite-size Lindblad gap. Both the length of the transient, and the degree of amplification increase with system size. We zoom in on those dynamically stable QBLs that possess non-trivial bulk topological invariants which, we deem *topologically metastable*. In the simplest cases, these invariants are the winding numbers of the bulk bands. The combined assumptions of dynamical metastability and non-trivial topology facilitate the emergence of pairs of operators consisting of one approximate zero mode and one approximate SG. These modes are *Hermitian,macroscopically separated*, and*canonically conjugate*. We call the members of these pairs *Majorana bosons* (MBs) due to their similarities with Majorana fermions. Consequentially, we identify the emergence of a manifold of long-lived *quasi-steady states* alongside MBs. We then explore the implications of a global number symmetry on topologically metastable systems, resulting in the

V. P. Flynn, *Effective Non-Hermiticity and Topology in Markovian Quadratic Bosonic Dynamics*, Springer Theses, https://doi.org/10.1007/978-3-031-52045-7_8

discovery of edge modes, which we call *Dirac bosons* (DBs), reminiscent of the Dirac fermion edge modes in topological insulators.

In this chapter,[1] we turn our focus to 1D bulk-translationally invariant QBLs and uncover *tight bosonic analogues* of the signatures of free-fermionic SPT phases that we deem *Majorana bosons* in the number-non-symmetric case and *Dirac bosons* in the number-symmetric case. Referring back to the story of QBHs, any attempt to uncover non-trivial zero-energy edge physics was met with one or more challenges associated to thermodynamical and dynamical instabilities. Ultimately, to uncover them, one must completely abandon a straightforward many-body picture (e.g., a ground state separated by bulk states by a many-body gap). Can we uncover them in a QBL setting while maintaining this picture? To answer this, we must define precisely what the QBL analogues of the hypothetical signatures and the many-body picture are. First, we propose that the signatures must be edge-localized modes that are either approximately (i) conserved (ZMs), or generate symmetries (Weyl SGs). The reasoning for considering both possibilities follows from the breakdown of Noether's theorem explained previously. By virtue of Sect. 7.1, we further expect (i) and (ii) to emerge in canonically conjugate pairs. Edge-localization leads us to consider open, or semi-infinite BCs. As for the many-body picture, we will replace the ground state with the steady state and substitute the many-body gap with the Lindblad gap. In particular, the existence of a steady state demands that we focus on *dynamically stable systems*. With the 'target' set, we can begin to determine which QBLs support such physics.

As we will see, we are able to determine a bulk criteria for the emergence of edge modes. Namely, we show that, if the bulk rapidity bands wind around the origin, then edge modes will emerge upon truncating the system. This introduces a complication: a rapidity band can only wind about the origin if it is allowed to enter the right-half complex plane. In this sense, the system must have a *bulk instability*. While this will lead to instabilities in the semi-infinite system, we argue that the finite-size OBC configuration can still retain stability, and, that one mechanism for ensuring this is to demand extreme *non-normality* of the dynamical matrix. When these conditions are met, one may fear that truncating to the finite size has eliminated the edge-modes altogether—thankfully, this is not true. We are lead to investigate the *pseudospectra*—essentially, the approximate spectra—of the dynamical matrix. Our key observation is that the (unstable) infinite-size spectra *imprints* itself into the pseudospectra of the finite-size system. Thus, any normal mode of the infinite-size system, when truncated, will behave as an *approximate* normal mode of the finite system. In particular, truncating unstable modes produces modes that *appear* unstable for a transient (whose duration we will argue increases with system size).

[1] All results from this section can be found in Refs. [1] and [2] which were both co-authored with Emilio Cobanera & Lorenza Viola.

Before assessing the fate of the infinite-size edge modes upon truncation, this observation leads us to define two novel dynamical phases of QBLs. The first, which we call the *anomalously relaxing phase*, is characterized by a discontinuous decrease of the Lindblad gap in the infinite-size limit. This occurs when the finite-size rapidity spectra remains gapped away from the bulk rapidity bands and leads to a distinctive *two-step* relaxation. The transient relaxation (as inferred by the evolution of generic observable expectation values) is governed by the smaller infinite-size gap, while the asymptotic relaxation rate is governed by the larger finite-size gap. The second dynamical phase is called *dynamical metastability* and is characterized by the discontinuous change of stability phase in the infinite-size limit. That is, dynamically metastable systems are dynamically stable for all finite sizes, but unstable *strictly* in the infinite-size limit. We will see that such systems exhibit *transient amplification* for an increasingly long transient.

With these more general considerations behind us, we will ultimately conclude that our systems of interest must be (i) dynamically metastable and (ii) have non-trivial band winding about zero. These two ingredients together make up what we call *topologically metastability*. Topologically metastable systems, by design, fit each of the desired criteria for tight bosonic analogues of fermionic SPT physics. In particular, the edge mode pairs, which we deem *Majorana bosons* (or *Dirac bosons*, in the case where the system possesses a weak number symmetry), are Hermitian, canonically conjugate, and edge-localized. However, there are some extremely striking differences. Generically, one element of the pair is a SG while the other is a ZM—that is, they are *split*, in the sense of Sect. 7.1. Moreover, they are not exact ZMs and SGs, instead they are approximate (ultimately stemming from their pseudospectral origin). As such, the ZM is stationary only during the transient associated to the topologically metastable phase. Similarly, the approximate symmetry only leaves the dynamics invariant for the same time-scale. However, since this transient increases with system size, we may take it to last longer than any meaningful timescale set by the system by simply increasing system size. We will see that one consequence of this is the emergence of a manifold of *quasi-steady states*. These states, which constitute the QBL-analogue of the degenerate ground states in non-interacting fermionic SPTs, exhibit diverging lifetimes and may be combined in non-trivial ways to generate long-lived (classically) *non-Gaussian* states.

The outline of this chapter is as follows. In Sect. 8.1, we initiate the search for SPT signatures in QBLs by spelling out exactly what behavior we attempt to uncover and exploring the consequences of these constraints. We argue that non-normality of the dynamical matrix is one such constraint and, as such, are forced to grapple with *pseudospectra*—a mathematical concept we introduce in detail in Sect. 8.1.2. Returning back to QBLs, we define anomalously relaxing and dynamically metastable systems in Sect. 8.2.1. We finally define topological metastability, Majorana bosons, Dirac bosons, and the aforementioned quasi-steady states in Sect. 8.3. Several results about the transient dynamics of topologically metastable systems follow.

8.1 Beginning the Search

8.1.1 The Need for Bulk Instabilities and Non-normality

The signatures of SPT physics in QFHs emerge as zero energy modes localized on boundaries separating regions of topologically inequivalent systems. In the simplest case, hard wall boundaries (OBCs) imposed on a topologically non-trivial translation-invariant system act as a separator between the topologically non-trivial bulk and the (necessarily) topologically trivial region outside of the system. In one-dimension, such modes manifest as edge-localized eigenstates of the OBC BdG Hamiltonian at exponentially small energies. As the system size increases, thus approaching the semi-infinite limit, these energies converge exactly zero.

With the fermionic story in mind, let us begin our search by looking for the simplest case in which a QBL possesses a zero rapidity mode in the semi-infinite limit. Mathematically, as long as bulk-translation symmetry is in place, the dynamical matrix of such a system will be a block-Toeplitz operator. Thankfully, the spectra of these operators have been completely characterized. Sticking to the language of QBLs, let us summarize the key features (see Appendix A for a more mathematical summary and Ref. [3] for a self-contained collection of results and proofs on this topic). Referring back to Sect. 6.2.4, the QBL in question will have $2d$ bulk rapidity bands $\{\lambda_n(k), \lambda_n^*(k)\}$, which are computed as eigenvalues of $-i\mathbf{g}(k)$, with $\mathbf{g}(k)$ the Bloch dynamical matrix. In the simplest case where the bands are independent from one-another, the SIBC spectrum is characterized in terms of the *winding numbers* of the bands. The winding number of the band $\lambda_n(k)$ about a point $\lambda \in \mathbb{C}$ not contained in any of the bands is defined as

$$\nu_n(\lambda) \equiv \frac{1}{2\pi i} \int_{-\pi}^{\pi} \frac{d}{dk} \ln(\lambda_m(k) - \lambda)\, dk. \tag{8.1}$$

The SIBC spectrum can then be characterized as containing (i) each of the bulk rapidity bands $\lambda_n(k)$ and (ii) every complex number λ, with $\nu_m(\lambda) \neq 0$ for at least one band index m. In the language of non-Hermitian physics, spectra of type (ii) are known as *point-gapped spectra* [4, 5]. The point-gapped modes are localized on the left (right) edge when the associated winding number is positive (negative). The signatures we are in search of must then be associated to a point-gapped zero rapidity in the semi-infinite limit. But, if a band winds around zero, it must necessarily enter the right-half complex plane. We conclude that *any system hosting such a mode must be unstable under bi-infinite and semi-infinite BCs*. In otherwords, bulk instability is unavoidable.[2]

[2] We remark that the general characterization of block-Toeplitz spectra involves the more sophisticated techniques of symbols, partial indices, Wiener-Hopf factorizations, and Fredholm invariants. In particular, partial indices may be thought of as a generalization of winding numbers for more complicated multiband systems. As such, the SIBC spectrum is precisely the BIBC bands joined

It seems QBLs have the same issue as QBHs: ZMs necessitate dynamical instability. Does our journey end here? Remarkably, the answer is *no*, and the reason is two-fold. First of all, while mathematically useful, the infinite-size limits we have considered are only idealizations of the true physical system. Any actual implementation of such a QBL will necessarily be finite, and thus we should be analyzing our system under OBCs. Secondly, unlike the fermionic case, the spectral properties of the finite size—*even in the limit as* $N \to \infty$—may be drastically different than the infinite-size limit. In fact, we have seen a concrete example of this phenomena in the BKC Hamiltonian of Sect. 3.3.3. The frequency bands of the BKC are ellipses in the complex plane. Consequently, the SIBC spectra consist of these ellipses and their interiors. However, the finite OBC spectra lies on the real axis and, even in the limit as $N \to \infty$, only fills a 1-dimensional line segment connecting the two foci of the ellipse. See also Fig. 8.1 for an example of this for a dissipative generalization of the BKC to be considered in Sect. 9.2. The key ingredient for this phenomena is *non-normality* of the dynamical matrix. Non-normality, i.e., failure of an operator to commute with its adjoint, is a stronger requirement than non-Hermiticity and is known to be responsible for a number of anomalous spectral behaviors in certain classes of matrices and operators [3, 7]. In particular, highly non-normal Toeplitz matrices are known to exhibit spectral discontinuities in the semi-infinite limit.[3]

The goal at this stage is to characterize those QBLs whose semi-infinite limits contain zero rapidity modes and whose finite OBC truncations are dynamically stable for all N. But then, what are the fate of the semi-infinite zero rapidity modes once these truncations are taken? As we have just discussed, non-normality generically means the semi-infinite spectra differ drastically from the finite spectra. Thus, a zero rapidity need not survive the truncation. Moreover, if it does survive, it will follows that the stability gap of the finite system will be zero. Referring back to Table 6.1, such a situation can lead to dynamical instabilities. Once again, however, we are saved. While the semi-infinite spectra and finite spectra are generally unrelated, the corresponding *pseudospectra* are not. Pseudospectra, which generalizes the notion of spectra by considering *approximate* eigenvalues and eigenvectors, behave far more predictably in the presence of extreme non-normality. Since non-normality is an unavoidable reality of our methodology, we are thus lead to consider this more general notion.

---

with all complex numbers for which there are non-trivial partial indices. While we will not address this more complicated case in this thesis (and instead refer to an upcoming work [6]), we conjecture that the need for bulk-instabilities persists.

[3] The degree of non-normality of a matrix can be measured in many different ways. Since normal matrices are precisely those that possess an orthonormal basis of eigenvectors, most measures concern themselves with the degree to which the eigenvectors fail to be orthonormal. One common measure is the condition number of the associated modal matrix. That is, if $\mathbf{X}$ is a diagonalizable matrix, one may infer the degree of non-normality from the quantity $\kappa = \|\mathbf{P}\| \left\|\mathbf{P}^{-1}\right\| \geq 1$, where $\mathbf{P}$ is the matrix whose columns are eigenvectors of $\mathbf{X}$ and $\|\cdot\|$ is any unitarily-invariant, submultiplicative matrix norm. In words, κ is the *condition number of* $\mathbf{P}$. If $\mathbf{X}$ is normal, $\mathbf{P}$ is unitary and so $\kappa = 1$. Highly non-normal matrices are then those with $\kappa \gg 1$ [7].

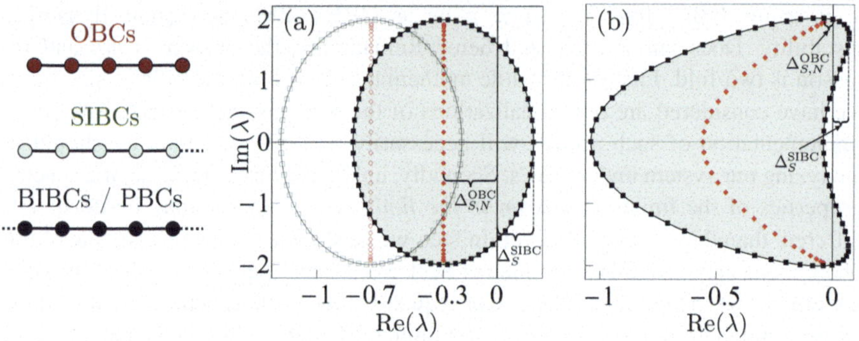

Fig. 8.1 Example of the disagreement between finite and semi-infinite spectra in the dissipative BKC of Sect. 9.2. Specifically, plotted are the rapidites for the chain under various BCs and system parameters along with the relevant stability gaps $\Delta_{S,N}^{\rm OBC}$ and $\Delta_S^{\rm SIBC}$. (**a**) The doubly-degenerate rapidities when $\mu = \Gamma = 0$. The filled (open) markers represent the topologically metastable (anomalously relaxing) regime, with $\kappa/\Delta = 0.6$ (1.4). The solid ellipses give the bulk spectrum, whereas the points on the ellipse are the rapidities for PBCs. The points on the vertical lines are the rapidities for OBCs. The shaded region denotes the semi-infinite spectrum. (**b**) The doubly-degenerate rapidity spectrum when $\Gamma = 0.12$ so that the winding around $\lambda = 0$ is zero. This is representative of non-topological dynamical metastability. In all cases, $J = 2$, $\Delta = 0.5$, and $\mu = 0$. Earlier versions of these figures first appeared in Ref. [1]

8.1.2 *Primer: The Pseudospectrum*

8.1.2.1 Definition and Dynamical Implications

We will cover the most important results related to the pseudospectra in this section and refer the interested reader to Appendix A for more details regarding the pseudospectra (and spectra) of block-Toeplitz matrices, and Ref. [7] for a comprehensive account of pseudospectra and its applications.

Consider a linear operator $\mathbf{X}$ acting on a complex vector space. In almost all instances, we will consider $n \times n$ matrices acting on $\mathbb{C}^n$. The spectrum $\sigma(\mathbf{X})$ of $\mathbf{X}$ is defined as the set of complex numbers λ such that $\mathbf{X} - \lambda\mathbb{1}$ is not invertible. In the finite dimensional case, these are precisely the eigenvalues of $\mathbf{X}$. With respect to a fixed operator norm $\|\cdot\|$, these correspond to singularities of the function $f(z) \equiv \|(\mathbf{X} - z\mathbb{1})^{-1}\|$, that is, the norm of the resolvent operator $\mathbf{R}(z) \equiv (\mathbf{X} - z\mathbb{1})^{-1}$. Again in the simplest case of finite dimension, the choice of norm is irrelevant since they are all equivalent (i.e., generate the same topology): if $f(z)$ diverges in one norm, then it diverges in all of them. Pseudospectra generalize this in the following way: Given an $\epsilon > 0$, the $(\epsilon, \|\cdot\|)$-*pseudospectrum* of $\mathbf{X}$ with respect to the norm $\|\cdot\|$ is

$$\sigma_{\epsilon,\|\cdot\|}(\mathbf{X}) \equiv \{\lambda \in \mathbb{C} : f(\lambda) = \|(\mathbf{X} - \lambda\mathbb{1})^{-1}\| > 1/\epsilon\}, \tag{8.2}$$

with the understanding that $f(\lambda) = \infty$ for any $\lambda \in \sigma(\mathbf{X})$ so that spectrum is always contained in the $(\epsilon, \|\cdot\|)$-pseudospectrum. If one or both of ϵ and $\|\cdot\|$ are understood, we will drop the prefix. We will also simply write $\sigma_{\epsilon,\|\cdot\|}(\mathbf{X}) = \sigma_\epsilon(\mathbf{X})$ when the norm is understood or of no particular importance. Elements of the pseudospectrum are called *pseudoeigenvalues*.

When the norm is induced by a vector norm, i.e., $\|\mathbf{X}\| \equiv \sup_{\vec{v}} \|\mathbf{X}\vec{v}\| / \|\vec{v}\|$, we have an equivalent (and more useful for our applications) definition, namely,

$$\sigma_{\epsilon,\|\cdot\|}(\mathbf{X}) = \{\lambda \in \mathbb{C} : \exists \vec{v} \in \mathbb{C}^n, \ \|\vec{v}\| = 1, \ \|(\mathbf{X} - \lambda \mathbb{1})\vec{v}\| < \epsilon\}. \tag{8.3}$$

The normalized vectors $\vec{v}$, with $\|(\mathbf{X} - \lambda)\vec{v}\| < \epsilon$ are called the $(\epsilon, \|\cdot\|)$-*pseudoeigenvectors* associated to the $(\epsilon, \|\cdot\|)$-eigenvalue λ.

The pseudospectrum becomes increasingly relevant as the operator $\mathbf{X}$ becomes highly nonnormal. One reason is the way it relates to perturbations. It follows from the definition that if $\lambda \in \sigma_{\epsilon,\|\cdot\|}(\mathbf{X})$, then there exists a perturbation $\mathbf{E}$ of size $\|\mathbf{E}\| < \epsilon$ such that $\lambda \in \sigma(\mathbf{X} + \mathbf{E})$ [7]. Hence

$$\sigma_{\epsilon,\|\cdot\|}(\mathbf{X}) = \bigcup_{\mathbf{E}:\ \|\mathbf{E}\| < \epsilon} \sigma(\mathbf{X} + \mathbf{E}) \tag{8.4}$$

Another reason can be seen simply by considering the case where the operator norm is induced by the usual vector 2-norm $\|\vec{v}\| = (\vec{v}^\dagger \vec{v})^{1/2}$. From Eq. (8.4), a perturbation of size ϵ can only shift the spectrum of a normal matrix by at-most ϵ. However, if $\mathbf{X}$ is nonnormal, small perturbations can dramatically modify the spectrum (one incarnation of this being the non-Hermitian skin-effect [8]). In sharp contrast, Eq. (8.4) illuminates the inherent *robustness* of pseudospectra. That is, consider perturbing a matrix $\mathbf{X}$ by a perturbation $\mathbf{E}$ of size $\delta > 0$. It follows from Eq. (8.4) that

$$\sigma_\epsilon(\mathbf{X} + \mathbf{E}) \subseteq \sigma_{\epsilon+\delta}(\mathbf{X}). \tag{8.5}$$

In words, a perturbation of size δ can only shift the pseudospectra by at most δ. Exact spectra enjoys no such robustness property in the absence of normality.[4]

If $\mathbf{X}$ is a highly non-normal matrix, the pseudospectra can dramatically influence the transient dynamics of a dynamical system generated by $\mathbf{X}$, e.g., $\dot{\vec{v}} = \mathbf{X}\vec{v}$. In particular, the pseudospectra bounds the maximal dilation of norm in the sense of

$$\sup_{t \geq 0} \left\| e^{t\mathbf{X}} \right\| \geq \frac{\alpha_\epsilon(\mathbf{X})}{\epsilon}, \quad \alpha_\epsilon(\mathbf{X}) \equiv \sup \operatorname{Re} \sigma_\epsilon(\mathbf{X}). \tag{8.6}$$

The quantity α_ϵ is called the *pseudospectral abcissa* and measures the extent to which the ϵ-pseudospectrum of $\mathbf{X}$ extends towards, or into, the right-half complex plane. The bound of Eq. (8.6) is particularly relevant for Hurwitz matrices. If $\alpha_\epsilon(\mathbf{X})$

[4] In the presence of normality, spectra enjoy this same robustness property. This fact is central to the standard theory of perturbation theory of self-adjoint operators, for example [9].

is positive and large compared to ϵ, $\|e^{t\mathbf{X}}\|$ will experience transient growth before asymptotically decaying to zero. That is, highly non-normal, but asymptotically stable, dynamical systems can appear unstable during a transient period. In fact, given a particular ϵ-pseudoeigenvector with ϵ-pseudoeigenvector λ, we have

$$\|e^{t\mathbf{X}}\vec{v} - e^{\lambda t}\vec{v}\| \leq \|(\mathbf{X} - \lambda)\vec{v}\|\, t + \mathcal{O}(t^2) < \epsilon t + \mathcal{O}(t^2), \tag{8.7}$$

so that $\vec{v}$ evolves like a normal mode with eigenfrequency λ for sufficiently small (set by ϵ) timescales, i.e., $\vec{v}(t) \simeq e^{\lambda t}\vec{v}(0)$. We call such modes *pseudonormal modes*. In the context of QBLs, we will also call the pseudoeigenvalues of the dynamical matrix *pseudorapidities*.

8.1.2.2 Pseudospectrum of Block-Toeplitz Operators and Matrices

We have seen how the spectrum of a block-Toeplitz operator and the finite block-Toeplitz matrix truncations can differ drastically, even when the truncations are arbitrarily large. Let us characterize this mathematically in the non-block case. Let $\mathbf{X}_N$ denote an $N \times N$ Toeplitz matrix and $\mathbf{X}_\infty \equiv \lim_{N\to\infty} \mathbf{X}_N$ denote the corresponding Toeplitz operator. Then, generically, we have

$$\lim_{N\to\infty} \sigma(\mathbf{X}_N) \neq \sigma(\mathbf{X}_\infty) = \sigma(\lim_{N\to\infty} \mathbf{X}_N). \tag{8.8}$$

Again, this is illustrated in Fig. 8.1 for a particular example to be introduced later. However, this spectral discontinuity does not plague the pseudospectrum. In fact,

$$\lim_{N\to\infty} \sigma_\epsilon(\mathbf{X}_N) = \sigma_\epsilon(\mathbf{X}_\infty) \Rightarrow \lim_{\epsilon\to 0}\lim_{N\to\infty} \sigma_\epsilon(\mathbf{X}_N) = \sigma(\mathbf{X}_\infty). \tag{8.9}$$

The left hand-side says that the pseudospectra of the finite truncations converges to the pseudospectra of the Toeplitz operator–it is much better behaved than the spectrum. It follows that, for every element λ of the infinite-size spectra, there is a pair (ϵ, N) such that $\lambda \in \sigma_\epsilon(\mathbf{X}_N)$. In this sense, the infinite-size spectra *imprints* itself in the pseudospectrum of the finite size Toeplitz matrices. These results generalize straightforwardly to block-Toeplitz matrices, see Appendix A.

8.2 Anomalous Transient Dynamics in QBLs

8.2.1 Anomalous Relaxation and Dynamical Metastability

Refocusing back to the discussion of Sect. 8.1.1, the types of QBLs we are attempting to characterize must exhibit two key properties: (i) the infinite-size limit are unstable, while all finite sizes are stable under OBCs; and (ii) the semi-infinite chain possesses a zero rapidity. The first property is rather peculiar from the dynamical perspective. These systems must change stability phase *discontinuously* in the infinite-size limit. Thinking to the observation that the semi-infinite spectra imprints itself in the pseudospectra of the finite chain (Eq. (8.9)), and that pseudospectra plays a key role in determining the transient behavior of dynamical systems with extremely non-normal generators (Eq. (8.7)), we expect that property (i) will engender dramatic consequences for the transient dynamics of the finite systems. Specifically, consider a normal mode associated to SIBC rapidity in the right-half complex plane—i.e., a dynamically unstable mode. If we truncate this mode to fit on a finite OBC chain, it can no longer be unstable (as we are assuming OBCs are stable). However, this truncated mode provides a *pseudonormal mode with pseudorapidity in the right-half plane*. Thus, the early dynamics will appear unstable according to Eq. (8.7). Physically, the truncated mode behaves in an unstable manner until it evolves long enough to detect the stabilizing presence of the two boundaries. See the concluding discussion of the BKC analysis in Sect. 3.3.3 and, more specifically, Eq. (3.58) for an example of this phenomenon.

To be more concrete, consider a fixed bulk-translation invariant QBL under finite N OBCs as well as SIBCs. Let $\Delta_{S,N}^{\text{OBC}}$ and Δ_S^{SIBC} denote the corresponding stability gaps. Note that Δ_S^{SIBC} is also the *bulk* stability gap in the sense that $\Delta_S^{\text{SIBC}} = \Delta_S^{\text{BIBC}}$. In general, we will always have

$$\Delta_{S,\infty}^{\text{OBC}} \equiv \lim_{N\to\infty} \Delta_{S,N}^{\text{OBC}} \leq \Delta_S^{\text{SIBC}}. \tag{8.10}$$

Now, suppose the dynamical matrices of interest are extremely non-normal so that a strict inequality $\Delta_{S,\infty}^{\text{OBC}} < \Delta_S^{\text{SIBC}}$ is possible. When this happens, there are two notable cases.

(i) *All gaps $\Delta_{S,N}^{\text{OBC}}$, $\Delta_{S,\infty}^{\text{OBC}}$, and Δ_S^{SIBC} are negative.* It follows that all relevant configurations are dynamically stable, but the asymptotic relaxation rate (which is set by the Lindblad gap) exhibits a discontinuity when N is taken to infinity. We then say that the OBC chain is in an *anomalously relaxing phase*. Such a phase is characterized by an increasingly long (with system size) transient time whereby the system possesses pseudonormal modes decaying at rates much slower than the rate set by the finite-size stability gap, i.e., the asymptotic decay rate. The transient is followed by asymptotic decay whose rate is indeed set by the finite-size stability gap. We conjecture that this mechanism is responsible for the anomalously long relaxation dynamics found in dissipative systems exhibiting the Liouvillian skin effect [10, 11]. This phenomena also bears close resemblence to the spectral separation-induced metastability of Ref. [12]. In our case, however, the spectral separation is between the finite- and infinite-size stability gaps, rather than between eigenvalues of a fixed Lindbladian.

(ii) *The finite size gaps $\Delta_{S,N}^{\text{OBC}}$ and their limit $\Delta_{S,\infty}^{\text{OBC}}$ are negative, while the semi-infinite gap Δ_S^{SIBC} is positive.* This case corresponds to the aforementioned discontinuous change of the stability phase in the infinite-size limit. We then call the OBC chain *dynamically metastable*, or just *metastable* for short. This phase is characterized by an increasingly long transient time whereby the pseudonormal modes whose pseudoeigenvalues are in the right half plane amplify exponentially. Once this transient concludes, all modes decay asymptotically with rate set by the finite-size stability gap. The terminology "dynamical metastability" refers to the metastable amplifying transient dynamical phase that eventually gives way to the necessarily stable asymptotic dynamics. Note that this is, at the current stage of understanding, a distinct notion of metastability from that known to arise in Markovian open quantum systems with large intra-spectral gaps [12–14]. However, both situations exhibit delayed relaxation to the true steady state.

In Sect. 9.2, we will uncover examples of both of these phenomena in a dissipative BKC (DBKC). Referring back to Fig. 8.1a, the transparent rapidities correspond to the system being in an anomalously relaxing phase while the solid rapidities correspond to a dynamically metastable phase. Generally speaking, both phenomena are characterized by a distinct separation of transient and asymptotic dynamics. In this sense, the dynamics have a two-step nature as depicted in Fig. 8.2a. In that figure, the mean trajectory of the expectation value $|\langle x_N \rangle (t)|$, averaged over 250 random initial conditions, is plotted for the DBKC in both an anomalously relaxing phase and a dynamically metastable phase. While the distinct two-step behavior is apparent in both cases, the dynamically metastable phase shows a much sharper separation between transient and asymptotic dynamics. In fact, the early time behavior shows a distinctive *amplification*. Such transient amplification is characteristic of dynamically metastable systems and is associated with the imprinting of the unstable SIBC rapidities into the pseudorapidities of the finite system. Finally, we refer to these behaviors as anomalous because they cannot be predicted from the *exact* spectral properties.

8.2.2 Divergence of the Transient Timescale in Dynamically Metastable Chains

Let us zoom in on dynamically metastable systems. In this case, we assert that the timescale of transient amplification actually *diverges in system size*. This can be seen by means of the *linear mixing time* $t_{\text{lin}}(\delta)$, which we define as the shortest time it takes for the quantity

$$d_{\text{lin}}(t) \equiv \sup_{\rho} \frac{\|\vec{m}_\rho(t) - \vec{m}_{\text{ss}}\|}{\|\vec{m}_\rho(0) - \vec{m}_{\text{ss}}\|}, \quad t \geq 0, \tag{8.11}$$

defined in terms of the mean vectors $\vec{m}_\rho(t) = \text{tr}[\Phi(t)\rho]$ and $\vec{m}_{\text{ss}} = \text{tr}[\Phi(t)\rho_{\text{ss}}] = \text{tr}[\Phi(0)\rho_{\text{ss}}]$, to fall, and remain, below a prescribed accuracy $\delta > 0$. Physically, $d_{\text{lin}}(t)$ measures the worst-case relative distance between the mean vector and the steady state at time t as compared to the initial separation at time 0. By design, $d_{\text{lin}}(0) = 1$ and $\lim_{t\to\infty} d_{\text{lin}}(t) = 0$. However, $d_{\text{lin}}(t)$ can rise *above* 1 during the

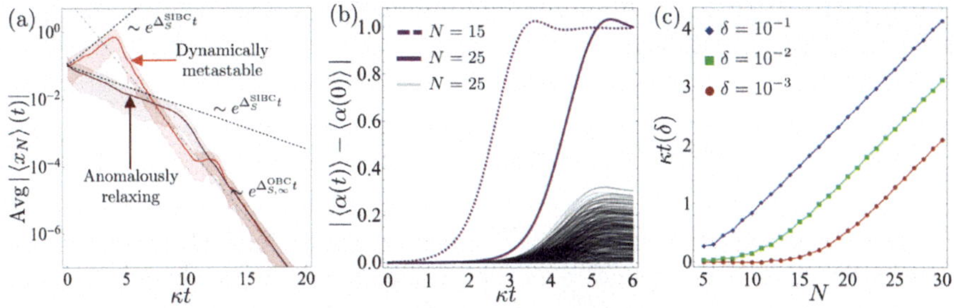

Fig. 8.2 (**a**) Illustration of anomalously relaxing vs. dynamically metastable dynamics. The solid lines are the trajectory of $|\langle x_N\rangle(t)|$ in the DBKC (Sect. 9.2), averaged over 250 initial conditions. The filled regions are $\pm$ one standard deviation from the mean. The black dashed lines are the dynamics predicted from the SIBC stability $\Delta_S^{\rm SIBC}$ gap while the gray dashed lines are the dynamics expected from the finite size stability games, $\Delta_{S,\infty}^{\rm OBC} = \lim_{N\to\infty}\Delta_{S,N}^{\rm OBC}$. In both cases, $\mu = \Gamma = 0$, and $N = 25$. The dynamically metastable (anomalously relaxing) curve corresponds to $\kappa/\Delta = 0.6$ (1.4). (**b**) Thin gray curves: expectation values of 250 randomly sampled linear observables in the quasi-steady state $\rho_\theta(t)$ generated by the left-localized MB γ_L^s (Eq. (9.17)) with $\|\theta\vec{\gamma}^s\| = 1$, in the DBKC, with $\kappa/\Delta = 0.6$, $N = 25$, $\|\vec{\alpha}\| = 1$. Thick purple curves: The upper bound in Eq. (8.34) for $N = 15$ (dashed) and $N = 25$ (solid). (**c**) The time $t(\delta)$ it takes for the aforementioned upper bound to exceed accuracy δ as a function of N. In all plots, $J = 2$ and $\Delta = 0.5$. Earlier versions of (**b**) and (**c**) first appeared in Ref. [1]

transient. In fact, it becomes arbitrarily large at intermediate times as N increases. Ultimately, we propose $t_{\rm lin}(\delta)$ as a proxy for the mixing time of the QBL, cf. Eq. (6.9). Let us concretely cement our claims in the form of a theorem.

Theorem 8.2.1 *Consider a dynamically metastable, bulk-translation invariant QBL. Then both (i) the maximum value of $d_{lin}(t)$ and (ii) the linear mixing time $t_{lin}(\delta)$, with δ sufficiently small, diverge in system size.*

Proof The quantity $d_{\rm lin}(t)$, which we will denote by $d_{\rm lin}(t, N)$ for a finite (N site) system under OBCs, may be calculated in terms of the OBC dynamical matrix, which we will denote by $\mathbf{G}_N^{\rm OBC}$. Firstly, observe that

$$\vec{m}_\rho(t) = {\rm tr}[\Phi(t)\rho] = {\rm tr}[e^{-i\mathbf{G}_N^{\rm OBC}t}\Phi(0)\rho] = e^{-i\mathbf{G}_N^{\rm OBC}t}{\rm tr}[\Phi(0)\rho] = e^{-i\mathbf{G}_N^{\rm OBC}t}\vec{m}_\rho(0). \tag{8.12}$$

Since we assume the system is dynamically stable (and more specifically, that the stability gap is strictly negative), we have $\vec{m}_{\rm ss} = 0$ (see Eq. (6.28)). From these two observations, we have

$$d_{\rm lin}(t, N) \equiv \sup_\rho \frac{\|\vec{m}_\rho(t)\|}{\|\vec{m}_\rho(0)\|} = \sup_{\vec{m}(0)} \frac{\|e^{-i\mathbf{G}_N^{\rm OBC}t}\vec{m}_\rho(0)\|}{\|\vec{m}_\rho(0)\|} = \|e^{-i\mathbf{G}_N^{\rm OBC}t}\|, \tag{8.13}$$

where the norm on the right hand-side is the operator norm induced by the vector 2-norm.

Equation (8.6) allows us to lower bound the maximum value of $d_{\text{lin}}(t)$ in terms of the pseudospectral abscissa:

$$d_{\text{lin}}^{\max}(N) \equiv \sup_{t\geq 0} d_{\text{lin}}(t, N) = \sup_{t\geq 0} \|e^{-i\mathbf{G}_N^{\text{OBC}} t}\| \geq \frac{\alpha_\epsilon(-\mathbf{G}_N^{\text{OBC}})}{\epsilon}. \tag{8.14}$$

The pseudospectral abscissa of $-i\mathbf{G}_N^{\text{OBC}}$ may be understood physically as the *pseudospectral stability gap*, i.e., the largest real part of the ϵ-pseudorapidities. For this reason, let us denote it by $\Delta_{S,N,\epsilon}^{\text{OBC}} \equiv \alpha_\epsilon(-\mathbf{G}_N^{\text{OBC}})$. We similarly define $\Delta_{S,\epsilon}^{\text{SIBC}}$ as the SIBC pseudospectral stability gap. In this notation,

$$d_{\text{lin}}^{\max}(N) \geq \frac{\Delta_{S,N,\epsilon}^{\text{OBC}}}{\epsilon}. \tag{8.15}$$

Referring to Eq. (8.9) (and its block-Toeplitz generalization, Eq. (A.6)), the finite size OBC pseudospectra converges to the infinite-size SIBC pseudospectra. In particular,

$$\lim_{N\to\infty} \Delta_{S,N,\epsilon}^{\text{OBC}} = \Delta_{S,\epsilon}^{\text{SIBC}} > 0, \tag{8.16}$$

where the inequality follows from the assumption of dynamical metastability. Thus,

$$\lim_{N\to\infty} d_{\text{lin}}^{\max}(N) \geq \frac{\Delta_{S,\epsilon}^{\text{SIBC}}}{\epsilon}.$$

Taking $\epsilon \to 0$ leaves the left hand-side invariant, while the fact that $\lim_{\epsilon\to 0} \Delta_{S,\epsilon}^{\text{SIBC}} = \Delta_S^{\text{SIBC}} > 0$ implies that the right hand-side diverges since. Thus, $\lim_{N\to\infty} d_{\text{lin}}^{\max}(N) > 0$, establishing the first claim.

The second claim follows by leveraging Eq. (A.3), which asserts $d_{\text{lin}}(t, N) < e^{\Omega t}$ for some system size-independent constant Ω. Now, let $t_{\max}(N) \equiv \arg\max d_{\text{lin}}(t, N)$, i.e., $d_{\text{lin}}(t_{\max}(N), N) = d_{\text{lin}}^{\max}(N)$. Thus, $d_{\text{lin}}^{\max}(N) < e^{\Omega t_{\max}(N)}$ from which it follows that

$$t_{\max}(N) > \frac{\ln d_{\text{lin}}^{\max}(N)}{\Omega}. \tag{8.17}$$

Divergence of the right hand-side ensures $t_{\max}(N)$ diverges. Thus, if $\delta < d_{\text{lin}}^{\max}(N)$ (if it is not, then $t_{\text{lin}}(\delta) = 0$ trivially), we have $t_{\text{lin}}(\delta) \geq t_{\max}(N) \to \infty$, completing the proof. □

8.3 Topological Metastability

With the tools of the last section in hand, we can finally pin down precisely those QBLs that are candidates for bosonic SPT physics. They are those QBLs that are

(i) dynamically metastable; and (ii) have a point-gapped zero rapidity about which at least one rapidity band winds. We refer to this class of QBLs as *topologically dynamically metastable*, or *topologically metastable*, for short.

We propose that these QBLs are candidates for tight bosonic analogues of fermionic SPT phases for several reasons. First, dynamical metastability ensures that the finite chain always possesses a steady state that can serve as the open system-generalization of the ground state. Second, we retain two distinct gap conditions, each of which constitutes a dissipative analogue to the many-body gap condition of fermionic SPTs: (1) the zero-rapidity is point-gapped in the infinite-size limit; and (2) the finite system has a Lindblad gap bounded away from zero as $N \to \infty$. This second feature, in particular, guarantees uniqueness of the steady state. Thirdly, the bulk has a non-trivial invariant in the form of the band winding number. Finally, we claim that topologically metastable systems host non-trivial edge modes analogous to those found in topologically non-trivial free fermionic systems. Let us now substantiate this final claim.

8.3.1 Majorana Bosons and Quasi-Steady States

Fix a topologically metastable QBL and denote the finite-size OBC dynamical matrix by $\mathbf{G}_N^{\text{OBC}}$. From the definition of topological metastability, we may immediately conclude that $\mathbf{G}_N^{\text{OBC}}$ and $\widetilde{\mathbf{G}}_N^{\text{OBC}}$ each possess at least one ϵ_N-pseudoeigenvector with zero ϵ_N-pseudoeigenvalue, and $\epsilon_N \to 0$ as $N \to \infty$. Let us call these two pseudoeigenvectors $\vec{v}_N^s$ and $\vec{v}_N^z$, respectively. Per usual, the property $\mathbf{G}^* = -\boldsymbol{\tau}_1\mathbf{G}^*\boldsymbol{\tau}_1$ allows us to ensure $\vec{v}_N^s = -\boldsymbol{\tau}_1\vec{v}_N^{s\,*}$ and similarly for $\vec{v}_N^z$ Generically, we will have $\vec{v}_N^{z\,\dagger}\boldsymbol{\tau}_3\vec{v}_N^s = ih_N$ for some[5] $h_N \in \mathbb{R} \setminus \{0\}$. From here we define $\vec{\gamma}^s \equiv A_s(N)\vec{v}_N^s$ and $\vec{\gamma}^z \equiv A_z(N)\vec{v}_N^z$, where $A_s(N)$ and $A_z(N)$ are any two real (possibly N-dependent) constants satisfying $A_z(N)A_s(N) = 1/h_N$, so that $\vec{\gamma}_N^{z\,\dagger}\boldsymbol{\tau}_3\vec{\gamma}_N^s = i$. Finally, note that the associated linear forms $\gamma_N^s \equiv \widehat{\vec{\gamma}_N^s}$ and $\gamma_N^z \equiv \widehat{\vec{\gamma}_N^z}$ are Hermitian, and have commutator equal to i (which is possible thanks to Theorem 7.1.4). We deem the pair (γ_N^z, γ_N^s) *Majorana bosons* (MBs). Henceforth, we will drop the N subscript on the MBs and their Nambu representations.

MBs enjoy a number of remarkable properties derived from their pseudospectral and topological origins.

(i) An MB pair consists of at least one approximate ZM γ^z and one generator of an approximate Weyl symmetry γ^s. Both are necessarily Hermitian. That is MBs are a particular instance of Noether modes.

[5] Following a similar proof to Theorem 7.1.4, one can find such a pair with $h_N \neq 0$ as long as the matrix $\mathbf{G}_N^{\text{OBC}} - \vec{\alpha}\vec{v}_N^{s\,\dagger}$, where $\vec{\alpha} \equiv \mathbf{G}_N^{\text{OBC}}\vec{v}_N^s$ (equivalently, $\widetilde{\mathbf{G}}_N^{\text{OBC}} - \vec{\beta}\vec{v}_N^{z\,\dagger}$, where $\vec{\beta} = \widetilde{\mathbf{G}}_N^{\text{OBC}}\vec{v}_N^z$) hosts Jordan chains of length at most one at the zero eigenvalue. 'Generically' here then refers to the fact that matrices are generically diagonalizable, and thus all Jordan chains are generically of length one.

(ii) The pair can generically be normalized to satisfy canonical commutation relations while maintaining their roles as approximate ZMs and SGs.
(iii) One member of the pair must be exponentially localized on the left half of the chain, while the other is localized on the right. This follows because, due to the adjoint relationship between $\widetilde{\mathbf{G}}$ and $\mathbf{G}$, the winding numbers that manifest γ^z and γ^s necessarily have opposite sign. This engenders the stated localization properties (see Theorem A.1.3).[6]
(iv) Combining (i)–(iii) allows us to construct a spatially split bosonic degree of freedom $\gamma^z + i\gamma^s$ whose quadrature components are the MBs. In the case where the MBs are non-split (in the sense of Sect. 7.1), this creates a long-lived bosonic excitation in the system.
(v) The pair (γ^z, γ^s) is robust against a large class of perturbations. Specifically, suppose the pair arises from the ϵ-pseudospectrum of the dynamical matrix. If the system is perturbed in such a way that the dynamical matrix is perturbed by a matrix of size $\delta > 0$, then there will be a pair of MBs arising from the $\epsilon + \delta$ pseudospectrum of the perturbed dynamical matrix.

Let us explore points (i) and (ii) in a simple, but rather typical , instance of MBs:

$$\gamma^z = \mathcal{M}_z(N)\sum_{j=1}^{N} \delta^{j-1} x_j, \quad \gamma^s = \mathcal{M}_s(N)\sum_{j=1}^{N} \delta^{N-j} p_j, \tag{8.18}$$

with $\mathcal{M}_{z,s}$ normalization constants[7] to be determined and δ real with $|\delta| < 1$. The approximate ZM and SG conditions then typically take on a form similar to

$$\mathcal{L}^\star(\gamma^z) = \mathcal{M}_z(N)\delta^N \chi, \tag{8.19}$$

$$\mathcal{L}^\star([\gamma^s, A]) - [\gamma^s, \mathcal{L}^\star(A)] = \mathcal{M}_s(N)\delta^N [\xi, A], \tag{8.20}$$

while the MBs are algebraically related according to

$$[\gamma^z, \gamma^s] = i\mathcal{M}_z(N)\mathcal{M}_s(N)N\delta^{N-1}. \tag{8.21}$$

Here, χ and ξ represent (typically localized) Hermitian linear forms whose coefficients in the Nambu basis are system-size independent. The goal of the normalization is to choose $\mathcal{M}_z(N)$ and $\mathcal{M}_s(N)$ such that (i) the right hand-sides of Eqs. (8.19) and (8.20) go to zero as $N \to \infty$ and (ii) the right hand-side of Eq. (8.21) is i. That

[6] The implicit assumption is that these MBs arise from winding of the rapidity bands. It is possible that more exotic MBs could arise without explicit band winding (see the discussion of Sect. A.2), thus rendering the localization properties more difficult to describe.

[7] In the notation used at the beginning of this section, $\mathcal{M}_s(N) = A_s(N) f_N(\delta)$ and $\mathcal{M}_z(N) = A_z(N)|\delta| f_N(\delta)$, with $f_N(x) = \sqrt{(1-x^2)/(1-x^{2N})}$, which converges to $\sqrt{1-x^2}$ as $N \to \infty$ for $|x| < 1$.

is, we want *canonically conjugate modes that provide asymptotically exact ZMs and SGs*. Mathematically, these conditions are

$$\text{(i)} \lim_{N\to\infty} \mathcal{M}_z(N)\delta^N = \lim_{N\to\infty} \mathcal{M}_s(N)\delta^N = 0, \quad \text{(ii)} \mathcal{M}_z(N)\mathcal{M}_s(N)N\delta^{N-1} = 1. \tag{8.22}$$

Remarkably, a scheme satisfying both (i) and (ii) always exists. The simplest, and most natural choice is

$$\mathcal{M}_z(N) = \mathcal{M}_s(N) = \frac{\delta^{-(N-1)/2}}{\sqrt{N}}. \tag{8.23}$$

With this, condition (ii) is clearly satisfied. As for (i),

$$\lim_{N\to\infty} \mathcal{M}_z(N)\delta^N = \lim_{N\to\infty} \frac{\delta^{(N+1)/2}}{\sqrt{N}} = 0, \tag{8.24}$$

and similarly for $\mathcal{M}_s(N)$. This scheme, which we call *symmetric normalization*, will be employed throughout the upcoming examples. Clearly, this choice is non-unique, and another may be adopted if certain features are desired over others. For example, if we wish for γ^z to have bounded coefficients in the x_j basis, then we may take $\mathcal{M}_z(N) = 1$ and $\mathcal{M}_s(N) = \delta^{-(N-1)}/N$. Importantly, the exponential profile of the MBs is *independent of the choice of normalization*.

With the symmetric normalization scheme adopted, we find ourselves with canonically conjugate operators (γ^z, γ^s) that become a pair consisting of an exact ZM and an exact SG, respectively, in the infinite-size limit. To what extent, however, do their roles as approximate ZMs and SGs manifest in the finite system? Starting with γ^z, from Eqs. (8.19) and (6.23), it follows that the Nambu representation $\vec{\gamma}^z$ satisfies

$$\|\widetilde{\mathbf{G}}_N^{\text{OBC}}\vec{\gamma}^z\| = \mathcal{M}_z(N)\delta^N \, \|\vec{\chi}\|, \tag{8.25}$$

where $\vec{\chi}$ is the Nambu representation of the operator χ on the right hand-side of Eq. (8.19), i.e., $\chi = \widehat{\vec{\chi}}$. Since the coefficients of χ are assumed to be independent of N, we will set $\|\vec{\chi}\| = 1$ for simplicity. It follows that γ^z is an approximate ZM with accuracy $\epsilon > 0$, where ϵ is any quantity satisfying

$$\epsilon > \epsilon_N \equiv \|\widetilde{\mathbf{G}}_N^{\text{OBC}}\vec{\gamma}^z\| = \mathcal{M}_z(N)\delta^N = \frac{\delta^{(N+1)/2}}{\sqrt{N}}. \tag{8.26}$$

The lifetime of γ^z can then be estimated via the quantity

$$\|\vec{\gamma}^z(t) - \vec{\gamma}^z(0)\| = \|\int_0^t \frac{d}{dt}\vec{\gamma}^z(t)\,dt\|, \tag{8.27}$$

with $\vec{\gamma}^z(0) = \vec{\gamma}^z$. Using Eq. (6.24), we have

$$\|\vec{\gamma}^z(t) - \vec{\gamma}^z(0)\| = \| \int_0^t i\widetilde{\mathbf{G}}_N^{\text{OBC}} \vec{\gamma}^z(t)\, dt \| = \| \int_0^t e^{i\widetilde{\mathbf{G}}_N^{\text{OBC}} t} \widetilde{\mathbf{G}}_N^{\text{OBC}} \vec{\gamma}^z(0)\, dt \| \tag{8.28}$$

We can then upper bound this by

$$\begin{aligned} \|\vec{\gamma}^z(t) - \vec{\gamma}^z(0)\| &\le \int_0^t \|e^{i\widetilde{\mathbf{G}}_N^{\text{OBC}} t}\| \|\widetilde{\mathbf{G}}_N^{\text{OBC}} \vec{\gamma}^z\|\, dt = \epsilon_N \int_0^t \|e^{i\widetilde{\mathbf{G}}_N^{\text{OBC}} t}\|\, dt \\ &\le \epsilon_N t \sup_{\tau \in [0,t]} \|e^{i\widetilde{\mathbf{G}}_N^{\text{OBC}} t}\|. \end{aligned} \tag{8.29}$$

Once again employing Eq. (A.3), the supremum on the right hand-side is bounded above by $e^{\Omega t}$ for some N-independent constant Ω. We ultimately arrive at the bound

$$\|\vec{\gamma}^z(t) - \vec{\gamma}^z(0)\| \le \epsilon_N t e^{\Omega t} = \frac{\delta^{(N+1)/2}}{\sqrt{N}} t e^{\Omega t}, \tag{8.30}$$

which vanishes as $N \to \infty$. That is, the lifetime of γ^z diverges as N increases.

As for γ^s, we can illuminate one particular implication of its existence by leveraging the uniqueness of the state-state under OBCs. Denoting said steady state by ρ_{ss}, we introduce the family of Weyl-displaced Gaussian states,

$$\rho_\theta \equiv e^{i\theta\gamma^s} \rho_{\text{ss}} e^{-i\theta\gamma^s}, \quad \theta \in \mathbb{R}. \tag{8.31}$$

Because γ^s generates an approximate symmetry, the states $\rho(\theta)$ are *quasi-steady*, in the sense that $\dot{\rho}_\theta(0) \sim 0 + \mathcal{O}(\theta\epsilon_N)$, with ϵ_N the same as in Eq. (8.26). Unlike steady states of a QBL, these quasi-steady states can possess nonzero mean vectors:

$$\vec{m}_\theta(0) = \text{tr}[\Phi\rho_\theta(0)] = i\theta\vec{\gamma}^s. \tag{8.32}$$

Since $\vec{\gamma}^s$ is edge-localized and a pseudonormal mode of the dynamical matrix, the quasi-steady mean vectors are exponentially localized on one edge of the chain and are long lived. In fact, leveraging an almost identical procedure to the one used to demonstrate the diverging lifetime of γ^z, we obtain the bound

$$\frac{\|\vec{m}_\theta(t) - \vec{m}_\theta(0)\|}{\|\vec{m}_\theta(0)\|} \le \epsilon t e^{\Omega t}, \tag{8.33}$$

where ϵ is such that the normalized vector $\vec{v}^s \equiv \vec{\gamma}^s / \|\vec{\gamma}^s\|$ is an ϵ-pseudoeigenvector of $\mathbf{G}$, with ϵ-pseudoeigenvalue 0. It follows that $\epsilon \to 0$ as $N \to \infty$ so that the mean-vector is long-lived. This long-livedness implies quasi-stationary evolution in certain observable expectation vlaues. Explicitly, if $\alpha = \widehat{\vec{\alpha}}$ is an arbitrary linear form, then

$$\frac{|\langle \alpha(t)\rangle - \langle \alpha(0)\rangle|}{\|\vec{\alpha}\| \, \|\vec{m}_\theta(0)\|} = \frac{|\vec{\alpha}^\dagger \boldsymbol{\tau}_3 (\vec{m}_\theta(t) - \vec{m}(0)|)|}{\|\vec{\alpha}\| \, \|\vec{m}_\theta(0)\|} \leq \frac{\|\vec{m}_\theta(t) - \vec{m}_\theta(0)\|}{\|\vec{m}_\theta(0)\|} \leq \epsilon t e^{\Omega t}, \tag{8.34}$$

where we have used the Cauchy-Schwarz inequality to bound the numerator by the quantity in Eq. (8.33). Long-livedness of the quasi-steady state mean vector and randomly sampled observable expectation values in a to-be-considered example are shown in Fig. 8.2b–d.

On the other hand, being only Weyl-displaced from the steady state, it follows that ρ_θ shares a covariance matrix (and hence all even moments) with ρ_{ss}. Interestingly, these states can be used to construct long-lived classical non-Gaussian states. Namely, any convex linear combination of the ρ_θ's will be long-lived and possess a non-Gaussian, but still nonnegative, Wigner function. In Sect. 9.3, we will also explore the possibility of constructing *non-classical* states utilizing these quasi-steady states.

Remarkably, the existence of any state whose mean vector is long-lived (in the above sense) implies that the system is topologically metastable. This can be seen by contradiction. Let $\mathbf{G}_N^{\text{O}}$ be the dynamical matrix for a bulk-translationally invariant system of size N under OBCs. The absence of topological metastability implies there exists a system-size-independent constant V such that,

$$\left\| \mathbf{G}_N^{\text{O}} \vec{v}_N \right\| > V \, \|\vec{v}_N\|, \quad \forall \vec{v}_N \in \mathbb{C}^{2N} \setminus \{0\}, \quad \forall N.$$

That is, $0 \notin \sigma_\epsilon(\mathbf{G}_N^{\text{OBC}})$ for all $\epsilon < V$ and for all N. In fact, we can take V to be the $N \to \infty$ limit of the minimal singular-value of $\mathbf{G}_N^{\text{OBC}}$. Consider an initial state $\rho(0)$, with mean vector $\vec{m}_\rho(t) \equiv \text{tr}[\Phi \rho(t)]$, satisfying $\vec{m}_\rho(0) \neq 0$. Since $\vec{m}_\rho(t) = e^{-i\mathbf{G}_N^{\text{O}} t} \vec{m}_\rho(0)$, we have, for sufficiently short t,

$$D(t) \equiv \frac{\left\| \vec{m}_\rho(t) - \vec{m}_\rho(0) \right\|}{\left\| \vec{m}_\rho(0) \right\|} \simeq \frac{\left\| \mathbf{G}_N^{\text{OBC}} \vec{m}_\rho(0) \right\|}{\left\| \vec{m}_\rho(0) \right\|} t > Vt.$$

The quantity $D(t)$, which measures the separation of $\vec{m}_\rho(t)$ from its initial value, immediately evolves away from zero with a nonzero "velocity" bounded below by V, for all N.

8.3.2 The Impact of Number Symmetry: Dirac Bosons

Being Hermitian, the topological modes in the above analysis bear closest resemblance to the Majorana fermions of topological superconductivity. Thus, one may suspect this same analysis fails to produce modes analogous of the topological edge-modes in topological insulators. In order to see why this is false, we will consider those QBLs with number symmetry to constitute the natural dissipative-bosonic analogues of insulators.

Number-symmetric QBLs are those QBLs that possess a weak U(1) symmetry, $a_j \to e^{i\phi} a_j$, generated by the total number operator $\sum_j a_j^\dagger a_j$. The corresponding Bogoliubov transformation is $\Phi \mapsto e^{i\tau_3 \theta}\Phi$. From this characterization, number symmetry implies, in particular, that $[\mathbf{G}, \boldsymbol{\tau}_3] = 0$. Hence, the dynamical matrix may be written as

$$\mathbf{G} = \mathbf{K} \otimes \begin{bmatrix} 1 & 0 \\ 0 & 0 \end{bmatrix} + (-\mathbf{K}^*) \otimes \begin{bmatrix} 0 & 0 \\ 0 & 1 \end{bmatrix}, \tag{8.35}$$

with $\mathbf{K}$ an arbitrary $N \times N$ complex matrix. Just as the spectrum of $\mathbf{G}$ is determined by that of $\mathbf{K}$, so is the pseudospectrum. In particular, if the system is topologically metastable, both $\mathbf{G}$ and $\mathbf{K}$ possess approximate kernel vectors.

To be more concrete, let us suppose the system is topologically metastable. This affords us at least one MB pair (γ_1^z, γ_1^s). Expanding in the bosonic basis yields

$$\gamma_1^z = \sum_{j=1}^{N} u_j^* a_j + u_j a^\dagger, \quad \gamma_1^s = \sum_{j=1}^{N} v_j^* a_j + v_j a_j^\dagger$$

where u_j and v_j are related to the Nambu vectors via $\vec{\gamma}_1^z = [u_1, -u_1^*, \ldots, u_N, -u_N^*]^T$ and $\vec{\gamma}_1^z = [v_1, -v_1^*, \ldots, v_N, -v_N^*]^T$. The HWRs imply $\text{Im} \sum_{j=1}^N u_j^* v_j = 1/2$ since

$$i = [\gamma_1^z, \gamma_1^s] = \sum_{j=1}^{N} u_j^* v_j - v_j^* u_j = 2i \,\text{Im} \sum_{j=1}^{N} u_j^* v_j. \tag{8.36}$$

Number symmetry implies that an approximate (or exact) ZM or SG remains so after a phase rotation $a_j \mapsto e^{i\phi} a_j$. In particular, we may rotate each of our MBs to construct a second linearly independent MB pair. Specifically, if we fix $\phi = \pi/2$, then we immediately find another linearly independent MB pair, namely,

$$\gamma_2^z = i \sum_{j=1}^{N} u_j^* a_j - u_j a^\dagger, \quad \gamma_2^s = i \sum_{j=1}^{N} v_j^* a_j - v_j a_j^\dagger.$$

We have not yet encountered anything resembling the edge modes of topological insulators. For that, we need to utilize both MB pairs. Explicitly, consider the operators

$$\alpha \equiv \frac{1}{2\sqrt{C^z}}(\gamma_1^z - i\gamma_2^z) = \frac{1}{\sqrt{C^z}} \sum_{j=1}^{N} u_j^* a_j,$$

$$\beta \equiv \frac{1}{2\sqrt{C^s}}(\gamma_1^s - i\gamma_2^s) = \frac{1}{\sqrt{C^s}} \sum_{j=1}^{N} v_j^* a_j, \tag{8.37}$$

where $C^z = \sum_j |u_j|^2$ and $C^s = \sum_j |v_j|^2$ are positive real numbers chosen to ensure α and β satisfy the following properties:

- They are bosonic: $[\alpha, \alpha^\dagger] = [\beta, \beta^\dagger] = 1_F$;
- They are algebraically related via $[\alpha, \beta] = 0$;
- They are edge localized according to the localization of the constituent MBs;
- The bosonic mode α is approximately conserved while the real and imaginary quadratures of β generate two (non-commuting) approximate symmetries.

Algebraically speaking, we have constructed edge modes on which the number symmetry acts trivially. The necessary and sufficient ingredients for this construction are topological metastability and number symmetry, and so this construction applies broadly.

The normalization of these operators is far more straightforward than it is in the number-non-symmetric case. To see this, consider a left-localized, approximately conserved, bosonic operator α, i.e., $u_j \propto \delta^{j-1}$, with $|\delta| < 1$ so that

$$\alpha = \mathcal{M}(N) \sum_{j=1}^{N} \delta^{j-1} a_j, \tag{8.38}$$

where $\mathcal{M}(N)$ is the normalization constant. Taking δ positive and real without loss of generality, the normalization constant is

$$\mathcal{M}(N) = \sqrt{\frac{1-\delta^2}{1-\delta^{2N}}}$$

which, unlike in the case of the MBs, converges to a finite value as $N \to \infty$. Namely, $\lim_{N\to\infty} \mathcal{M}(N) = \sqrt{(1-\delta^2)}$. With this, the exact ZM in the $N \to \infty$ limit is

$$\alpha = \sqrt{1-\delta^2} \sum_{j=1}^{\infty} \delta^{j-1} a_j. \tag{8.39}$$

While, in the generic case, the exact expression for α will be more complicated than Eq. (8.38), this argument demonstrates the unambiguous normalizability of the modes in the infinite-size limit.

We conclude the general analysis by considering the possibility of a purely dissipative, number-symmetric chain that exhibits dynamical metastability (e.g., a number symmetric analogue to the PDC to be discussed in Sect. 9.1). The number symmetry property $[\mathbf{G}, \boldsymbol{\tau}_3] = 0$ manifests in the purely dissipative case $\mathbf{G} = -i\boldsymbol{\tau}_3\mathcal{F}(\mathbf{M})$ as $[\mathcal{F}(\mathbf{M}), \boldsymbol{\tau}_3] = 0$. Thus, $\mathbf{G}^\dagger = (-i\boldsymbol{\tau}_3\mathcal{F}(\mathbf{M}))^\dagger = i\boldsymbol{\tau}_3\mathcal{F}(\mathbf{M}) = -\mathbf{G}$. That is, a purely dissipative number symmetric chain must have an anti-Hermitian (thus, normal) dynamical matrix $\mathbf{G}$. It follows that no such model can exhibit dynamical metastability.

8.4 Relationships to Existing Work

Before moving to explicit examples, it is important to place our contributions in the context of existing work. Firstly, concepts of metastability motivated by classical statistical physics have been extended to Markovian systems, and studied in great detail [12, 14]. In essence, this form of metastability is characterized by multi-step relaxation, mandated by the presence of large gaps in the Lindblad spectrum. While our *dynamically* metastable systems possess no such spectral gaps, it turns out they do exhibit *pseudospectral gaps*. That is, the nontrivial pseudospectra (determined by the spectra under semi-infinite BCs) remains gapped away from the exact spectrum in such a way to mandate anomalous transient dynamics (see Fig. 8.2a). In fact, pseudospectra has been conjectured [15] to also play a role in the recent discoveries of anomalous relaxation dynamics in Markovian systems exhibiting a non-Hermitian skin-effect [10, 11]. They have further been successfully applied to study anomalous dynamics in random quantum circuits [16–18] and explicitly non-Hermitian Hamiltonians [15, 19].

As hinted to in Chap. 1, a related branch of research known as *topological amplification* [20–25], while not motivated by the many-body physics of topological free fermions, shares notable points of contact with our analysis. The standard approach to topological amplifiers employs input-output theory. Central to the input-output treatment is the *susceptibility matrix*, or Green's function, say, $\chi(\omega)$, which connects incoming fields at frequency ω to outgoing fields at frequency ω. For photonic systems on N modes, whose coherent and incoherent dynamics can be cast in the form of a Lindblad master equation, this susceptibility matrix takes the form $\chi(\omega) = i(\omega\mathbb{1}_{2N} - \mathbf{G})^{-1}$, where $\mathbf{G}$ is (in essence) the dynamical matrix we have studied extensively. It is then known that topological amplification can only take place if $\chi^{-1}(\omega)$ "winds" around the origin, in a suitable sense. In our language this is, in the simplest case, equivalent to nontrivial winding of the rapidity bands about the point $i\omega$. According to our pseudospectral approach, this implies that there exist pseudonormal modes with pseudoeigenvalues $i\omega + \lambda$, with $\lambda > 0$. Such pseudonormal modes must necessarily amplify in the transient, and hence contribute to gain in the output signal. Moreover, we see that when, these systems amplify zero-frequency ($\omega = 0$) signals, they *must* possess MBs and all of the associated non-trivial transient dynamics they entail. To summarize, in our framework, general topological amplifiers are classified as dynamically metastable, while those that amplify zero-frequency input signals are classified as topologically metastable. We further observe that the introduction and application of the "doubled matrix" approach in Refs. [21, 22] can be connected naturally to pseudospectral theory (see Ref. [15], for instance).

References

1. V.P. Flynn, E. Cobanera, L. Viola, Phys. Rev. Lett.**127**, 245701 (2021)
2. V.P. Flynn, E. Cobanera, L. Viola, Phys. Rev. B **108**, 214312 (2023)
3. A. Böttcher, B. Silbermann, *Introduction to Large Truncated Toeplitz Matrices* (Springer, Berlin, 2012)
4. K. Kawabata, K. Shiozaki, M. Ueda, M. Sato, Phys. Rev. X**9**, 041015 (2019)
5. Y. Ashida, Z. Gong, M. Ueda, Adv. Phys. **69**, 249 (2020)
6. V.P. Flynn, E. Cobanera, L. Viola, Upcoming paper covering more mathematically sophisticated instances of topological metastability (2024)
7. L.N. Trefethen, M. Embree, *Spectra and Pseudospectra: The Behavior of Nonormal Matrices and Operators* (Princeton University Press, Princeton, 2005)
8. S. Yao, Z. Wang, Phys. Rev. Lett. **121**, 086803 (2018)
9. T. Kato, *Perturbation Theory for Linear Operators*, vol. 132 (Springer, Berlin, 2013)
10. T. Haga, M. Nakagawa, R. Hamazaki, M. Ueda, Phys. Rev. Lett. **127**, 070402 (2021)
11. T. Mori, T. Shirai, Phys. Rev. Lett. **125**, 230604 (2020)
12. K. Macieszczak, M. Guţă, I. Lesanovsky, J.P. Garrahan, Phys. Rev. Lett. **116**, 240404 (2016)
13. L.M. Vasiloiu, F. Carollo, J.P. Garrahan, Phys. Rev. B **98**, 094308 (2018)
14. K. Macieszczak, arXiv:2104.05095 (quant-ph) (2021)
15. N. Okuma, M. Sato, Phys. Rev. B **103**, 085428 (2021)
16. J. Bensa, M. Žnidarič, Phys. Rev. X **11**, 031019 (2021)
17. M. Žnidarič, Phys. Rev. Res. **4**, 033041 (2022)
18. J. Bensa, M. Žnidarič, Phys. Rev. A **107**, 022604 (2023)
19. G.Z., Y. Ashida, K. Kawabata, K. Takasan, S. Higashikawa, M. Ueda, Phys. Rev. X **8**, 031079 (2018)
20. J.C. Budich, E.J. Bergholtz, Phys. Rev. Lett. **125**, 180403 (2020)
21. D. Porras, S. Fernández-Lorenzo, Phys. Rev. Lett. **122**, 143901 (2019)
22. T. Ramos, J.J. García-Ripoll, D. Porras, Phys. Rev. A **103**, 033513 (2021)
23. C.C. Wanjura, M. Brunelli, A. Nunnenkamp, Nat. Commun. **11**, 3149 (2020)
24. C.C. Wanjura, M. Brunelli, A. Nunnenkamp, Phys. Rev. Lett. **127**, 213601 (2021)
25. M. Brunelli, C.C. Wanjura, A. Nunnenkamp, arXiv:2207.12427v2 [quant-ph] (2022)

Chapter 9
The Realm of Possibilities

Abstract This chapter is devoted to dissecting a series of four models that exemplify the general results of Chap. 8. The first model is a purely dissipative QBL derived by applying the first recipe of Chap. 7 to the fermionic Kitaev chain (FKC) Hamiltonian. The resulting QBL is topologically metastable in certain parameter regimes that overlap with the topological phase diagram of the FKC and, most importantly, sustains an MB pair that derive from the Majorana edge modes of the FKC. Interestingly, these provide an example of "non-split" MBs, i.e., MBs that are both ZMs *and* SGs. In this sense, these are the *tightest* bosonic analogues of Majorana fermions. The second model is a dissipative version of the BKC first explored in Chap. 3. By completely characterizing the five-parameter topological phase diagram, we uncover three relevant dynamical phases: an anomalously relaxing one, a non-topological dynamically metastable one, and a topologically metastable one. We explore the dynamical features of each phase in detail and, in particular, compute MBs that arise in the topologically metastable phases. The third model once again describes a dissipative BKC. However, in this model, the dissipator is constructed following the second recipe of Chap. 7, and thus, the resultant QBL possesses a unique, pure steady-state. Additionally, we find that this QBL has a metastable regime. This allows us to explore the interplay between MBs and pure steady states. In particular, we analytically compute the quasi-steady states and uncover surprisingly non-trivial parity dynamics of certain cat-state superpositions. The final model is a number-symmetric QBL that possesses a topologically metastable phase. We explicitly construct the DBs predicted in Chap. 8. We conclude with an analysis of certain multitime correlation functions in QBLs. We find that, generically, topological metastability may be characterized by the existence of long-live two-time quantum correlation functions between the macroscopically separated MB partners, in addition to the emergence of divergent zero-frequency power-spectral peaks. We further use two-time correlation functions to distinguish the unique features of the models of interest.

V. P. Flynn, *Effective Non-Hermiticity and Topology in Markovian Quadratic Bosonic Dynamics*, Springer Theses, https://doi.org/10.1007/978-3-031-52045-7_9

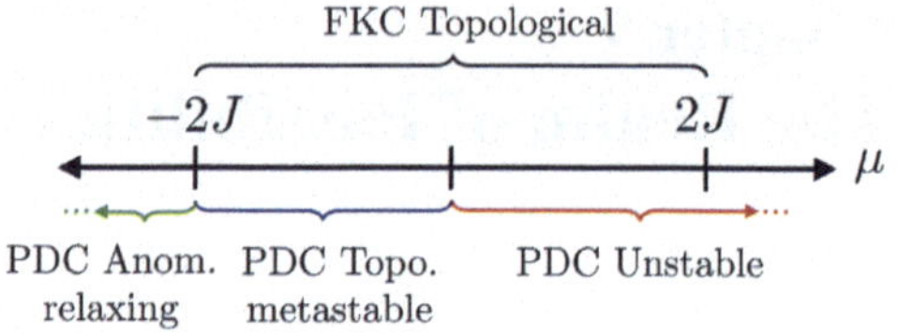

Fig. 9.1 Top: the topological phase diagram of the FKC Hamiltonian. Bottom: the topological stability diagram of the PDC for the same parameters

In this chapter,[1] we introduce four models that best exemplify the novel aspects of dynamical (both non-topological and topological) metastability and anomalous relaxation dynamics and, in addition, present an in-depth analysis of multitime correlation functions in them. Let us summarize the key physical ingredients for each model, and most important takeaways from each model.

The first model, which we will call the *purely dissipative topologically metastable chain*, or simply the purely dissipative chain (PDC), will provide us with, in a sense, the *tightest* bosonic analogues of Majorana edge fermions. As the name suggest, the chain is purely dissipative ($H = 0$) and, in fact, is built according to the procedure detailed in Sect. 7.2.1 with the fermionic Kitaev chain Hamiltonian used as the input. The model exhibits topological metastability for parameters that correspond to *half* of the topological region of the FKC Hamiltonian. The other half of the topological region maps to a dynamically unstable QBL (see also Fig. 9.1). The Majorana bosons emerging in the topological metastable regime are direct descendants of the Majorana fermions in the original fermionic Hamiltonian. Since the overall model is purely dissipating, each Majorana boson plays the role of an approximate ZM as well as an approximate SG - they are *non-split*, in the language of Sect. 7.1.

The second model, which we will call the dissipative BKC (DBKC), consists of the BKC Hamiltonian with the uniform DPA term (Eq. (5.20)) paired with both uniform on-site, and next-nearest neighbor (NNN), loss. The full model, which depends on five parameters, exhibits each of the three dynamical phases of interest: an anomalously relaxing phase, a non-topological dynamically metastable phase, and a topologically metastable phase. We will explore the transient dynamics in each of these three phases, and explicitly construct two Majorana pairs in the topologically metastable regime. In sense, these Majorana boson pairs are the most *generic*, as they are split. Thus, we can explore the consequences of the breakdown of Noether's theorem via these edge modes. We further are able to adjust the winding numbers of the two rapidity bands of the model in such a way to vary the total number of MB pairs.

The third model, which we will call the DBKC with a pure steady state, consists of the BKC Hamiltonian with a dissipator derived from the procedure detailed in Sect. 7.2.2. Physically, the dissipator consists of uniform on-site loss, but in a squeezed basis. We find that the dynamical matrix of this QBL coincides with that

[1] All results from this section can be found in Refs.[1] and [2] which were both co-authored with Emilio Cobanera & Lorenza Viola.

of the DBKC in a certain parameter regime. As such, we are able to explore the interplay between a pure steady state and topological metastability. We determine the form of the (pure) quasi-steady states and determine that they are squeezed coherent states. Thanks to their simple form, we are able to analytical study their relaxation dynamics and find an sharp distinction between the relaxation times depending on which edge symmetry displacement they are built from. Finally, we combine these states into "cat"-states and analytically determine their parity dynamics. Again, depending on which edge symmetry is used, the parity undergoes either relatively tame, or extremely dramatic, evolution.

The final model, which we will call the dissipative number-symmetric chain (DNS), consists of a tight-binding Hamiltonian subject to both on-site gain and loss. We find that this model, which has a weak total number symmetry, exhibits a topologically metastable phase. Within this phase, we are able to analytically construct the DBs that we predicted must arise in topologically metastable systems with total number symmetry.

Following the detailed analysis of each model, we propose a class of observable signatures of (topological and non-topological) dynamical metastability. Namely, two-time correlation functions and their associated power spectra. By focusing on correlations between linear forms, we uncover a state-independent notion of quantum correlations. These correlations, and their power spectra, are then shown to behave in distinctive ways depending on the dynamical phase of the QBL. In particular, we show that topologically metastable systems will exhibit long-lived quantum correlations between the macroscopically separated, edge-localized MBs, in addition to divergent zero frequency power spectral peaks. In contrast, non-topological dynamically metastable systmes are shown to have divergent peaks away from zero-frequency, while non-metastable systems lack any such divergent peaks. We further develop the theory of two-time quantum correlations in QBLs to distinguish split and non-split MBs, as well as MBs from DBs. All predictions are confirmed within the previously studied four models.

The outline for this chapter is as follows. In Sects. 9.1, 9.2, 9.3, and 9.4, we introduce, and analyze, the PDC, the DBKC, the pure steady state DBKC, and the DNS chain. In Sect. 9.5, we discuss aspects of two-time correlations and specifically apply them to distinguish various incarnations of dynamical and topological metastability in QBLs.

9.1 A Purely Dissipative Topologically Metastable Chain

To construct our first model, we will apply the procedure of Sect. 7.2.1 to the fermionic Kitaev chain Hamiltonian (Eq. (5.10)). The first step is to identify the FKC BdG Hamiltonian. Under OBCs, it is determined to be

$$\mathbf{H}_{\text{FKC}} = \mathbb{1}_N \otimes \mathbf{h}_0 + \mathbf{T} \otimes \mathbf{h}_1 + \mathbf{T}^\dagger \otimes \mathbf{h}_1^\dagger, \tag{9.1}$$

$$\mathbf{h}_0 \equiv \begin{bmatrix} -\mu & 0 \\ 0 & \mu \end{bmatrix}, \quad \mathbf{h}_1 \equiv \begin{bmatrix} -J & \Delta \\ -\Delta & J \end{bmatrix}. \tag{9.2}$$

The construction of Sect. 7.2.1 produces a purely dissipative QBL defined (solely) by its GKS matrix, $\mathbf{M} = \mathbf{H}_{\text{FKC}} + \mathbf{B}$, with $\mathbf{B}$ any bosonic matrix ensuring $\mathbf{M} \geq 0$. We define the purely dissipative topologically metastable chain, or simply the purely dissipative chain (PDC), by taking $\mathbf{B} = \alpha \mathbb{1}_{2N}$, with α any real quantity satisfying $\alpha \geq |\min \sigma(\mathbf{H}_{\text{FKC}})|$. All of the features we will focus on are independent of the specific chosen value of α. The physical dissipator contains five different contributions: $\mathcal{D}_{\text{PDC}} \equiv \mathcal{D}_{-,0} + \mathcal{D}_{+,0} + \mathcal{D}_{-,1} + \mathcal{D}_{+,1} + \mathcal{D}_{p,1}$ with

$$\mathcal{D}_{-,0} \equiv (\alpha - \mu) \sum_{j=1}^{N} \mathcal{D}[a_j, a_j^\dagger],$$

$$\mathcal{D}_{+,0} \equiv (\alpha + \mu) \sum_{j=1}^{N} \mathcal{D}[a_j^\dagger, a_j],$$

$$\mathcal{D}_{-,1} \equiv -J \sum_{j=1}^{N} \mathcal{D}[a_j, a_{j+1}^\dagger] + \mathcal{D}[a_{j+1}, a_j^\dagger],$$

$$\mathcal{D}_{+,1} \equiv J \sum_{j=1}^{N} \mathcal{D}[a_j^\dagger, a_{j+1}] + \mathcal{D}[a_{j+1}^\dagger, a_j],$$

$$\mathcal{D}_{p,1} \equiv \Delta \sum_{j=1}^{N} \mathcal{D}[a_j, a_{j+1}] - \mathcal{D}[a_{j+1}, a_j] - (a \leftrightarrow a^\dagger).$$

The first and second term correspond to onsite damping and pumping of strengths $\alpha - \mu$ and $\alpha + \mu$, respectively. The third and fourth encode nearest- neighbor damping and pumping ("dissipative hopping"), each of strength $|J|$. The final term encodes the nearest-neighbor dissipative pairing of strength Δ. Importantly, this final term breaks the U(1) number symmetry and allows for topological metastability to arise (recall the final remark of Sect. 8.3.2).

Before embarking on a detailed analysis of the model, we ask: does the dissipator "detect" the topological phase transition of the underlying FKC? The answer is "Yes" and can be understood by diagonalizing the GKS matrix $\mathbf{M}_{\text{PDC}}$. Since $\mathbf{M}_{\text{PDC}}$ differs from the BdG Hamiltonian of the FKC $\mathbf{H}_{\text{FKC}}$ by only a constant shift, its eigenvectors are the same as those of the FKC while the eigenvalues are those of $\mathbf{H}_{\text{FKC}}$ shifted by α. It follows that

$$\mathbf{M}_{\text{PDC}} = \sum_{n=1}^{2N} (\epsilon_n + \alpha) \vec{\psi}_n \vec{\psi}_n^\dagger, \quad \mathbf{H}_{\text{FKC}} \vec{\psi}_n = \epsilon_n \vec{\psi}_n, \quad \vec{\psi}_n^\dagger \vec{\psi}_m = \delta_{nm}. \tag{9.3}$$

This allows us to diagonalize the dissipator according to

$$\mathcal{D}_{\mathrm{PDC}} = \sum_{n=1}^{2N} \mathcal{D}[L_n], \quad L_n \equiv \vec{\psi}_n^{\dagger} \Phi = \sum_{j=1}^{2N} \left(\vec{\psi}_n\right)_j^* \Phi_j. \tag{9.4}$$

These diagonal Lindblad operators L_n are defined by the eigenvectors of $\mathbf{H}_{\mathrm{FKC}}$. The FKC topological phase transition is then detected as follows. In the trivial phase, $|\mu/2J| > 1$, the eigenvectors $\vec{\psi}_n$ are "in the bulk", that is, they are delocalized standing waves spread uniformly throughout the chain. It follows that the L_n have the same spatial distribution. In the non-trivial phase $|\mu/2J| < 1$, two edge-localized eigenvectors emerge, say $\vec{\psi}_1$ and $\vec{\psi}_2$. It follows that the Lindblad operators L_1 and L_2 are edge-localized. The topological phase transition of the FKC mandates a dramatic change in the spatial distribution of the Lindblad operators. We then ask, how does this dramatic change reflect in the actual dissipative dynamics of the bosonic system?

The dynamical matrix of this model is given by $\mathbf{G}_{\mathrm{PDC}} = -i\boldsymbol{\tau}_3\mathcal{F}(\mathbf{M}_{\mathrm{PDC}}) = -i\boldsymbol{\tau}_3\mathbf{H}_{\mathrm{FKC}}$. Consequently, the Bloch dynamical matrix may be computed from the Bloch Hamiltonian of the FKC (Eq. (5.11)), i.e.,

$$\mathbf{g}(k) = -i\boldsymbol{\sigma}_3\mathbf{h}_{\mathrm{FKC}}(k) = i(\mu + 2J\cos(k))\mathbb{1}_2 + 2\Delta\sin(k)\boldsymbol{\sigma}_1 \tag{9.5}$$

The two rapidity bands are

$$\lambda_\pm(k) = \mu + 2J\cos(k) \pm 2i\Delta\sin(k), \tag{9.6}$$

and bulk stability gap is $\Delta_S^{\mathrm{BIBC}} = \Delta_S^{\mathrm{SIBC}} = \mu + 2|J|$. The bulk is (un)stable if $\mu + 2|J| < 0$ (> 0). Bulk stability is then excluded if the underlying FKC Hamiltonian is topological. In fact, one may verify that the rapidity bands wind about the origin if and only if $|\mu/2J| < 1$. Thus, if topological metastability is to emerge, it must be in the same parameter regime as the non-trivial phase of the FKC.

To diagnose topological metastability, we require the OBC rapidity spectrum. It turns out that this may be determined analytically (see Appendix B.3) and is given by

$$\lambda_n = \mu + 2i\sqrt{\Delta^2 - J^2}\cos\left(\frac{n\pi}{N+1}\right), \qquad n = 0, \ldots, 2N-1. \tag{9.7}$$

Henceforth, we will take $\Delta \geq J > 0$ so that the OBC stability gap is $\Delta_{S,N}^{\mathrm{OBC}} = \mu$. In particular, $\Delta_{S,N}^{\mathrm{OBC}} \to \mu$ for $N \to \infty$. We conclude that the chain is topologically metastable whenever $\mu < 0$ and $|\mu/2J| < 1$. This corresponds to "half" of the topological phase diagram of the FKC. For $\mu < 0$ and $|\mu/2J| > 1$, the system is anomalously relaxing, since $|\Delta_{S,N}^{\mathrm{OBC}} - \Delta_S^{\mathrm{SIBC}}| = 2|J| \neq 0$. See Fig. 9.1 for a superimposition of the FKC topological phase diagram and the PDC's stability phase diagram.

While we may generically compute MBs numerically, it is possible to determine analytical expressions for the special case $J = \Delta$. The unnormalized MB pair reads

$$\gamma_L = \sum_{j=1}^{N} \delta^{j-1} x_j, \quad \gamma_R \equiv \sum_{j=1}^{N} \delta^{N-j} p_j, \quad \delta \equiv \frac{-\mu}{2J}. \tag{9.8}$$

Immediately, we see that these modes have the exact same spatial distribution as the Majorana fermions in the FKC, Eqs. (5.13). This is to be expected by the general arguments of Sect. 7.2.1 which, in particular, guarantee that localized approximate ZMs of the fermionic Hamiltonian manifest as localized approximate ZMs of the corresponding QBL. General arguments about purely dissipative, topologically metastable systems further imply that if γ_L or γ_R is an approximate ZM, then it is also an approximate SG. We verify this explicitly:

$$\mathcal{L}^\star(\gamma_L) = -2J\delta^N x_N, \quad \mathcal{L}^\star([\gamma_L, A]) - [\gamma_L, \mathcal{L}^\star(A)] = -2iJ\delta^N [x_N, A], \ \forall A,$$

$$\mathcal{L}^\star(\gamma_R) = -2J\delta^N p_1, \quad \mathcal{L}^\star([\gamma_R, A]) - [\gamma_R, \mathcal{L}^\star(A)] = -2iJ\delta^N [p_1, A], \ \forall A.$$

The two equations on the left verify that each member of the pair is an approximate ZM, while the two on the right verify that each is also an approximate SG: they are non-split. Finally, the unnormalized MB algebra is $[\gamma_L, \gamma_R] = iN\delta^{N-1}$. Referring back to Sect. 8.3.1, we see that the PDC hosts exactly the non-split incarnations of the "typical" MBs previously analyzed. In particular, we may normalize them , $\gamma_{L,R} \mapsto \mathcal{M}(N)\gamma_{L,R}$, with $\mathcal{M}(N) = \delta^{-(N-1)/2}/\sqrt{N}$, so that (γ_L, γ_R) satisfy all of these desired criteria. In particular, the robustness property can be analytically verified (see Appendix B.4).

9.2 A Dissipative BKC

The previous model represents the simplest bosonic extension of Majorana fermions. However, the chain itself is atypical in the sense that a vanishing system Hamiltonian is rather non-generic. More generally, one may expect both coherent and incoherent processes encoded by a non-zero Hamiltonian and a non-zero dissipator, respectively. If such a system is topologically metastable, the MBs will be generically *non-split*. That is, the breakdown of Noether's theorem (in the sense of Sect. 6.1.2), and its partial restoration (in the sense of Sect. 7.1), are to be expected.

For an example of such a system, we will take the system Hamiltonian to be the BKC Hamiltonian with hopping and pairing plus uniform degenerate parametric amplification, i.e., Eq. (5.20). We will focus on the case where the hopping J and pairing Δ are related via $J \geq \Delta > 0$. Appended to this system will be two damping mechanisms, via the dissipator

$$\mathcal{D}_{\text{DBKC}} \equiv \sum_{j=1}^{N} \left(\kappa - \frac{(2\Gamma)^2}{\kappa} \right) \mathcal{D}[a_j] + \kappa \mathcal{D}[a_j + \frac{2\Gamma}{\kappa} a_{j+2}], \tag{9.9}$$

where $\kappa \geq 0$ is a uniform onsite damping rate and $\Gamma \geq 0$ is a next-nearest neighbor (NNN) damping rate. BCs are enforced by taking $a_{j+N} = 0$ (a_j) for OBCs (PBCs). One may verify that the GKS matrix is guaranteed to be positive-semidefinite if $\kappa \geq 2\Gamma$. We will refer to the overall five parameter model simply as a *dissipative bosonic Kitaev chain* (DBKC).

The Bloch dynamical matrix for this system is determined to be

$$\mathbf{g}_{\text{DBKC}}(k) \equiv -i(\kappa - iJ\sin(k) + 2\Gamma\cos(2k))\mathbb{1}_2 + i(\mu + \Delta\cos(k))\boldsymbol{\sigma}_1, \tag{9.10}$$

from which the rapidity bands follow

$$\lambda_{\pm}(k) \equiv -(\kappa \pm \mu) \mp \Delta\cos(k) - iJ\sin(k) - 2\Gamma\cos(2k). \tag{9.11}$$

For convenience, we will explore the properties of this system in two specific parameter regimes: (i) $\mu = 0$ and (ii) $\Gamma = 0$.

9.2.1 The Parameter Regime $\mu = 0$

When $\mu = 0$, Eq. (9.11) reveals that the two bands are degenerate. For this degenerate case, we refer to Fig. 8.1 for example rapidity bands and finite-system rapidities. With respect to the elliptical bands of the BKC Hamiltonian, the onsite damping introduces an overall shift (we will formalize this in the next section) while the NNN damping introduces more complex curvature. Computing $\max_k \text{Re}[\lambda_{\pm}(k)]$ yields the bulk stability gap

$$\Delta_S^{\text{BIBC}} = \Delta_S^{\text{SIBC}} = \begin{cases} -\kappa + \Delta - 2\Gamma, & \Gamma/\Delta < 1/8 \\ -\kappa + \Delta^2/16\Gamma + 2\Gamma & \Gamma/\Delta > 1/8 \end{cases} \tag{9.12}$$

Interestingly, the region where $\Delta_S^{\text{BIBC}} > 0$ is divided into two distinct sectors: one in which the two bands wind around the origin and one where neither do. This is illustrated in Fig. 9.2a. This means that, depending on the sign of the OBC stability gap $\Delta_{S,N}^{\text{OBC}}$, this parameter regime can potentially correspond to a topologically metastable phase and a non-topological metastable phase. Unfortunately, however, analytical determination of $\Delta_{S,N}^{\text{OBC}}$ is difficult due to the presence of the NNN term.

Numerical determination of the OBC rapidity spectrum (shown for one choice of parameters in Fig. 8.1) reveals that, generally, $\Delta_{S,N}^{\text{OBC}}$ is negative bounded away from the bulk stability gap as $N \to \infty$. In the case where the bulk stability gap is negative, this means the system is in an anomalously relaxing phase. When the

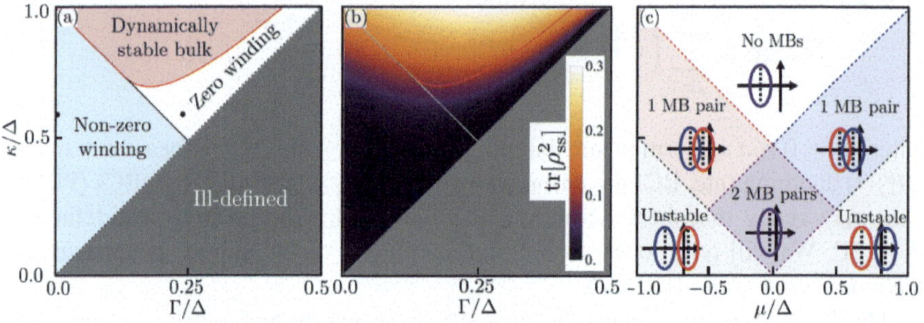

Fig. 9.2 (**a**) The topological phase diagram of the DBKC with $J \geq \Delta > 0$ and $\mu = 0$. The "Non-zero winding" region indicates the parameter regime where the rapidity bands wind around the origin. The "Zero winding" region indicates the parameter regime where neither band winds around the origin. If the OBC chain is dynamically stable in either of these two regions, then it is dynamically metastable, additionally being topologically metastable in the former case. The "Dynamically stable bulk" region is where the rapidity bands lie in the left-half plane. If the OBC chain is dynamically stable here, then it is anomalously relaxing. The "Ill-defined" region is where **M** is no longer positive-semidefinite. The black dots indicate the position of two representative parameter choices used in later figures. (**b**) For later reference, the steady state purity under OBCs in the same phase space as (**a**) with $J = 2$, $\Delta = 0.5$, and $N = 25$. (**c**) The topological stability phase diagram of the DBKC under OBCs with $\Gamma = 0$. The inset figures show representative rapidity band structure in each region. The number of bands winding around the origin determines the number of MB pairs. Earlier versions of (**a**) and (**c**) first appeared in Ref. [1]

bulk stability gap is positive, there are particular choices of parameters where both topological and non-topological dynamical metastability are present. Since the case $\Gamma = 0$, which we will consider in the next section, offers the cleanest platform for studying the topological case, we will focus on the non-topological dynamical metastability when $\Gamma > 0$. One example of such a parameter regime is the one used to generate Fig. 8.1b, which, corresponds to the rightmost point of Fig. 9.2a, i.e., $\Gamma/\Delta = 0.12$ and $\kappa/\Delta = 0.6$.

9.2.2 The Parameter Regime $\Gamma = 0$

When $\Gamma = 0$, the system is simply the BKC with uniform DPA of strength μ subjected to uniform on-site loss of rate κ. From Eq. (9.11), the rapidity bands consist of two ellipses in the complex plane. The major and minor axes are J and Δ, respectively. The center of the ellipse $\lambda_\pm$ is at the point $-\kappa \mp \mu$ and the corresponding bulk stability gap follows as

$$\Delta_S^{\mathrm{BIBC}} = \Delta_S^{\mathrm{SIBC}} = -\kappa + \Delta + |\mu|. \tag{9.13}$$

Thus, bulk stability requires $\kappa \geq \Delta + |\mu|$, i.e., sufficiently strong loss to overcome the coherent amplification mechanisms. The band $\lambda_\pm$ winds about the origin whenever $-\kappa \mp \mu < 0 < \Delta \mp \mu + \Delta$. For the simple case $\mu = 0$, this corresponds to $\kappa/\Delta < 1$. If $\mu \neq 0$, then the winding numbers of the bands are generally independent. In particular, an appropriate choice of μ can ensure one band winds around the origin while the other does not. Notably, the chiralities of the band are opposite: λ_+ winds counterclockwise while λ_- winds clockwise. Thus, if both bands wind about the origin, then the corresponding winding numbers are of opposite sign.

To diagnose topological metastability, we require the OBC rapidities. Thankfully, uniform onsite damping corresponds to a very simple GKS matrix:

$$\mathbf{M} = \kappa \bigoplus_{j=1}^{N} \begin{bmatrix} 1 & 0 \\ 0 & 0 \end{bmatrix}, \quad \mathcal{F}(\mathbf{M}) = \kappa \boldsymbol{\tau}_3 \tag{9.14}$$

Thus, the dynamical matrix is generally given by the BKC (with the uniform DPA term) Hamiltonian dynamical matrix, plus $-i\boldsymbol{\tau}_3\mathcal{F}(\mathbf{M}) = -i\kappa\mathbb{1}_{2N}$. Consequently, the rapidities (under any BCs) are given by $\lambda_n = -\kappa + i\omega_n$, with ω_n the normal mode frequencies of the Hamiltonian. For our model,

$$\lambda_n^{\pm} = -\kappa \mp \mu + i\sqrt{J^2 - \Delta^2}\cos\left(\frac{n\pi}{N+1}\right), \quad n = 1, \ldots, N. \tag{9.15}$$

The OBC stability gap is then $\Delta_{S,N}^{\text{OBC}} = -\kappa + |\mu|$, which differs from the bulk stability gap by a system-size independent constant Δ. With this, and the band winding conditions, we determine the $\Gamma = 0$ DBKC to be topologically metastable whenever $-\kappa + |\mu| < 0$ and at least one of $|(\kappa \pm \mu)/\Delta| < 1$. The second condition corresponds to the band $\lambda_\pm$ having nonzero winding. Otherwise, the system is anomalously relaxing.

Per the general arguments of Sect. 8.3.1, each non-trivial band will contribute a pair of MBs. Analogously to the previous model, we can compute them exactly in the special case $J = \Delta$. The two unnormalized MB pairs are now given by (γ_L^z, γ_R^s) and (γ_L^s, γ_R^z) where,

$$\gamma_L^z \equiv \sum_{j=1}^{N} \delta_-^{j-1} x_j, \qquad \gamma_R^s \equiv \sum_{j=1}^{N} \delta_-^{N-j} p_j \tag{9.16}$$

$$\gamma_L^s \equiv \sum_{j=1}^{N} \delta_+^{j-1} x_j, \qquad \gamma_R^z \equiv \sum_{j=1}^{N} \delta_+^{N-j} p_j \tag{9.17}$$

with $\delta_\pm = -(\mu \pm \kappa)/J$. The relevant non-vanishing commutators are $[\gamma_L^z, \gamma_R^s] = iN\delta_-^{N-1}$ and $[\gamma_L^s, \gamma_R^z] = iN\delta_+^{N-1}$. The MB's roles as approximate ZMs and SGs (indicated by z and s superscripts, respectively) follow from

$$\mathcal{L}^\star(\gamma_L^z) = -J\delta_-^N x_N, \quad \mathcal{L}^\star([\gamma_R^s, A]) - [\gamma_R^s, \mathcal{L}^\star(A)] = J\delta_-^N [p_1, A], \ \forall A.$$

$$\mathcal{L}^\star(\gamma_R^z) = J\delta_+^N p_1, \quad \mathcal{L}^\star([\gamma_L^s, A]) - [\gamma_L^s, \mathcal{L}^\star(A)] = -J\delta_+^N [x_N, A], \ \forall A.$$

Whenever δ_+ or δ_- has modulus less than one, the corresponding z/s mode is an approximate ZM / SG. This modulus condition, which also ensures exponential edge-localization, is precisely tied to the band winding condition: $|\delta_\pm| < 1$ if and only if $\lambda_\pm(k)$ winds around the origin. In particular, the number of MB pairs can be 0, 1, or 2. The difference in band chirality is reflected in which member of the ZM-SG pair is localized on which side. The $\Gamma = 0$ topological phase diagram is shown in Fig. 9.2c. Finally, we remark that the fact that there are regimes where the total winding number is non-vanishing (specifically, when only one of the $\delta_\pm$ has modulus less than 1) explicitly requires dissipation and is interesting from the perspective of Fredholm operator theory [3].

In Sect. 8.3.1, we made several predictions regarding the dynamical behavior of the quasi-steady states in topologically metastable systems. Since the DBKC offers us a rather generic instance of such a behavior, let's test these predictions. Consider specifically the quasi-steady states generated by the left approximate SG, i.e., $\rho_\theta^L \equiv e^{i\theta\gamma_L^s}\rho_{\rm ss}e^{-i\theta\gamma_L^s}$. Equations (8.33) and (8.34) assert that both the mean vector, as well as the expectation values of arbitrary linear observables, will remain within an ever-decreasing distance of their initial value for an ever-increasing amount of time as the system size is increased. This behavior is directly verified in Fig. 8.2b, c. Figure 8.2b specifically shows both the left hand-side of Eq. (8.33) as well as the left hand-side of Eq. (8.34) for randomly sampled choices of the operator α. Clearly, the increasing of system size leads to a larger suppression of $\vec{m}_\theta(t) - \vec{m}(0)$ for longer times. Consequently, the expectation values of observables follow suit. Figure 8.2c further bolsters this point, by demonstrating the dependence of $t(\delta)$, the time it takes for the left hand-side of Eq. (8.33) to exceed a prescribed accuracy $\delta > 0$, on system size. Asymptotically, $t(\delta)$ appears to increase linearly with N, regardless of δ. This linear behavior suggests that the topologically metastable transient time may be closely related to the time it takes for the state to 'detect' the presence of both boundaries. Specifically, the left-localized quasi-steady state needs to evolve long enough for the boundary on the right side to be detected. A standard Lieb-Robinson-type argument then leads to the expectation that this time should scale linearly with system size. In particular, as $N \to \infty$, the time it takes to detect the right boundary diverges and the quasi-steady state becomes truly steady. This is consistent with the undersanding that the semi-infinite rapidity spectrum, which contains zero, *imprints* itself into the pseudospectra of the finite system.

9.3 A Dissipative BKC with a Pure Steady State

9.3.1 Constructing the Model

The concluding analysis of the previous model shed light on the behavior of the quasi-steady states in a generic instance of a topologically metastable system. However, unlike the ground states of topological QFHs, steady states of Lindbladians are generally mixed states. For QBLs, the purity can be directly computed using the covariance matrix according to Eq. (2.16). It turns out that the steady states of all models considered thus far are non-pure (e.g., see Fig. 9.2b for the DBKC steady state purity under OBCs). This has prevented us from exploring the interplay between topological metastability and pure steady states (so-called "dark states" in quantum optics). The more general procedure detailed in Sect. 7.2.2 allows us, given a dynamically stable Hamiltonian with quasiparticle vacuum $|\widetilde{0}\rangle$, to engineer a dissipator that relaxes any initial condition to the state $|\widetilde{0}\rangle$. Such a stabilization procedure is especially interesting when the vacuum state possesses certain nontrivial properties such as non-zero squeezing, as is the case with the BKC. Can topological metastability arise in such a system? The answer is Yes, as we will now show.

Consider the Lindbladian defined via the BKC Hamiltonian (with $\mu = 0$ for simplicity) and the site-local dissipator

$$\mathcal{D} = 2\kappa \sum_{j=1}^{N} \mathcal{D}[\beta_j(r)], \tag{9.18}$$

where $\kappa \geq 0$ and we have introduced squeezed bosonic degrees of freedom

$$\beta_j(r) \equiv \cosh(jr)a_j - \sinh(jr)a_j^\dagger. \tag{9.19}$$

The squeezing parameter r is the same as defined in Sect. 3.3.3, i.e., $e^{2r} = (J + \Delta)/(J - \Delta)$. This represents a local dissipative process of the squeezed degrees of freedom with constant loss rate 2κ. A number of remarks are in order.

(1) We claim that the dissipator in Eq. (9.18) is precisely the output of the procedure detailed in Sect. 7.2.2 when applied to the BKC Hamiltonian. To see this, we will utilize Eq. (6.6) to implicitly determine the GKS matrix via the "parent Hamiltonian" $\sum_{j=1}^{N} L_j^\dagger L_j$ (in the langauge of Ref. [4]) with the choice of Lindblad operators $L_j \equiv \sqrt{2\kappa}\,\beta_j(r)$. We proceed by noting that $\beta_j(r) = (\mathbf{R}(r)\Phi)_j$ in terms of the duality transformation $\mathbf{R}(r)$ of Sect. 4.2.2. The parent Hamiltonian is then

$$\sum_{j=1}^{N} L_j^\dagger L_j = 2\kappa \sum_{j=1}^{N} \alpha_j^\dagger(r)\beta_j(r) = 2\kappa \sum_{j=1}^{N} \sum_{k,\ell=1}^{2N} \mathbf{R}(r)^*_{jk}\mathbf{R}_{j\ell}(r)\Phi_k^\dagger \Phi_\ell \tag{9.20}$$

The sum over j can be simplified by utilizing a number of facts. First, we will leverage three properties of $\mathbf{R}(r)$ (recall Chap. 4): (i) Hermiticity, (ii) $\mathbf{R}^2(r) = \mathbf{S}(r)$, and (iii) $\mathbf{R}(r)\boldsymbol{\tau}_3\mathbf{R}(r) = \boldsymbol{\tau}_3$. Second, note that a sum from $j = 1$ to N can be extended to a sum from $j = 1$ to $2N$ by inserting a coefficient $(\mathbb{1}_{2N} + \boldsymbol{\tau}_3)_{jj}/2$, which is 1 for $j \leq N$ and 0 for $j > N$. Bringing these together reveals

$$\begin{aligned}\sum_{j=1}^{N} \mathbf{R}(r)^*_{jk}\mathbf{R}_{j\ell}(r) &= \sum_{j=1}^{N} \mathbf{R}(r)_{kj}\mathbf{R}_{j\ell}(r) = \frac{1}{2}\sum_{j=1}^{2N} \mathbf{R}(r)_{kj}(\mathbb{1}_{2N} + \boldsymbol{\tau}_3)_{jj}\mathbf{R}_{j\ell}(r)\\ &= \frac{1}{2}\sum_{j=1}^{2N} \mathbf{R}(r)_{kj}\mathbf{R}(r)_{j\ell} + \mathbf{R}(r)_{kj}(\boldsymbol{\tau}_3)_{jj}\mathbf{R}_{j\ell}(r)\\ &= \frac{1}{2}(\mathbf{S}(r) + \boldsymbol{\tau}_3)_{k\ell}\,.\end{aligned}$$

The "parent Hamiltonian" is then

$$\sum_{j=1}^{N} L_j^\dagger L_j = \kappa \sum_{k,\ell=1}^{2N} (\mathbf{S}(r) + \boldsymbol{\tau}_3)_{k\ell}\, \Phi_k^\dagger \Phi_\ell, \tag{9.21}$$

from which we infer $\mathbf{M} = \kappa\,(\mathbf{S}(r) + \boldsymbol{\tau}_3)$, i.e., the GKS matrix output from the recipe of Sect. 7.2.2. Following Eq. (7.26) the dissipator has a particularly simple representation in terms of the quasiparticles of the BKC Hamiltonian under OBCs, Eq. (3.37). Explicitly, these quasi-particles are 'standing-waves' in the squeezed basis $\beta_j(r)$,

$$\psi_n = \sqrt{\frac{2}{N+1}} \sum_{j=1}^{N} i^{-j} \sin\left(\frac{n\pi j}{N+1}\right) \beta_j(r). \tag{9.22}$$

From here, the dissipator is given by

$$\mathcal{D} = 2\kappa \sum_{n=1}^{N} \mathcal{D}[\psi_n]. \tag{9.23}$$

As expected, the chain is a set of N decoupled dissipative harmonic oscillators with mode-varying frequencies ω_n and uniform damping κ in the quasiparticle basis.[2] Equivalently, we have an alternative choice of Lindblad operators $L_n =$

[2] The fact that the dissipator more than one diagonal representation, Eqs. (9.18) and (9.23), results from a large degeneracy in the GKS matrix spectrum. In particular, $\mathbf{M}$ has an N-fold degenerate 0 eigenvalue. Different diagonal representation of $\mathcal{D}$ are derived from different choices of bases for the 0 eigenvalue eigenspace.

$\sqrt{2\kappa}\,\psi_n$. Ultimately, the steady state is $\rho_{ss} = |\widetilde{0}\rangle\langle\widetilde{0}|$ in terms of the BKC quasiparticle vacuum $|\widetilde{0}\rangle$ of Eq. (3.39). We further note that dissipators diagonal in the normal mode basis of the system Hamiltonian often arise in microscopic derivations of quadratic master equations [5].

(2) The dissipator is not bulk-translationally invariant. Explicitly, the squeezing of the mode $\beta_j(r)$ increases with j. Remarkably, however, the dynamical matrix $\mathbf{G} = \mathbf{G}\big|_{\kappa=0} - i\kappa\mathbb{1}_{2N}$ *is* translationally invariant, up to BCs. This is one particular incarnation of the "restricted symmetry" phenomena detailed in Sect. 7.2.2. We then say this model has a *restricted translational symmetry*. Furthermore, the dynamical matrix of this model is precisely equivalent to that of the DBKC with $\Gamma = \mu = 0$ considered previously. This allows for a straightforward computation of rapidities and pseudospectra. In particular, the chain is topologically metastable for $|\kappa/\Delta| < 1$ and is anomalously relaxing for $|\kappa/\Delta| > 1$. There are two MB pairs in the topologically metastable regime.

Since the steady state is the quasi-particle vacuum of $H_{\rm BKC}$, it is necessarily a function of the Hamiltonian parameters J and Δ only (in addition to system size N). In particular, it is insensitive to κ, and hence to the topology of the rapidity bands. This reveals a remarkable fact about topological metastability: its consequences need not be reflected in the structure of the steady state at all. Topological metastability is truly a *transient* phenomena.

9.3.2 Relaxation Dynamics of the Quasi-Steady States

Quasi-steady states of topoloigcally metastable chains are characterized as Weyl displacements, in the direction of the edge-SG, of the steady state. Let us explore the properties of these states given that our steady state is pure. Let γ^s be one of two edge-symmetries γ_L^s or γ_R^s of the DBKC. Following Sect. 8.3.1, we construct the quasi-steady states $\rho_\theta \equiv e^{i\theta\gamma^s}|\widetilde{0}\rangle\langle\widetilde{0}|e^{-i\theta\gamma^s}$, which are also pure. In fact, the states $|\vec{\alpha}(\theta), \{\psi_\mu\}\rangle \equiv e^{i\theta\gamma^s}|\widetilde{0}\rangle$, which are generically squeezed coherent states with respect to the physical degrees of freedom $\{a_j\}$, can be interpreted as coherent states with respect to the squeezed bosonic normal modes $\{\psi_\mu\}$. Explicitly,

$$\psi_\mu\,|\vec{\alpha}(\theta), \{\psi_\mu\}\rangle = \alpha_\mu(\theta)\,|\vec{\alpha}(\theta), \{\psi_\mu\}\rangle\,. \tag{9.24}$$

The complex numbers $\alpha_\mu(\theta)$ encode the extent to which the normal modes commute with the SG, i.e., $[\psi_\mu, \gamma^s] = i\theta\vec{\psi}_\mu^\dagger\boldsymbol{\tau}_3\vec{\gamma}^s\,1_F = \alpha_\mu(\theta)1_F$ where $\vec{\psi}_\mu$ and $\vec{\gamma}^s$ are the Nambu representations of ψ_μ and γ^s, respectively. As in the general case, the mean vector $\vec{m}(t) = \langle\widetilde{0}|\Phi(t)|\widetilde{0}\rangle = i\theta\vec{\gamma}^s(t)$ is long-lived in the sense of Fig. 8.2b. Due to the correspondence between dynamical matrices, this is precisely the same mean vector that the *mixed* quasi-steady states of the DBKC give rise to.

Identifying the quasi-steady states as coherent states evolving under decoupled dissipative harmonic dynamics (in the normal-mode basis) affords us the ability to compute their dynamics exactly [6, 7]. In particular, we can analyze their relaxation dynamics analytically. As they evolve, the states remain coherent, i.e., $\rho_\theta(t) = |\vec{\alpha}(\theta, t), \{\psi_\mu\}\rangle \langle \vec{\alpha}(\theta, t), \{\psi_\mu\}|$, with amplitudes

$$\alpha_\mu(\theta, t) = \alpha_\mu(\theta) e^{-(\kappa + i\omega_\mu)t} = \langle \psi_\mu \rangle (t). \tag{9.25}$$

Almost paradoxically, the amplitudes relax to equilibrium exponentially fast. This paradox is resolved by noting that these amplitudes correspond to the expectation values of ψ_μ, which, due to squeezing, have exponentially large coefficients when expressed in the basis of the physical relevant degrees of freedom $\{a_j\}$. This explains why the normal-mode amplitudes decay exponentially, while the mean vector has an increasingly long lifetime. Although this property remains consistent regardless of whether the left-localized or right-localized SG is chosen, the specific way in which the relaxation dynamics occur, at the level of the many-body states, may differ.

Consider the trace distance between two quantum states $T(\rho, \sigma) = (1/2) \|\rho - \sigma\|_1$, which provides a measure of distinguishability between ρ and σ. Thanks to our knowledge of the exact dynamics, we can study the relaxation time of the quasi-steady states via the distance from equilibrium

$$T(\rho_\theta(t), \rho_{\mathrm{ss}}) = \sqrt{1 - |\langle \vec{\alpha}(\theta, t), \{\psi_\mu\}|\widetilde{0}\rangle|^2} = \sqrt{1 - \exp\left(-\|\vec{\alpha}(\theta, 0)\|^2 e^{-2\kappa t}\right)},$$

where $\vec{\alpha}(\theta, t)$ is either $\vec{\alpha}_L(\theta, 0)$ or $\vec{\alpha}_R(\theta, 0)$ depending which SG γ_L^s or γ_R^s is used to generate the quasi-steady states. We then define the relaxation time to be the time $t_{\mathrm{rel}}(\delta)$ such that the relative distance from equilibrium $T(\rho_\theta(t), \rho_{\mathrm{ss}})/T(\rho(0), \rho_{\mathrm{ss}})$ falls (and remains) below a prescribed accuracy $\delta > 0$. A straightforward calculation yields

$$\kappa t_{\mathrm{rel}}(\delta) = \frac{1}{2} \ln \left[\frac{\|\vec{\alpha}(\theta, 0)\|^2}{\ln \left[\left(1 - \delta(1 - e^{-\|\alpha(\vec{\theta}, 0)\|^2})\right)^{-1} \right]} \right]. \tag{9.26}$$

The system-size scaling of $t_{\mathrm{rel}}(\delta)$ is thus explicitly tied to the system-size scaling of the norm of the initial amplitude vectors $\vec{\alpha}(\theta, 0)$. Remarkably, the two manifolds of quasi-steady states display dramatically different relaxation dynamics, as inferred from their relaxation times. We find, numerically, that $\|\vec{\alpha}_L(\theta, 0)\|$ increases exponentially with N, while $\|\vec{\alpha}_R(\theta, 0)\|$ decreases exponentially. The consequences of this observation are clearly displayed in Fig. 9.3: The left-localized quasi-steady state displays a increasingly long relaxation time, while the right-localized shows just the opposite. The asymmetry in the dynamics may be understood by noting that, while the dynamical matrix is translationally invariant (up to boundaries), the full generator is not. Specifically, the Lindblad operators $L_j = \sqrt{2\kappa}\, \beta_j(r)$ are right localized.

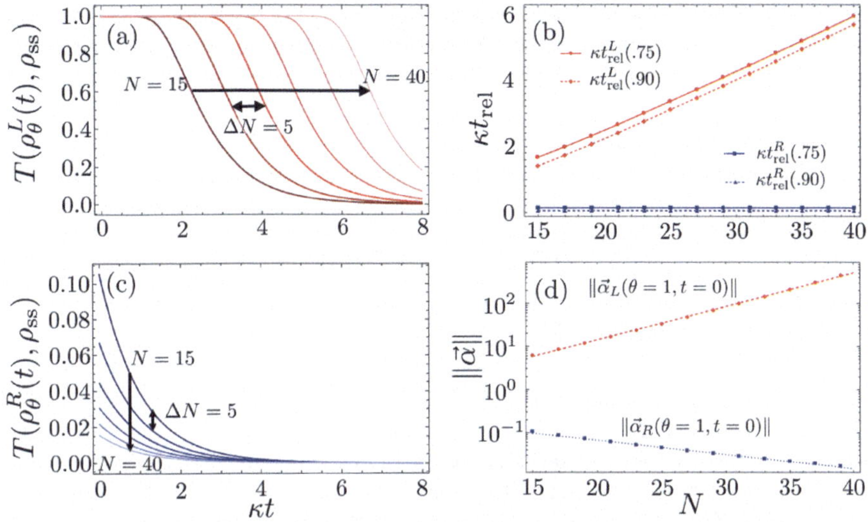

Fig. 9.3 (**a**) The relaxation dynamics of the left-localized quasi-steady state as N increases. (**b**) The same as in (**a**), but for the right-localized quasi-steady state. (**c**) The relaxation times $t_{\rm rel}(\delta)$ for the left (red) and right (blue) localized quasi-steady states for accuracies $\delta = 0.75$ (solid) and $\delta = 0.90$ (dashed). (**d**) The scaling of the amplitude vector norms for the left (red) and right(blue) localized quasi-steady states. In all cases, we take $\theta = 1$, symmetrically normalize the MBs, and set $J = 2$, $\Delta = 0.5$, and $\kappa = 0.3$

The boundedness of the relaxation times for the right-localized state does not contradict our claim of long-livedness. Per the general theory, the lifetime of the *physically accessible* degrees of freedom $\vec{m}^R_\theta(t)$ remain indistinguishable from their initial value $\vec{m}^R_\theta(0)$ (again, in the sense of Fig. 8.2b). On the other hand, the quasi-steady state and the true steady state have overlap exponentially close to one as N increases. The resolution comes by noting that the macroscopic expectation values $\vec{m}^R_\theta(t)$ are obtained from the exponentially small normal mode amplitudes $\vec{\alpha}_R(\theta, t)$ via the squeezing transformation taking a_j to ψ_n. This transformation dilates the exponentially small amplitudes into the physical amplitudes. In the same vein, the same transformation squeezes the exponentially large normal mode amplitudes $\vec{\alpha}_L(\theta, t)$ into the physical amplitudes $\vec{m}^L_\theta(\theta, t)$.

9.3.3 *Transient Odd-Parity Behavior*

To conclude our analysis of this model, we pose the question: Do the pure quasi-steady states span a subspace of long-lived states? That is, do linear combinations of these pure states provide long-lived states themselves? This question is interesting from two perspectives. Firstly, from the perspective of continuous variable quantum information, linear combinations of coherent states provide the simplest realization of bosonic cat-codes [8, 9]. Thus, the ability to potentially dissipatively prepare and sustain such states is a powerful one. Secondly, from the perspectives of condensed-matter physics and quantum optics, bosonic ground states and steady states are intrinsically averse to odd parity. Specifically, if a QBH (QBL) has a unique ground (steady) state, then it must have even bosonic parity (see Secs. 2.3 and 6.2.1). As we have seen in Sect. 5.2.2, this eliminates the possibility for the parity switching behavior characteristic of topological superconductor, in QBHs. Moreover, negative expectation values of parity signify quantum non-Gaussianity (recall Eq. (2.41)).

In general, the question is difficult to answer. However, we can make interesting statements about particularly relevant linear combinations. Concretely, let's focus our attention to the "cat states"

$$|\mathcal{C}_\phi(\vec{\alpha})\rangle = \mathcal{N}_\phi(\vec{\alpha})(|\vec{\alpha}(\theta), \{\psi_\mu\}\rangle + e^{i\phi}\,|-\vec{\alpha}(\theta), \{\psi_\mu\}\rangle), \tag{9.27}$$

where $\mathcal{N}_\phi(\vec{\alpha})$ is a normalization constant. Being generally non-Gaussian, these states are not simply characterized by their first and second moments. Moreover, computation of overlaps becomes drastically more difficult than the previous case. Instead of a direct analysis of relaxation times, let us focus explicitly on parity dynamics,

$$\langle P\rangle_{\phi,\vec{\alpha}}\,(t) = \langle\mathcal{C}_\phi(\vec{\alpha})|P(t)|\mathcal{C}_\phi(\vec{\alpha})\rangle\,, \tag{9.28}$$

with P the bosonic parity operator, Eq. (2.13). Once again, our simple characterization of the dynamics affords us an exact analytical result:

$$\langle P\rangle_{\phi,\vec{\alpha}}\,(t) = \frac{e^{-2\|\vec{\alpha}\|^2} + \cos(\phi)e^{-2\|\vec{\alpha}\|^2(1-e^{-2\kappa t})}}{1+\cos(\phi)e^{-2\|\vec{\alpha}\|^2}}. \tag{9.29}$$

We explain how to derive this formula in the Appendix B.5. Here we will focus on its physical implications. Again, which quasi-steady state manifold is chosen dramatically affects the displayed behavior. If $\vec{\alpha} = \vec{\alpha}_L(\theta, 0)$, the parity shows a dramatic dependence on both N and ϕ, as seen in Fig. 9.4. Unless $\phi = \pm\pi/2$, the parity drops extremely fast to zero, remains zero for a transient time that scales linearly with system size, and then eventually increases to its asymptotic value of 1. The extremely fast initial drop to zero corresponds to a singularity in the derivative,

$$\langle\dot{P}\rangle_{\phi,\vec{\alpha}}\,(0) = -4\,\|\vec{\alpha}_L(\theta,0)\|^2\left(\frac{e^{-2\|\vec{\alpha}\|^2} - \cos(\phi)}{1+e^{-2\|\vec{\alpha}\|^2}\cos(\phi)}\right), \tag{9.30}$$

as N (and hence $\|\vec{\alpha}\|$) goes to infinity. The transient state of zero parity that follows is interesting, as it represents a long-lived period where the measurement statistics of parity are split evenly between the $+1$ and -1 outcomes. While the system does not sustain a state of odd parity deterministically, it does, in fact, sustain a state equally distributed between even and odd parity sectors.

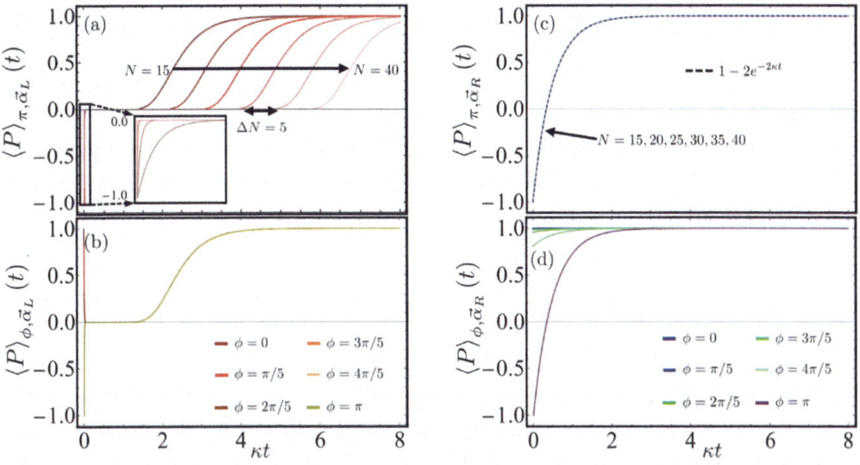

Fig. 9.4 (**a**) The parity of the odd-parity cat state formed from the left-localized quasi-steady states for increasing N. Inset: the very short time dynamics of the parity. (**b**) The parity of the cat state formed from the left-localized quasi-steady states for fixed $N = 25$ and varying ϕ. (**c**) The parity of the odd-parity cat state formed from the right-localized quasi-steady states for increasing N. The dashed line indicates the $N \to \infty$ limit. (**d**) The parity of the cat state formed from the right-localized quasi-steady states for fixed $N = 25$ and varying ϕ. In all cases, we take $\theta = 1$, symmetrically normalize the MBs, and set $J = 2$, $\Delta = 0.5$, and $\kappa = 0.3$

In sharp contrast, if $\vec{\alpha} = \vec{\alpha}_R(\theta, 0)$, the parity dynamics are much more well-behaved. The exponentially small norms of $\vec{\alpha}_R(\theta, 0)$ ensure that, for sufficiently large, N,

$$\lim_{N\to\infty} \langle P \rangle_{\phi,\vec{\alpha}_R}(t) = \begin{cases} 1, & \phi \neq \pi, \\ 1 - 2e^{-2\kappa t}, & \phi = \pi. \end{cases} \tag{9.31}$$

Unless $\phi = \pi$, the parity approaches 1 for all t as N increases. Moreover, when $\phi = \pi$, the parity is indistinguishable from $1 - e^{-2\kappa t}$ for sufficiently large N. Regardless, unlike their left-localized partners, these right-localized cat states fail to support any semblance of an odd-parity state for any meaningful amount of time.

9.4 Dirac Bosons in a Number-Symmetric Dissipative Chain

Let us explore the interplay between number symmetry and topological metastability in a concrete example. The dynamical matrix of a number symmetric QBL has the general form Eq. (8.35), where $\mathbf{K}$ is an arbitrary $N \times N$ complex matrix. Topological metastability can then be engineered through an appropriate choice of $\mathbf{K}$. For concreteness, we consider the example

$$\mathbf{K} = -i\kappa \mathbb{1}_N + J_L \mathbf{S} + J_R \mathbf{S}^\dagger, \tag{9.32}$$

with $\kappa \geq 0$, $J_L, J_R \in \mathbb{R}$, and $\mathbf{S}$ the usual BC-dependent shift operator. We identify this matrix as that of the Hatano-Nelson asymmetric hopping model [10], with an identity shift that will ultimately serve to stabilize the QBL. For convenience, we define $J_\pm = (J_L \pm J_R)/2$. The Hamiltonian can be unambiguously determined from $\mathbf{G}$ and is given by

$$H_{\text{NS}} = \frac{J_+}{2} \sum_{j=1}^{N} \left(a_j^\dagger a_{j+1} + a_{j+1}^\dagger a_j \right). \tag{9.33}$$

Per usual, the QBL is not fully determined until we specify $\mathcal{B}(\mathbf{M})$. Moreover, the second necessary condition for the U(1) symmetry is that $\boldsymbol{\tau}_3 \mathcal{B}(\mathbf{M})$ commutes with $\boldsymbol{\tau}_3$. Together, $[\mathbf{G}, \boldsymbol{\tau}_3] = 0$ and $[\boldsymbol{\tau}_3 \mathcal{B}(\mathbf{M}), \boldsymbol{\tau}_3] = 0$ are necessary and sufficient for the U(1) symmetry. We specify $\mathcal{B}(\mathbf{M})$ implicitly by defining the dissipator $\mathcal{D}_{\text{NS}} = \mathcal{D}_{-,0} + \mathcal{D}_{+,0} + \mathcal{D}_{-,1}$ with

$$\mathcal{D}_{-,0} = 2\kappa_- \sum_{j=1}^{N} \mathcal{D}[a_j], \quad \mathcal{D}_{+,0} = 2\kappa_+ \sum_{j=1}^{N} \mathcal{D}[a_j^\dagger], \tag{9.34}$$

$$\mathcal{D}_{-,1} = 2i J_- \sum_{j=1}^{N} \mathcal{D}[a_j, a_{j+1}^\dagger] - \mathcal{D}[a_{j+1}, a_j^\dagger]. \tag{9.35}$$

Here, $2\kappa_- \geq 0$ and $2\kappa_+ \geq 0$ are the onsite loss and gain rates, respectively, while $2J_-$ takes the role of the nearest-neighbor loss rate. The GKS matrix will be positive-semidefinite for OBCs and PBCs, and for all N, if $\kappa_- \geq 2|J_-|$. This QBL has a dynamical matrix specified by Eq. (9.32) if we further identify $\kappa \equiv \kappa_- - \kappa_+$. We refer to this model as the dissipative number-symmetric (DNS) chain. The steady state behavior of a related model has been considered in Ref. [11].

The rapidities can be obtained straightforwardly from the well-known Hatano-Nelson spectrum and closely resemble that of the DBKC. For BIBCs, the bands are given by $\{\lambda(k), \lambda(k)^*\}$, with

$$\lambda(k) = -\kappa + 2J_- \sin(k) + i2J_+ \cos(k). \tag{9.36}$$

The bands trace out an ellipse centered at $-\kappa$ in the complex plane. Winding about the origin requires $2|J_-| > \kappa$. For OBCs, the eigenvalues are given by λ_m, $m = 0, \ldots, 2N-1$, where

$$\lambda_m = -\kappa + 2i\sqrt{J_+^2 - J_-^2} \cos\left(\frac{m\pi}{N+1}\right). \tag{9.37}$$

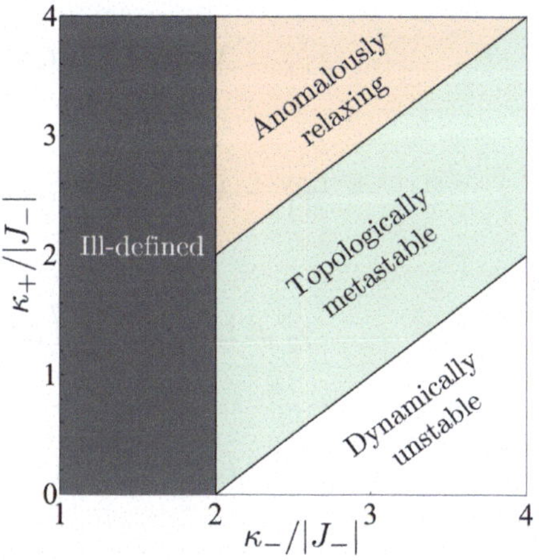

Fig. 9.5 The stability phase diagram for the DNS under OBCs with $|J_+| > |J_-|$. The "Ill-defined" region corresponds to the parameter regime where $\mathbf{M}$ is no-longer positive-semidefinite

To simplify the discussion (and fix the OBC Lindblad gap $\Delta_{\mathcal{L}} = \kappa$, for all N), we focus on the case $|J_+| \geq |J_-|$. Combining the GKS matrix positivity condition and the rapidity band winding condition, we identify a topologically metastable regime whenever $\kappa_-/|J_-| \geq 2$ and $0 \leq \kappa/2|J_-| \leq a1$. The OBC stability phase diagram is shown in Fig. 9.5.

In the special case $J_R = 0$, the pseudoeigenvectors of $\mathbf{K}$ (and $\mathbf{K}^\dagger$) with zero pseudoeigenvalue may be computed analytically. Such pseudoeigenvectors can be used to build approximate kernel vectors of $\mathbf{G}$ and $\widetilde{\mathbf{G}} = \mathbf{G}^\dagger$, which correspond to approximate SGs and ZMs, respectively. Specifically, consider the bosonic modes

$$\alpha \equiv \mathcal{M}(N) \sum_{j=1}^{N} (i\delta)^{N-j} a_j, \quad \beta \equiv \mathcal{M}(N) \sum_{j=1}^{N} (-i\delta)^{j-1} a_j, \tag{9.38}$$

where $\delta \equiv -\kappa/2J_-$ and $\mathcal{M}(N) \equiv \sqrt{(1-\delta^2)/(1-\delta^{2N})}$. First and foremost, we have $[\alpha, \alpha^\dagger] = [\beta, \beta^\dagger] = 1_{\mathcal{F}}$ and $[\alpha, \beta] = 0$. Second, these operators satisfy

$$\mathcal{L}^\star(\alpha) = -\kappa \mathcal{M}(N) \, (-i\delta)^{N-1} a_1, \tag{9.39}$$

$$\mathcal{L}^\star([\beta, A]) - [\beta, \mathcal{L}^\star(A)] = -\kappa \mathcal{M}(N) \, (i\delta)^{N-1} [a_N, A], \quad \forall A. \tag{9.40}$$

Utilizing $\mathcal{L}^\star(A^\dagger) = [\mathcal{L}^\star(A)]^\dagger$ yields similar expressions for $\alpha^\dagger$ and $\beta^\dagger$. For $|\delta| < 1$, the right hand-side of each equation goes to zero as $N \to \infty$. Physically, this means α is a bosonic approximate ZM, while the real and imaginary quadratures of β generate approximate symmetries. That is, α and β are the Dirac bosons of Sect. 8.3.2.

We may trace these two bosonic modes back to two pairs of MBs using Eqs. (8.37). Specifically, we have

$$\gamma_1^z = \mathcal{M}_z(N) \sum_{j=1}^{N} \left((i\delta)^{N-j} a_j + (-i\delta)^{N-j} a_j^\dagger \right),$$

$$\gamma_1^s = \mathcal{M}_s(N) \sum_{j=1}^{N} \left((-i\delta)^{j-1} a_j + (i\delta)^{j-1} a_j^\dagger \right), \tag{9.41}$$

$$\gamma_2^z = i\mathcal{M}_z(N) \sum_{j=1}^{N} \left((i\delta)^{N-j} a_j - (-i\delta)^{N-j} a_j^\dagger \right),$$

$$\gamma_2^s = i\mathcal{M}_s(N) \sum_{j=1}^{N} \left((-i\delta)^{j-1} a_j - (i\delta)^{j-1} a_j^\dagger \right), \tag{9.42}$$

for some normalization constants $\mathcal{M}_z(N)$ and $\mathcal{M}_s(N)$ chosen to ensure $[\gamma_j^z, \gamma_j^s] = i1_{\mathcal{F}}$ for $j = 1, 2$. The relevant equations of motion follow from Eqs. (9.39)–(9.40), that is

$$\mathcal{L}^\star(\gamma_1^z) = -\kappa\mathcal{M}_z(N) \left((-i\delta)^{N-1} a_1 + (i\delta)^{N-1} a_1^\dagger \right), \tag{9.43}$$

$$\mathcal{L}^\star(\gamma_2^z) = -i\kappa\mathcal{M}_z(N) \left((-i\delta)^{N-1} a_1 - (i\delta)^{N-1} a_1^\dagger \right), \tag{9.44}$$

$$\mathcal{L}^\star([\gamma_1^s, A]) - [\gamma_1^s, \mathcal{L}^\star(A)] = -\kappa\mathcal{M}_s(N)[(i\delta)^{N-1} a_N + (-i\delta)^{N-1} a_N^\dagger, A], \tag{9.45}$$

$$\mathcal{L}^\star([\gamma_2^s, A]) - [\gamma_2^s, \mathcal{L}^\star(A)] = -i\kappa\mathcal{M}_s(N)[(i\delta)^{N-1} a_N - (-i\delta)^{N-1} a_N^\dagger, A]. \tag{9.46}$$

Following the discussion of Sect. 8.3.1, there exists normalization schemes that ensure the right hand-sides of these four equations vanish as $N \to \infty$, while keeping each pair canonically conjugate (e.g., the symmetric normalization scheme).

We remark that these pairs of MBs are not simply the real and imaginary quadratures of α and β, i.e., the operators $x_\alpha \equiv (\alpha + \alpha^\dagger)/\sqrt{2}$, $p_\alpha \equiv i(\alpha - \alpha^\dagger)/\sqrt{2}$, and similarly for β. Instead, the MBs are proportional to these quadratures (e.g., $\gamma_1^z = \sqrt{2}\mathcal{M}_z(N) x_\alpha$ and $\gamma_2^z = \sqrt{2}\mathcal{M}_s(N) p_\alpha$). The proportionality constants ensure that the macroscopically separated pairs (γ_1^z, γ_1^s) and (γ_2^z, γ_2^s) are canonically conjugate. We further remark that each edge supports two of the same type Noether mode: the left edge supports two SGs, while the right edge supports two ZMs. This is to be contrasted with the previous models, which all featured at most one of each type on each edge.

9.5 Observable Signatures

In a traditional equilibrium statistical mechanics setting, quantum phase transitions are accompanied by critical behavior, i.e., the presence of long range (typically algebraically decaying) correlation functions. More specifically, these correlation functions are evaluated at equal times and are of the form $C(j, r, t) = \langle A_j(t)A_{j+r}(t)\rangle$ taken with respect to some equilibrium state (e.g., the ground state). The equilibrium assumptions allows one to take $C(j, r, t) = C(j, r, 0)$, so that time-dependence is removed. On the one hand, it is natural to expect that correlation functions can detect the dynamical phases present in our QBLs. On the other hand, haphazardly transplanting this concept into our setting would be inappropriate. For our systems, the only 'equilibrium state' (in the sense of having trivial dynamics) is the steady state. However, dynamical (both topological and non-topological) metastability is a *transient* phenomenon and thus need not have any impact on the steady state. In fact, the pure steady state of the model in Sect. 9.3 completely lacked any dependence on the dissipative parameter κ, and thus, was completely decoupled from the topological phase diagram. For these reasons, we will instead investigate *multitime correlation functions*. This will allow us to probe the dynamical features of correlations in our system which, nominally, should be privy to the metastable phases of interest.

Beyond their general interest for characterizing Markovian systems [12–14], specific instances of multitime correlations arise frequently in the quantum optical settings in the form of *coherence functions*. For example, first order coherence functions $g_{ij}^{(1)}(\tau)$ are proportional to the two-time correlation function between creation and annihilation operators, i.e., $\langle a_i^\dagger(t)a_j(t+\tau)\rangle$. Such quantities are experimentally accessible via photon counting and interference experiments [15, 16]. Thus, it is conceivable that the specific correlations we will consider could be probed in certain experimental settings.

9.5.1 Two-Time Correlation Functions and Power Spectra

Under the assumptions of the quantum regression theorem [12], the two-time correlation functions for operators A and B in a state ρ is given by

$$C_{A,B}^{+}(t,\tau) = \langle A(t+\tau)B(t)\rangle = \mathrm{tr}[A(\tau)B(0)\rho(t)], \;\; t,\tau \geq 0, \tag{9.47}$$

where $A(\tau) = e^{\mathcal{L}^\star\tau}(A)$ and $\rho(t) = e^{\mathcal{L}t}(\rho)$. Such quantities are related to statistical correlations between measurements of two different observables at two different time t and $t+\tau$ [12]. If instead the measurement of B comes first, one should instead consider

$$C^{-}_{A,B}(t,\tau) = \langle A(t)B(t+\tau)\rangle = \mathrm{tr}[A(0)B(\tau)\rho(t)] = (C^{+}_{B,A}(t,\tau))^*, \quad t,\tau \geq 0. \tag{9.48}$$

In the case of a unique steady state ρ_{ss}, we consider the steady state two-time correlation functions $\lim_{t\to\infty} C^{\pm}_{A,B}(t,\tau)$, which we can express compactly as

$$C^{ss}_{A,B}(\tau) = \begin{cases} \mathrm{tr}[A(\tau)B(0)\rho_{ss}], & \tau \geq 0, \\ \mathrm{tr}[A(0)B(|\tau|)\rho_{ss}], & \tau < 0. \end{cases} \tag{9.49}$$

As it may be more practically significant in certain situations, we additionally define the (two-sided[3]) steady state power spectrum

$$S^{ss}_{A,B}(\omega) \equiv \int_{-\infty}^{\infty} e^{i\omega\tau} C_{A,B}(\tau)\, d\tau. \tag{9.50}$$

Long-lived correlations are then revealed through large power-spectral peaks at zero frequency. To more appropriately captures the relative decay of correlations, we define the normalized correlation functions and power spectrum as

$$\widetilde{C}^{ss}_{A,B}(\tau) = \frac{C^{ss}_{A,B}(\tau)}{C^{ss}_{A,B}(0)}, \quad \widetilde{S}^{ss}_{A,B}(\omega) = \frac{S^{ss}_{A,B}(\omega)}{C^{ss}_{A,B}(0)}. \tag{9.51}$$

For QBLs, we will focus entirely on the case where A and B are linear forms, i.e., $A = \alpha = \widehat{\vec{\alpha}}$ and $B = \beta = \widehat{\vec{\beta}}^{\dagger}$, with $\vec{\alpha}, \vec{\beta} \in \mathbb{C}^{2N}$. In this case, the unnormalized correlation function and power spectra take on a simple closed form

$$C^{ss}_{\alpha,\beta^\dagger}(\tau) = \begin{cases} \vec{\alpha}^\dagger \boldsymbol{\tau}_3 e^{-i\mathbf{G}\tau} \mathbf{Q}_{ss} \boldsymbol{\tau}_3 \vec{\beta}, & \tau \geq 0. \\ \vec{\alpha}^\dagger \boldsymbol{\tau}_3 \mathbf{Q}_{ss} e^{i\mathbf{G}^\dagger \tau} \boldsymbol{\tau}_3 \vec{\beta}, & \tau < 0. \end{cases} \tag{9.52}$$

$$S^{ss}_{\alpha,\beta^\dagger}(\omega) = \vec{\alpha}\boldsymbol{\tau}_3[\boldsymbol{\chi}(\omega)\mathbf{Q}_{ss} + \mathbf{Q}_{ss}\boldsymbol{\chi}^\dagger(\omega)]\boldsymbol{\tau}_3\vec{\beta}, \tag{9.53}$$

where $\boldsymbol{\chi}(\omega) \equiv i(\omega\mathbb{1}_{2N} - \mathbf{G})^{-1}$ is the susceptibility matrix. Mathematically, $\boldsymbol{\chi}(\omega)$ is resolvent of $-i\mathbf{G}$ evaluated at $-i\omega$.

The restriction to linear forms further yields a state-independent notion of *quantum correlation functions*. Note that

$$\alpha(\tau)\beta^\dagger(0) = \frac{1}{2}\{\alpha(\tau), \beta^\dagger(0)\} + \frac{1}{2}[\alpha(\tau), \beta^\dagger(0)]$$

[3] Instead, one may consider the one-sided power spectrum, i.e., the one-sided ($\tau \geq 0$) Fourier transform of the steady state correlation function. The distinction is irrelevant for our applications.

$$= \frac{1}{2}\{\alpha(\tau), \beta^\dagger(0)\} + \frac{1}{2}\vec{\alpha}^\dagger e^{-i\widetilde{\mathbf{G}}^\dagger \tau} \boldsymbol{\tau}_3 \vec{\beta}\, 1_{\mathcal{F}}.$$

Since quantum states have unit trace, we have

$$C^{+}_{\alpha,\beta^\dagger}(t,\tau) = C^{+,\text{cl}}_{\alpha,\beta^\dagger}(t,\tau) + \frac{1}{2}\vec{\alpha}^\dagger e^{-i\widetilde{\mathbf{G}}^\dagger \tau} \boldsymbol{\tau}_3 \vec{\beta}, \tag{9.54}$$

where $C^{+,\text{cl}}_{\alpha,\beta^\dagger}(t,\tau) = \frac{1}{2}\text{tr}[\{\alpha(\tau), \beta^\dagger(0)\}\rho(t)]$ is the *classical (symmetrized) correlation function*. The nomenclature indicates the fact that, for classical degrees of freedom, the commutator term would vanish, leaving only the classical correlations. The quantum correlations between linear forms are then given by

$$C^{+,\text{qu}}_{\alpha,\beta^\dagger}(\tau) = \frac{1}{2}\vec{\alpha}^\dagger e^{-i\widetilde{\mathbf{G}}^\dagger \tau} \boldsymbol{\tau}_3 \vec{\beta}, \tag{9.55}$$

whose state- and t-independence follows from the state-independent nature of $\langle[\alpha(\tau), \beta^\dagger(0)]\rangle$ when α and β are linear forms. We define $C^{-,\text{cl}}_{\alpha,\beta^\dagger}(t,\tau)$ and $C^{-,\text{qu}}_{\alpha,\beta^\dagger}(\tau)$ analogously. In the case where α and β are observables (Hermitian), we have

$$\text{Im}\, C^{\pm}_{\alpha,\beta}(t,\tau) = -iC^{\pm,\text{qu}}_{\alpha,\beta}(\tau), \tag{9.56}$$

that is, the quantum correlation function is the (negative of the) imaginary part of the full two-time correlation function.

We can drop the $\pm$ notation by defining

$$C^{\text{qu}}_{\alpha,\beta^\dagger}(\tau) = \begin{cases} C^{+,\text{qu}}_{\alpha,\beta^\dagger}(\tau) & \tau \geq 0, \\ C^{-,\text{qu}}_{\alpha,\beta^\dagger}(|\tau|) & \tau < 0, \end{cases} \tag{9.57}$$

Finally, we may define the quantum power spectrum as

$$S^{\text{qu}}_{\alpha,\beta^\dagger}(\omega) = \int_{-\infty}^{\infty} e^{i\omega\tau} C^{\text{qu}}_{\alpha,\beta^\dagger}(\tau)\, d\tau = \vec{\alpha}^\dagger \left(\boldsymbol{\tau}_3 \boldsymbol{\chi}(\omega) + \boldsymbol{\chi}^\dagger(\omega)\boldsymbol{\tau}_3\right)\vec{\beta}, \tag{9.58}$$

which is again, a state-independent quantity thanks to our restriction to quantum correlation functions between linear forms.

9.5.2 Signatures of Topological Dynamical Metastability

As we saw, dynamically metastable systems can be either topologically trivial or non-trivial. Starting from the susceptibility matrix, one can predict certain properties of the steady state power spectrum that can distinguish the two regimes. We will focus on the normalized power spectrum in order to capture the relative dynamics of correlations. This eliminates the influence of exponentially large steady state second moments $\mathbf{Q}_{ss}$ (e.g., occupation numbers), that may arise in systems displaying transient amplification.

Let $\boldsymbol{\chi}_N(\omega)$ denote the susceptibility matrix/resolvent of the dynamical matrix $-i\mathbf{G}_N$ for an open chain of length N. On one hand, if the chain is anmalously relaxing, then necessarily $\boldsymbol{\chi}_N(\omega)$ is bounded (in norm) for all ω. The reason is that the rapidity bands of anomalously relaxing systems are bound to the left-half of the complex plane and so $-i\omega$ is not in the SIBC spectrum. Thus, we have a system-size independent upper bound on $\|\chi_N(\omega)\|$. On the other hand, if the chain is dynamically metastable, then there is necessarily a subset of the imaginary axis contained within the SIBC rapidity spectrum. Equivalently, there are intervals on the imaginary axis about which the rapidity bands wind. Ultimately, the restriction of $\boldsymbol{\chi}_N(\omega)$ to these intervals will necessarily grow without bound as $N \to \infty$. Since topological metastability is characterized by the presence of zero in these non-trivial intervals, we conjecture that it generically elicits a peak of the power spectrum at zero frequency that grows without bound with system size. In contrast, there should be no such peak in a dynamically metastable system that is topologically trivial.

The distinctive behavior of $\boldsymbol{\chi}_N(\omega)$ in these regimes is exemplified in Fig. 9.6. In (a), the 2-norm of the susceptibility matrix converges for all values of ω considered. In (b) and (c) we see a divergence of the norm at frequencies ω such that $i\omega$ is contained in the non-trivial interior of the rapidity bands. The zero frequency behavior of $\|\boldsymbol{\chi}_N(\omega)\|_2$ is shown in (d). In particular, the topological regime is distinguished from both the non-topological dynamically metastable regime and the anomalously relaxing regime by an *exponential divergence* in system size. We will see that this manifests directly in certain quantum power spectra, see Fig. 9.9.

Two-time *steady state* correlations may also detect various dynamical regimes of interest. This is exemplified in Fig. 9.7 whereby the x_N-x_N (normalized) power spectrum is shown for the DBKC in the topologically metastable, non-topologically metastable, and anomalously relaxing phases. The crucial difference between the steady-state correlations and the behavior of the susceptibility matrix can be seen in the behavior of the zero-frequency peaks of the power spectrum. In Fig. 9.7b, we find that the steady state power spectral peak at zero frequency diverges *algebraically*, rather than exponentially, with system size in the topologically metastable phase, while the behavior in the other two regimes is unambiguously bounded (in fact, it is strongly suppressed in the non-toplogically metastable case).

9.5.3 Distinguishing Split and Non-split Majorana Bosons

Within the class of topologically metastable, number-non-conserving QBLs, models may be distinguished based on whether their MBs are split or non-split. Let (γ^z, γ^s) denote a split MB pair which satisfy $[\gamma^z, \gamma^s] = i$, and are localized on opposite sides of a chain. During the transient timescale ($t < t_N$ for some t_N

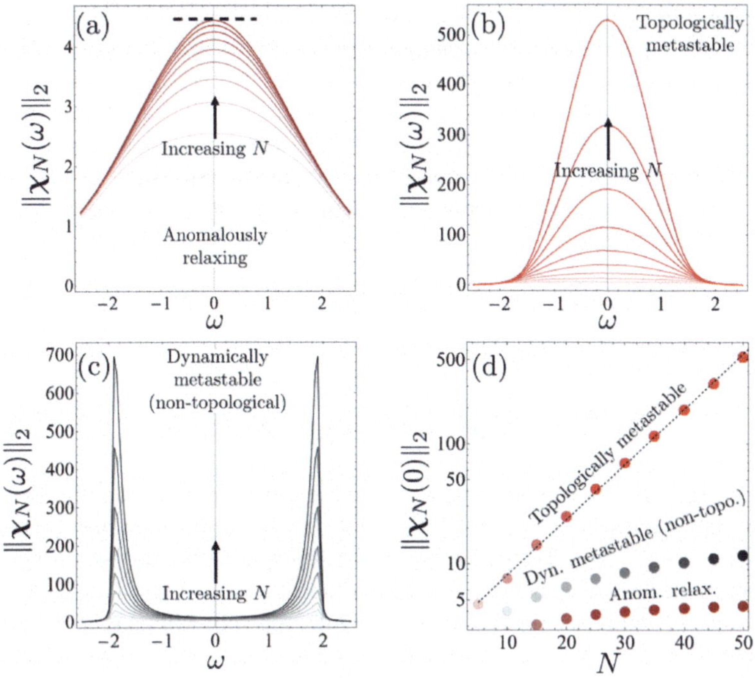

Fig. 9.6 (**a**)–(**d**) The 2-norm of the DBKC's OBC susceptibility matrix in the anomalously relaxing, topologically metastable, and non-topological dynamically metastable phases, respectively. (**a**) and (**b**) correspond to the regimes whose rapidities are the open and filled markers in Fig. 8.1a, respectively. (**c**) corresponds to the regime whose rapidities are shown in Fig. 8.1b. (**d**) shows the zero-frequency susceptibility norm in the three aforementioned regimes. The dashed line is a linear fit

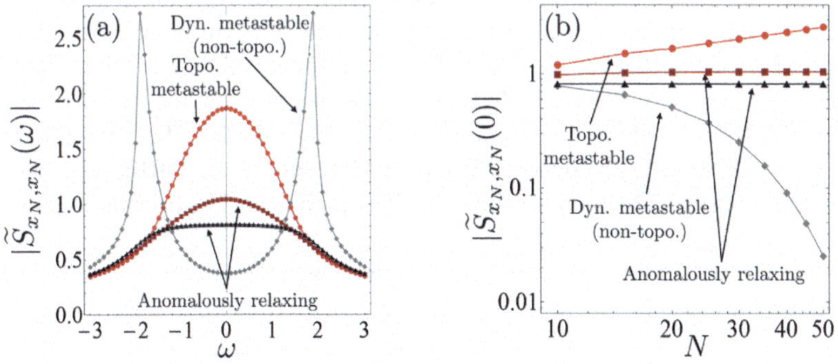

Fig. 9.7 (**a**) Modulus of the normalized power spectra (Eq. (9.51)) of the operators $\alpha = \beta = x_N$ for the DBKC in the topologically metastable phase (light red disks $\kappa/\Delta = 0.6$), non-topologically metastable phase (gray diamonds $\kappa/\Delta = 0.6$, with $\Gamma = 0.12$), and the anomalously relaxing phase (dark red squares $\kappa/\Delta = 1.4$ with $\Gamma = 0$ and black triangles $\kappa/\Delta = 1.4$ with $\Gamma = 0.12$). In all cases $J = 2$, $\Delta = 0.5$, $N = 25$. (**b**) Modulus of the zero-frequency component for the same parameters in (a) as a function of N. Earlier versions of these figures first appeared in Ref. [1]

increasing with system size N) we have $\gamma^z(t) \simeq \gamma^z(0)$. However, because the MBs are split, $\gamma^s(t)$ deviates meaningfully from $\gamma^s(0)$ over the same timescale. Now, consider the associated correlation function $C^{\mathrm{ss}}_{\gamma^z,\gamma^s}(\tau)$. Firstly, we note that canonical commutation implies the existence of a non-zero quantum correlation function at $\tau = 0$. Explicitly, $C^{\mathrm{qu}}_{\gamma^z,\gamma^s}(0) = i/2$. Remarkably, this persists in spite of the macroscopic spatial separation of the two modes. However, non-split MBs satisfy the same identity. To distinguish them, we must go beyond $\tau = 0$. For $0 < \tau < t_N$, we have

$$C^{\mathrm{ss}}_{\gamma^z,\gamma^s}(\tau) = \mathrm{tr}[\gamma^z(\tau)\gamma^s(0)\rho_{\mathrm{ss}}] \simeq \mathrm{tr}[\gamma^z(0)\gamma^s(0)\rho_{\mathrm{ss}}] = C^{\mathrm{ss}}_{\gamma^z,\gamma^s}(0). \tag{9.59}$$

On the other hand, for $\tau < 0$

$$C^{\mathrm{ss}}_{\gamma^z,\gamma^s}(\tau) = \mathrm{tr}[\gamma^z(0)\gamma^s(|\tau|)\rho_{\mathrm{ss}}] \not\simeq C^{\mathrm{ss}}_{\gamma^z,\gamma^s}(0). \tag{9.60}$$

If instead the MBs were non-split, we would additionally have that the SG γ^s is approximately conserved, i.e., $\gamma^s(t) \simeq \gamma^s(0)$ for $t < t_N$. It would then follow that $C^{\mathrm{ss}}_{\gamma^z,\gamma^s}(\tau) \simeq C^{\mathrm{ss}}_{\gamma^z,\gamma^s}(0)$. Theqrefore, split and non-split MBs may be distinguished by asymmetries in the associated correlation function around $\tau = 0$. Split MBs are approximately stationary for $0 < \tau < t_N$ while non-split MBs are approximately stationary for $-t_N < \tau < t_N$. Two remarks are in order. (i) While we have treated the full steady state correlation function explicitly, the same conclusions hold for both the classical and the (state-independent) quantum contributions. (ii) Due to the properties of MBs, the stationarity of the correlation functions in each case will become more pronounced as system size is increased.

To exemplify these distinctions, we focus on DBKC (with $\mu = \Gamma = 0$) and the PDC as representative examples of these two classes. Comparing the rapidity bands of the DBKC (Eq. (9.11)) with $\mu = \Gamma = 0$ and the rapidity bands of the PDC (Eq. (9.6)), we observe that there is an isomorphism between the two topological phase diagrams. First, to distinguish the two models, we will relabel the quantities J, Δ, and μ of the PDC as J_F, Δ_F, and μ_F, respectively.[4] If we then make the identification $J_F = \Delta/2$, $\Delta_F = J/2$, $\mu_F = -\kappa$, we find that the rapidity bands of each model are coincident - in particular, the topological phase diagrams coincide.[5] Let $(\gamma_L^{zz}, \gamma_R^s)$ and (γ_L^s, γ_R^z) denote the (split) MB pairs of the DBKC and (γ_L, γ_R) denote the (non-split) MB pair of the PDC. The parameter identification yields

$$\gamma_L = \gamma_L^z, \qquad \gamma_R = \gamma_R^s, \tag{9.61}$$

[4] The subscript F is to remind the reader that the PDC arises from the FKC.

[5] Note that this identification is not the same mapping as the isospectral mapping in Appendix B.3.

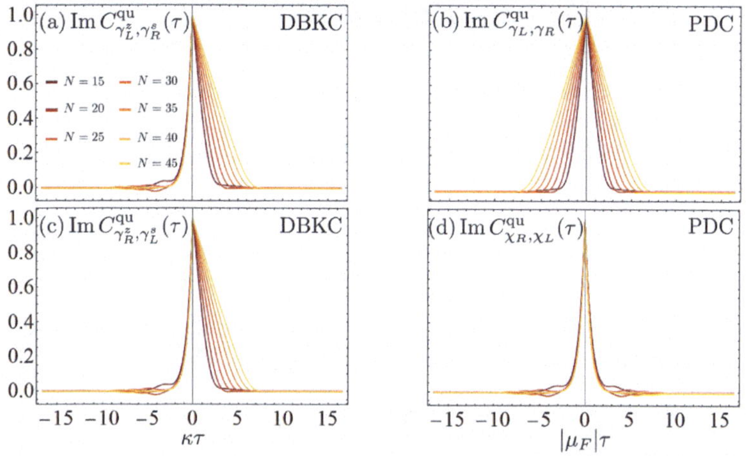

Fig. 9.8 (**a**) An MB correlation function for the DBKC. (**b**) An MB correlation function for the PDC. (**c**) A different MB correlation function for the BKC. (**d**) A correlation function for the same operators in (**c**) but for the FKC. In all cases, the modes are normalized so that canonical commutation relations hold at $\tau = 0$. Note that $\gamma^z_L = \gamma_L$, $\gamma^s_R = \gamma_R$, $\gamma^z_R = \chi_R$, and $\gamma^s_L = \chi_L$ at $\tau = 0$ when the parameter mapping discussed in the main text is applied

which may be directly verified in the case $J = \Delta$. In particular, the second MB pair of the DBKC (γ^s_L, γ^z_R) are not approximate ZMs, nor Weyl SGs, in the PDC. This has several implications for certain two-time correlation functions. Since the steady states of these two models may differ in meaningful ways, we can directly compare the state-independent quantum correlation function of the MBs. Our general analysis above predicts that the DBKC correlation function $C^{\mathrm{qu}}_{\gamma^z_L,\gamma^s_R}(\tau)$ will be asymmetric about zero and increasingly stationary in the positive τ direction as N increases. On the contrary, the FKC correlation function $C^{\mathrm{qu}}_{\gamma_c,\gamma_R}(\tau)$ should be symmetric and increasingly stationary in both the positive and negative τ direction as N increases. These predictions are verified in Fig. 9.8a, b.

A further distinction between these two models can be seen by focusing on the second pair of MBs in the DBKC. As previously noted, the operators γ^s_L and γ^z_R, when mapped to the PDC, are neither approximate ZMs nor Weyl SGs. To distinguish which model we are working in, let $\gamma^s_L \mapsto \chi_L$ and $\gamma^z_R \mapsto \chi_R$ denote the image of the DBKC's second MB pair in the PDC under the parameter identification. As argued above, the correlation function $C_{\gamma^z_R,\gamma^s_L}(\tau)$ will become more and more stationary for $\tau \geq 0$ as $N \to \infty$. On the contrary, no such argument applies to $C_{\chi_R,\chi_L}(\tau)$ and so we generally expect exponentially decaying correlations. This is verified in Fig. 9.8c, d.

9.5.4 Signatures of Dirac Edge Bosons

We have thus far focused on observable signatures of MBs. How, if at all, do these quantum correlations behave in the presence of number symmetry? That is, can we detect the Dirac bosons of Sect. 8.3.2? To begin, consider the elementary steady state correlator

$$C^{\text{ss}}_{i,j}(\tau) \equiv C^{\text{ss}}_{\Phi_i,\Phi_j^\dagger}(\tau). \tag{9.62}$$

This correlator is elementary in the sense that any correlation function of linear observables can be written as a linear combination of the above:

$$C^{\text{ss}}_{\alpha,\beta^\dagger} = \sum_{i,j} c^{i,j}_{\alpha,\beta^\dagger} C^{\text{ss}}_{i,j}(\tau), \tag{9.63}$$

with $c^{i,j}_{\alpha,\beta^\dagger}$ determined by the coefficients of Φ_i and $\Phi_j^\dagger$ in the definitions of α and $\beta^\dagger$. Explicitly, $c^{i,j}_{\alpha,\beta^\dagger} = \alpha_i \beta_j^*$, with α_i and β_j elements of $\vec{\alpha}$ and $\vec{\beta}$.

Number symmetry, combined with uniqueness of the steady state, immediately yields $[\rho_{\text{ss}}, \sum_j a_j^\dagger a_j] = 0$. This guarantees that the "off-diagonal" elementary correlators, i.e., correlators of the form $\langle a_i^\dagger(\tau) a_j^\dagger(0)\rangle_{\text{ss}}$ and their Hermitian conjugate counterparts, vanish. In fact, the off-diagonal, state-independent quantum correlation functions always vanish. This follows because number symmetry guarantees

$$a_i(\tau) = \sum_{j=1}^{N} d_{ij}(\tau) a_j(0),$$

for some time-dependent coefficients d_{ij}. This observation combined with canonical commutation relations ensures that $[a_i(\tau), a_j(0)] = 0$ for all τ. We remark that the equivalent statement in the quadrature basis is $C^{\text{qu}}_{x_j,p_i}(\tau) = C^{\text{qu}}_{x_i,p_j}(-\tau)$.

With this, we can characterize topologically metastable, number-symmetric chains by vanishing off-diagonal correlators and long-lived correlations / divergent-zero frequency power-spectral peaks. This behavior is reflected in Fig. 9.9. The quantum power spectra $S_{a_1^\dagger,a_N}(\omega)$ displays exponential divergence at zero frequency in both the DNS chain and the DBKC. However, the off-diagonal spectra $S_{a_1^\dagger,a_N^\dagger}(0)$ is exactly zero for the DNS chain and diverging exponentially for the DBKC.

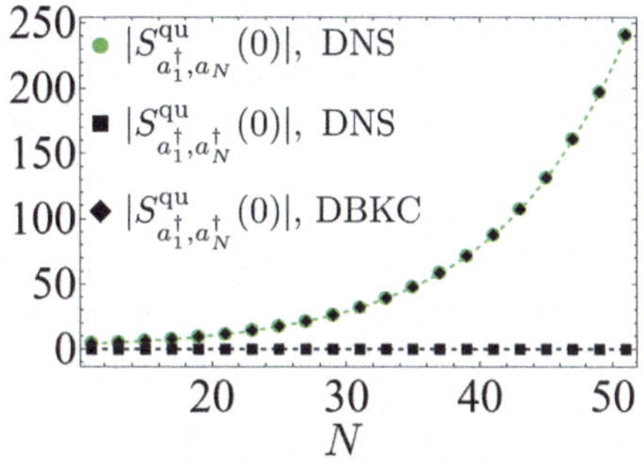

Fig. 9.9 Various quantum power spectra for the DNS chain and the DBKC in their respective topologically metastable regimes. The DNS shows exponential divergence of the zero frequency peak $S^{qu}_{a_1^\dagger,a_N}(0)$ (green circles) and the vanishing of the off-diagonal spectra $S^{qu}_{a_1^\dagger,a_N^\dagger}$ (black squares). To contrast, the DBKC exhibits exponential growth of the off-diagonal spectra (black diamonds). The parameters for the DNS are $J_+ = 1$, $J_- = 0.25$, and $\kappa = 0.3$, while the parameters for the DBKC are $J = 2$, $\Delta = 0.5$, $\mu = 0$, $\kappa = 0.3$, and $\Gamma = 0$. These choices ensure an isospectral relationship between the two models. An exponential fit is shown as a green dashed line while 0 is emphasized with a black dashed line

References

1. V.P. Flynn, E. Cobanera, L. Viola, Phys. Rev. Lett. **127**, 245701 (2021)
2. V.P. Flynn, E. Cobanera, L. Viola, Phys. Rev. B **108**, 214312 (2023)
3. V.P. Flynn, E. Cobanera, L. Viola, Upcoming paper covering more mathematically sophisticated instances of topological metastability (2024)
4. C.E. Bardyn, M.A. Baranov, C.V. Kraus, E. Rico, A. İmamoglu, P. Zoller, S. Diehl, New J. Phys. **15**, 085001 (2013)
5. A. D'Abbruzzo, D. Rossini, Phys. Rev. A **103**, 052209 (2021)
6. K. Fujii, Chap. 7 in Advances in Quantum Mechanics, ed. by P. Bracken (InTech, London, 2012). arXiv:1209.1437 (quant-ph)
7. H.J. Korsch, arXiv:1908.01187v2 (quant-ph) (2019)
8. P.T. Cochrane, G.J. Milburn, W.J. Munro, Phys. Rev. A **59**, 2631 (1999)
9. V.V. Albert, K. Noh, K. Duivenoorden, D.J. Young, R.T. Brierley, P. Reinhold, C. Vuillot, L. Li, C. Shen, S. M. Girvin, B.M. Terhal, L. Jiang, Phys. Rev. A **97**, 032346 (2018)
10. N. Hatano, D.R. Nelson, Phys. Rev. Lett. **77**, 570 (1996)
11. A. McDonald, R. Hanai, A.A. Clerk, Phys. Rev. B **105**, 064302 (2022)
12. C.W. Gardiner, P. Zoller, *Quantum Noise* (Springer, Heidelberg, 2000)
13. I.I. Arkhipov, A. Miranowicz, F. Minganti, F. Nori, Phys. Rev. A **102**, 033715 (2020)
14. L.M. Vasiloiu, F. Carollo, J.P. Garrahan, Phys. Rev. B **98**, 094308 (2018)
15. M.O. Scully, M.S. Zubairy, *Quantum Optics* (American Association of Physics Teachers, Maryland, 1999)
16. D.F. Walls, G.J. Milburn, *Quantum Optics* (Springer, Berlin, 2007)

Chapter 10
Summary and Outlook

Abstract We reflect on the two questions that broadly motivated the work in this thesis. Namely, 'What are the most salient consequences of the effective non-Hermiticity intrinsic to the equations of motion for closed, non-interacting bosonic systems?' and 'To what extent can closed, or open, non-interacting bosons manifest physics associated to SPT phases of non-interacting fermions?'. We concisely summarize the key results uncovered in the pursuit of answers to these questions, and discuss several future directions for this exciting line of research.

10.1 Summary of Key Results

Let us reflect on the two questions that broadly motivated the work in this thesis:

(1) What are the most salient consequences of the effective non-Hermiticity intrinsic to the equations of motion for closed, non-interacting bosonic systems?
(2) To what extent can closed, or open, non-interacting bosons manifest physics associated to SPT phases of non-interacting fermions?

To what extent have we addressed these questions? We will approach this in parts.

10.1.1 Effective Non-Hermiticity in Quadratic Bosonic Hamiltonians

We have seen that, unlike in the fermionic case, non-Hermiticity lies deeply within the equations of motion for non-interacting bosonic systems, even if the full many-body dynamics are *explicitly* Hermitian. While this may seem unintuitive on the surface, it ultimately engenders the physics of amplification and, more broadly, it provides a rich arena to explore the physics of instabilities. Notably, this provides us with the notion of a stability phase diagram for QBHs, that is, a separation of

V. P. Flynn, *Effective Non-Hermiticity and Topology in Markovian Quadratic Bosonic Dynamics*, Springer Theses, https://doi.org/10.1007/978-3-031-52045-7_10

parameter phase space into dynamically stable and dynamically unstable regimes. To understand these phase diagrams more precisely, it is useful to make use of further structural features of dynamical matrices associated to QBHs. In particular, the intrinsic *pseudo-Hermiticity* of these matrices makes applicable the tools of *Krein stability theory* to classify the transitions to instability. These transitions are signaled by the emergence of two distinct types of spectral degeneracies in the dynamical matrix spectrum: *exceptional points* and *Krein collisions*. Merging these notions with tools from non-Hermitian quantum mechanics, we arrive at our first three major contributions:

(i) Intrinsic to every QBH is an underlying *generalized PT-symmetry.*
(ii) Stability phase transitions in QBHs are directly attributable to the *spontaneous breaking* of this GPT symmetry.
(iii) Stability phase transitions are identifiable via the vanishing of a newly introduced type of phase rigidity, called the *Krein phase rigidity*.

Most importantly, these concepts (GPT symmetry and KPR, specifically) are not simply arrived at by transplanting concepts from non-Hermitian quantum mechanics into a bosonic setting. As we have argued, they arise naturally by considering the behavior of normal modes in the presence of stability phase transitions.

We exemplified the consequences of these results in three models: an elementary single-mode system, a two-mode cavity QED system, and the flagship bosonic Kitaev chain. Our analysis of the BKC, in particular, revealed several non-trivial aspects of both general QBHs, as well as the response of the BKC to changes in system size and BCs. Lessons from the BKC may be summarized as follows:

(iv) Any QBH exhibiting phase-dependent transport must necessarily be (a) odd under time-reversal symmetry; (b) invariant under certain squeezing transformations; (c) sitting at the *cusp on instability,* as signaled by a macroscopic number of Krein collisions in the normal mode spectrum.
(v) The BKC is dynamically stable under the *same BCs* that manifest Majorana fermion edge modes in the FKC. Increasing system-size, causes these these regions of stability to shrink, and thus, dramatically enhance dynamical sensitivity of the system to perturbations.
(vi) *Bosonic "shadows" of Majorana fermions* emerge in the BKC when subjected to uniform degenerate parametric amplification. These shadows are explicitly tied to *topological properties of the bulk.*

Up until this point, it was understood that number-non-conservation is necessary, but *not sufficient* for dynamical instabilities. We have been able to fill this conceptual gap by gaining a deeper understanding of the role played by number-non-conserving terms in *dynamically stable* QBHs. Specifically:

(vii) By leveraging results in the field of pseudo-Hermitian quantum mechanics, we have formulated an explicit *number-conservation-restoring duality transformation* for dynamically stable QBHs. Moreover, we have characterized the resulting transmutation of translation and quasiparticle-number symmetry.

(viii) This duality, which has a natural geometric interpretation as a metric in Nambu space, can be interpreted physically in terms of the vacuum covariance matrix.

(ix) This duality points to the possibility of realizing analog quantum simulations of genuinely non-Hermitian PT-symmetric Hamiltonians, *without* the need for loss, gain, or coherent driving.

As with with GPT-symmetry and KPR, we put the duality transformation to the test in two paradigmatic examples of interest: the gapped harmonic chain and the BKC. In the former, we provided a recipe for explicitly constructing the duality that generalizes to arbitrary single-band systems with pairing, explored the exponentially decaying nature of the dual hopping amplitudes and established a connection to the decay of ground-state correlations, and explored the viability of truncating the dual system in such a way to faithfully reproduce the band structure while explicitly restoring quasi-locality. In the case of the BKC, we found that the duality transformation comes in the form of a local squeezing transformation that ultimately allowed us to extend the analytically-known region of dynamical stability in the boundary parameter phase space. We also explored the behavior of both of these duality transformations in the vicinity of stability phase boundaries. The duality transformation proved useful for two further applications, the first of which being the following:

(x) The discrepancy between *pseudo-Hermitian Berry phases*, which lie at the core of defining topological invariants for bosonic systems, and the standard *Hermitian Berry phase* used, e.g., to define fermionic topological invariants, may be explicitly computed in terms of the duality transformation.

The second implication is best left to the following section, as it appears in the context of open bosonic systems.

10.1.2 Manifestations of SPT Physics in Quadratic Bosonic Lindbladians

The first step we took toward addressing question (2) was to examine the possibility for SPT physics in the closed-system setting. Here, we were met with two insurmountable hurdles: (i) there are no SPT phases of free bosonic matter (modeled using thermodynamically stable QBHs); and (ii) any instance of bosonic analogues to fermionic zero-energy edges states are intrinsically unstable, in a dynamical sense. These facts have forced us to forgo at least one of the assumptions we had made up until this point. We concluded that, if we are to retain the non-interacting and time-independent nature of our systems, we are forced to move beyond the closed-system settings. Unlike in the fermionic case, this move to open systems is absolutely necessary if we wish to uncover anything resembling SPT phases of free bosons. To retain the simplest possible dynamical description, we allowed our non-interacting bosonic systems to undergo a simple (quadratic) form of Markovian dissipation, and have modeled our open-system dynamics using quadratic bosonic Lindbladians.

To establish a convincing bosonic analogue to fermionic SPT physics, it is crucial to identify the characteristics of bosonic incarnations of the topologically-mandated ZMs central to topological free-fermions. To this end, we defined bosonic ZMs of QBLs and, due to the breakdown in Noether's theorem, the closely related Weyl SGs. These modes, which we unify under the umbrella of *Noether modes*, are generically independent from one-another. However, our first concrete result about QBLs establishes a fundamental correspondence:

(i) Under certain assumptions placed on the associated dynamical matrix, for each (approximate) ZM of a QBL, there exists a *canonically conjugate* (approximate) Weyl SG, and vice-versa.

In the context of open quantum systems, whereby there exists no simple relationship between conserved quantities (e.g., ZMs) and SGs, this result comes as a surprise. It offers, in a sense, a partial restoration of Noether's theorem for QBLs, at least within the Nambu space. We proceeded to present two recipes that would later provide essential for uncovering signatures of SPT systems. These may be summarized as:

(ii) Given a QFH with (approximate) ZMs, it is *always* possible to engineer a QBL possessing non-split (approximate) Noether modes whose spatial distribution is identical to that of the fermionic modes.
(iii) Given a dynamically stable QBH, it is possible,by leveraging our duality transformation, to engineer a purely dissipative Markovian generator that relaxes the system *uniquely to the quasiparticle vacuum of the Hamiltonian.*

With these fundamental results and recipes behind us, we were ready to identify a class of 1D QBLs capable of supporting bosonic signatures of SPT physics. Explicitly, we require that any such system must be dynamically stable (in order to retain an appropriate notion of robustness) and exhibit topologically mandated edge ZMs and SGs. Leveraging the known spectral properties of block-Toeplitz matrices and operators, we came to three conclusions. Systems satisfying our requirements must possess *non-trivial bulk topology* (in the form of rapidity band winding), *bulk instabilities*, and a *highly non-normal* dynamical matrix. This third requirement, in particular, forced us to move beyond the techniques of spectral analysis, and instead bring forth the tools of pseudospectral theory. With pseudospectra centered at the heart of our analysis, an important result about the relationship between finite- and infinite-size QBLs revealed itself, namely:

(iv) The rapidity spectrum of a bulk-translation invariant, semi-infinite QBL *imprints* itself into the rapidity pseudospectrum of its finite-size truncation.

One physical implication of this result is that normal modes of the semi-infinite system behave as approximate normal modes of the finite system for a timescale that scales (roughly) linearly with system size. That is, the transient lasts as long as it takes for the mode to "detect" the presence of both boundaries. While this may seem to be a rather innocuous fact, it becomes extremely relevant in the case where the semi-infinite rapidity spectrum (in particular, the stability gap) differs *dramatically* from that of the finite-size truncations - a possibility granted to us by non-normality. Along this line, we identified two novel dynamical phases for finite-size, 1D, bulk-translationally invariant QBLs. These are:

(v) An *anomalously relaxing phase* characterized by two-step relaxation dynamics. The dominant relaxation rate in the first (transient) step is set by the infinite-size Lindblad gap, while the dominant relaxation rate in the second (asymptotic) step corresponds to that of the finite-size system.

(vi) A *dynamically metastable phase* characterized by an transient amplification at a rate set by the positive infinite-size stability gap, followed by asymptotic relaxation at a rate set by the finite-size Lindblad gap.

Refocusing back to our search, we concluded that a system with all of our desired properties must be *dynamically metastable*. Combining this with the requirement of a non-trivial bulk-topology allowed us to finally pin down the precise systems of interest: *topologically metastable QBLs*. These systems support a number of key features, namely:

(vii) A *unique steady state* and a finite spectral gap are maintained for all finite system sizes. In particular, *dynamical stability* is present for all finite system sizes.

(viii) *Tight bosonic analogues of Majorana fermions*, which we deemed *Majorana bosons*, emerge localized on opposite ends of the chain. They consist of an approximate ZM and the generator of an approximate (Weyl) symmetry, and are canonically conjugate, despite macroscopic spatial separation. If, additionally, number symmetry is present, then the MBs can be arranged into bosonic degrees of freedom, which we deemed *Dirac bosons*.

(ix) A manifold of *degenerate quasi-steady states* manifest in the finite-size chains. Physically, such states are phase-space displacements of the unique steady state.

(x) The Majorana and Dirac bosons, as well as the quasi-steady states, persist in a transient dynamical regime whose duration diverges (at least linearly) with system size. Crucially, their existence elicits divergent zero-frequency peaks in certain power spectra and increasingly stationary behavior of quantum correlation functions.

These general features were explicitly demonstrated in four flagship models, the purely dissipative topologically metastable chain (derived using recipe (ii) above), a dissipative BKC, a dissipative BKC with a pure steady state (derived using recipe (iii) above), and the dissipative number-symmetric chain. Several additional lessons were learned from these examples. Two particularly notable ones were the following:

(xi) Non-split MBs, which may be thought of as the tightest analogues of Majorana fermions, and topological metastability more broadly, can be supported in a *purely dissipative* setting ($H = 0$).

(xii) The quasi-steady states of topologically metastable systems with pure steady states can be thought of as coherent states with respect to a particular basis and, when arranged into cat-state superpositions, can support long-lived regimes of indefinite bosonic parity.

10.2 Outlook

The research program we have presented here provides a vast array of potential next steps. Let us describe a small number of them. In terms of "fundamental" open questions, there is one that towers above the rest.

(1) **Is topological metastability a true SPT phase of non-interacting bosonic systems?**

At this stage, we have presented several pieces of evidence in the affirmative. However, there still remains major theoretical hurdles in the way of a concrete answer. Unlike for fermions, there is no agreed-upon set of classifying symmetries for bosonic systems. This is to be expected - up until this point, there has been no evidence for non-interacting bosonic SPTs, and thus, no reason to attempt classifying them. Moreover, this is not as simple as importing the fermionic classification scheme, or its extension to the (semiclassical) non-Hermitian and (fully quantum) open settings. As pointed out in Ref. [1], two of the three main classifying symmetries (specifically, particle-hole and chiral) symmetry lack any bosonic analogue. Simply put, these transformations cannot be implemented unitarily, or anti-unitarily, for bosonic particles. So, the first main challenge would be to identify a set of classifying symmetries that is appropriate in the bosonic context. A second, perhaps less severe, challenge involves properly defining the concept of a SPT phase for Markovian systems. After all, symmetries can be either weak, or strong, and need not correspond to conserved quantities in the continuous case. Luckily, however, there have been considerable efforts in this direction, albeit in a fermionic context [2, 3]. One final hurdle specifically applies to bulk considerations. When a finite-system undergoes a transition to a topologically metastable phase, it is unavoidable that the bulk becomes unstable. In particular, there is no bulk steady state.[1] Thus, it is not clear to what extent this can be thought of as dissipative phase transition in the bulk.

[1] The system may not even have a steady state prior to the transition. For example, consider the DBKC with $\Gamma \neq 0$ so that the finite-system is dynamically, but not topologically, metastable. In this configuration, the bulk lacks a steady state. Taking $\Gamma \to 0$ so that the finite-system becomes topologically metastable induces a bulk transition in which there is no steady state on either side.

There are numerous other questions falling under this "fundamental" category. For closed-systems, it would be interesting to explore the deformation of stability phase diagrams as a consequence of periodic (Floquet) driving. Inspired by the classical *Kapitza's pendulum* phenomenon, whereby an inverted oscillator is stabilized by means of periodically shifting the pivot point, it is conceivable to assume that periodically driving coupling constants in an unstable QBH could result in an overall stable dynamical evolution. In fact, the conceptual hurdles that ultimately lead us to consider open bosonic systems could have just as easily lead us to the realm of *driven* bosonic systems. The no-go theorems and ZM instability results of Chap. 5 explicitly assume time-independence. Thus, can they be circumvented by means of driving? This line of research would require making connections with the now booming field of Floquet topological matter. Perhaps not coincidentally, Floquet phases are also metastable, since driving is eventually tied to heating and trivial (infinite-temperature) steady states [4]. Interestingly, however, Floquet bosonic systems would include the extra twist of non-Hermiticity at the level of the relevant dynamical matrices. One concrete goal would be to stabilize the bosonic "shadows" of Majorana fermions found in the BKC by means of periodically driving certain Hamiltonian parameters, and thus, synthesize a periodically driven analogue of MBs. Additionally, the effects of non-periodic time-dependence (e.g., quenching a boundary) would be interesting to explore in a more systematic fashion. One existing result to this end involves the selective population of bosonic edge states via quenching [5].

Another avenue involves expanding the theory of metastability to include fermionic (or even spin) systems. Although the topological aspects of these systems are well-understood, it is conceivable that our framework of metastability can be useful in explaining anomalous relaxation phenomena [6, 7] and cutoff effects [8]. While dynamically metastable phases are not possible in fermionic systems (since the bulk is always dynamically stable [9–11]), anomalously relaxing phases are. Furthermore, exploring the potential splitting of ZMs and SGs in fermionic systems is also intriguing. As for bosonic systems, one natural extension of our framework is to higher dimensions. For instance, what is the dissipative bosonic equivalent of surface bands in two-dimensional topological insulators and superconductors, such as quantum Hall systems or the $p + ip$ superconductor?

As for "practical" open questions, most fall under the umbrella of the following:

(2) What are the experimentally accessible consequences and applications our results?

There exists a variety of different experimental platforms and proposals for realizing many coupled bosonic degrees of freedom. Some of the most promising are cavity- and circuit-QED platforms [12–25], microlasers and ring resonators [26, 27], optomechanical systems [28, 29], and vibronic lattices [30]. Arguably, the more exotic ingredients appearing in our topologically metastable models are the dissipative hopping and the (coherent and dissipative) bosonic pairing. Notably, Ref. [31] proposes a method for realizing dissipative hopping, while realizations of both coherent and dissipative pairing have been proposed by means of three-wave mixing with suitably tuned couplings to auxiliary modes [32–35]. With these platforms becoming available, various lines of investigation open up.

For closed systems, we conjecture there are applications of QBHs for quantum sensing. Concretely, we believe the answer to "Can the dynamical consequences of Krein collisions be leveraged to develop high-precision quantum sensors?" is in the affirmative. This conjecture is motivated by several factors. Firstly, like EPs, KCs populate the stability phase boundaries of QBHs. In particular, based on the many models we have studied, it appears that KCs typically arise at the locus of two EP-dominated phase boundaries. It is then reasonable to expect that systems with KCs respond dramatically to small perturbations. In fact, there is mathematical precedent for this: the splitting of a Krein-collided eigenvalue in response to a perturbation of strength $\epsilon \ll 1$ can scale like $\epsilon^{1/n} \geq \epsilon$ for some integer n (see Ch.9 of Ref [36], for instance). Moreover, we conjecture that KCs, or something closely related related to KCs, are the main source of enhanced sensing recently discovered in non-Hermitian topological models. First, in Ref. [37], a non-Hermitian quantum system featuring (i) the NHSE, (ii) non-EP spectral degeneracies, and (iii) exponentially small overlap between left and right eigenvectors, was demonstrated to have exponentially enhanced (in system size) sensitivity to certain perturbations. While property (ii) is clearly reminiscent to KCs (in that the degeneracies are 'diagonalizable'), property (iii) is actually equivalent to the near-vanishing of the KPR, at least in the case where the system is pseudo-Hermitian. Similarly, in Ref. [38], a non-Hermitian generalization of the BKC was shown to provide a similar degree of enhanced sensitivity (inferred from signal-to-noise ratio enhancements) to certain perturbations without explicit need for EPs. The essential requirement was that these perturbations break an intrinsic $\mathbb{Z}_2$-symmetry that, we believe, is precisely the symmetry ensuring the macroscopic number of KCs in the BKC. However, bridge between these phenomena and the realm of Krein stability theory and KPR has not yet been established. We believe these connections are a promising avenue for future research.

Moving to open systems, experimental detection of Majorana (or Dirac) bosons, or more broadly, topological metastability seems promising. Once the appropriate coherent and incoherent mechanisms are engineered, the unique properties of two-time correlation functions, and their power-spectra, discussed in Sect. 9.5, should offer a path towards detection. To this end, it would be particularly useful to establish further connections between these correlation functions and first-order coherence functions. Since coherence functions are experimentally accessible, at least in quantum-optical platforms, such a connection would provide specific experimental techniques for detection.

Beyond their detection, one potential application of MBs may be in the field of continuous variable quantum information processing. Drawing analogy with Majorana-based quantum computing proposals, we expect that MBs could find utility in the context of continuous-variable schemes based on Gottesman-Kitaev-Preskill (GKP) codes. Central to GKP-based computation (first proposed in Ref. [39] and realized experimentally in trapped ion [40] and a circuit-QED [41] platforms nearly two decades later) is a pair of canonically conjugate quadratures.

From here, the GKP code is built from fixed displacement operators within the associated phase space and the logical states are built from the (ideally) infinitely squeezed eigenstates of these operators. In our topological metastability paradigm, we are provided with two canonically conjugate quadratures that have an additional non-trivial property: they are *macroscopically separated in space*. This fact, which arises due to their topological origin, provides us with a degree of robustness that may provide utility for such applications. Notably, in the non-split case, the MBs generate (orthogonal) phase-space displacements that leaves the overall dynamics invariant to an arbitrarily high degree of precision (as set by the system size).

It is our hope to continue to investigate and address a large fraction of these fundamental and practical questions in the future.

References

1. Q.-R. Xu, V.P. Flynn, A. Alase, E. Cobanera, L. Viola, G. Ortiz, Phys. Rev. B **102**, 125127 (2020)
2. S. Lieu, M. McGinley, N.R. Cooper, Phys. Rev. Lett. **124**, 040401 (2020)
3. A. Altland, M. Fleischhauer, S. Diehl, Phys. Rev. X **11**, 021037 (2021)
4. D.A. Abanin, W. De Roeck, F. Huveneers, Phys. Rev. Lett. **115**, 256803 (2015)
5. B. Galilo, D.K.K. Lee, R. Barnett, Phys. Rev. Lett. **115**, 245302 (2015)
6. T. Haga, M. Nakagawa, R. Hamazaki, M. Ueda, Phys. Rev. Lett. **127**, 070402 (2021)
7. T. Mori, T. Shirai, Phys. Rev. Lett. **125**, 230604 (2020)
8. E. Vernier, SciPost Phys. **9**, 049 (2020)
9. T. Prosen, New J. Phys. **10**, 043026 (2008)
10. T. Prosen, J. Stat. Mech. P07020 (2010)
11. T. Barthel, Y. Zhang, J. Stat. Mech. 113101 (2022)
12. V. Peano, M. Houde, C. Brendel, F. Marquardt, A.A. Clerk, Nat. Commun. **7**, 1 (2016)
13. A. McDonald, T. Pereg-Barnea, A.A. Clerk, Phys. Rev. X **8**, 041031 (2018)
14. J. Koch, A.A. Houck, K. Le Hur, S.M. Girvin, Phys. Rev. A **82**, 043811 (2010)
15. A. Kamal, J. Clarke, M.H. Devoret, Nat. Phys. **7**, 311 (2011)
16. B. Abdo, K. Sliwa, L. Frunzio, M. Devoret, Phys. Rev. X **3**, 031001 (2013)
17. K.M. Sliwa, M. Hatridge, A. Narla, S. Shankar, L. Frunzio, R.J. Schoelkopf, M.H. Devoret, Phys. Rev. X **5**, 041020 (2015)
18. A. Metelmann, A.A. Clerk, Phys. Rev. X **5**, 021025 (2015)
19. B.M. Anderson, R. Ma, C. Owens, D.I. Schuster, J. Simon, Phys. Rev. X **6**, 041043 (2016)
20. P. Roushan, C. Neill, A. Megrant, Y. Chen, R. Babbush, R. Barends, B. Campbell, Z. Chen, B. Chiaro, A. Dunsworth, A. Fowler, E. Jeffrey, J. Kelly, E. Lucero, J. Mutus, P.J.J. O'Malley, M. Neeley, C. Quintana, D. Sank, A. Vainsencher, J. Wenner, T. White, E. Kapit, H. Never, J. Martinis, Nat. Phys. **13**, 146 (2017)
21. M. Fitzpatrick, N.M. Sundaresan, A.C.Y. Li, J. Koch, A.A. Houck, Phys. Rev. X **7**, 011016 (2017)
22. C. Owens, A. LaChapelle, B. Saxberg, B.M. Anderson, R. Ma, J. Simon, D.I. Schuster, Phys. Rev. A **97**, 013818 (2018)
23. Y. Zhang, B.J. Lester, Y.Y. Gao, L. Jiang, R.J. Schoelkopf, S.M. Girvin, Phys. Rev. A **99**, 012314 (2019)
24. E. Kim, X. Zhang, V.S. Ferreira, J. Banker, J.K. Iverson, A. Sipahigil, M. Bello, A. González-Tudela, M. Mirhosseini, O. Painter, Phys. Rev. X **11**, 011015 (2021)
25. Y. Gong, L. Guo, S. Wong, A.J. Bennett, S. Soon Oh, Sci. Rep. **11**, 1055 (2021)

26. H. Zhao, P. Miao, M.H. Teimourpour, S. Malzard, R. El-Ganainy, H. Schomerus, L. Feng, Nat. Commun. **9**, 981 (2018)
27. M. Parto, S. Wittek, H. Hodaei, G. Harari, M.A. Bandres, J. Ren, M.C. Rechtsman, M. Segev, D.N. Christodoulides, M. Khajavikhan, Phys. Rev. Lett. **120**, 113901 (2018)
28. D. Malz, L.D. Tóth, N.R. Bernier, A.K. Feofanov, T.J. Kippenberg, A. Nunnenkamp, Phys. Rev. Lett. **120**, 023601 (2018)
29. A. Youssefi, S. Kono, A. Bancora, M. Chegnizadeh, J. Pan, T. Vovk, T.J. Kippenberg, Nature **612**, 666 (2022)
30. V. Peano, C. Brendel, M. Schmidt, F. Marquardt, Phys. Rev. X **5**, 031011 (2015)
31. D. Porras, S. Fernández-Lorenzo, Phys. Rev. Lett. **122**, 143901 (2019)
32. B. Abdo, A. Kamal, M. Devoret, Phys. Rev. B **87**, 014508 (2013)
33. N.E. Frattini, U. Vool, S. Shankar, A. Narla, K.M. Sliwa, M.H. Devoret, Appl. Phys. Lett. **110**, 222603 (2017)
34. V.V. Sivak, N.E. Frattini, V.R. Joshi, A. Lingenfelter, S. Shankar, M.H. Devoret, Phys. Rev. Appl. **11**, 054060 (2019)
35. A. Pocklington, Y. Wang, A.A. Clerk, arXiv:2210.09252 [quant-ph] (2022)
36. I. Gohberg, P. Lancaster, L. Rodman, *Indefinite Linear Algebra and Applications* (Springer, Berlin, 2005)
37. J.C. Budich, E.J. Bergholtz, Phys. Rev. Lett. **125**, 180403 (2020)
38. A. McDonald, A.A. Clerk, Nat. Commun. **11**, 5382 (2020)
39. D. Gottesman, A. Kitaev, J. Preskill, Phys. Rev. A **64**, 012310 (2001)
40. C. Flühmann, T.L. Nguyen, M. Marinelli, V. Negnevitsky, K. Mehta, J.P. Home, Nature **566**, 513 (2019)
41. P. Campagne-Ibarcq, A. Eickbusch, S. Touzard, E. Zayls-Geller, N.E. Frattini, V.V. Sivak, P. Reinhold, S. Puri, S. Shankar, R.J. Schoelkopf, L. Frunzio, M. Mirrahimi, M.H. Devoret, Nature **584**, 368 (2020)

Appendix A
Spectra and Pseudospectra of Block-Toeplitz Matrices and Operators

The focus on 1D lattice models throughout this thesis required us to study the spectral and pseudospectral properties of certain classes of matrices and operators. The four most commonly encountered classes are block-Toeplitz matrices, block-circulant matrices, block-Toeplitz operators, and block-Laurent operators. Each of these correspond to various configurations of BCs, see Table 2.1. In addition, we have considered *corner-modified* block-Toeplitz matrices which arise when one considers lattice models subject to arbitrary BCs. In this appendix, we collect the essential mathematical properties of these objects which may be useful for interested readers to gain a deeper, self-contained understanding. We will also provide a summary of the techniques developed in Ref. [1] for casting corner-modified block-Toeplitz matrices into Jordan canonical form and apply these techniques to the BKC Hamiltonian subject to generalized BCs, i.e., Eq. (3.25).

The general form of a $dN \times dN$ *block-Toeplitz matrix* is[1]

$$\mathbf{X}_N = \begin{bmatrix} \mathbf{x}_0 & \mathbf{x}_{-1} & & \cdots & \mathbf{x}_{1-N} \\ \mathbf{x}_1 & \mathbf{x}_0 & & & \vdots \\ & \ddots & \ddots & \ddots & \\ \vdots & & & \mathbf{x}_0 & \mathbf{x}_{-1} \\ \mathbf{x}_{N-1} & \cdots & & \mathbf{x}_1 & \mathbf{x}_0 \end{bmatrix}, \quad \mathbf{x}_j \in \mathbb{C}^{d\times d}.$$

If, in addition, $\mathbf{x}_j = \mathbf{x}_{N-j}$ for $1 \leq j \leq N-1$, then the matrix is called *circulant*. In physical language, circulant matrices correspond to PBCs for lattice models. The symbol associated to $\mathbf{X}_N$ is the matrix-valued function

[1] Note that, in the context of bosonic systems, we adopt the convention $\mathbf{g}_j = \mathbf{x}_{-j}$. This choice stems from conventions established in previous works [1, 2].

V. P. Flynn, *Effective Non-Hermiticity and Topology in Markovian Quadratic Bosonic Dynamics*, Springer Theses, https://doi.org/10.1007/978-3-031-52045-7

$$\mathbf{x}(z) = \sum_{j=1-N}^{N-1} \mathbf{x}_j z^j, \quad z \in \mathbb{C}.$$

One can regard $\mathbf{X}_N$ as the result of truncating an infinite matrix. There are two natural possibilities for the infinite matrices themselves. They are

$$\mathbf{X}^{(\mathrm{T})} = \begin{bmatrix} \mathbf{x}_0 & \mathbf{x}_{-1} & \cdots \\ \mathbf{x}_1 & \mathbf{x}_0 & \ddots \\ \vdots & \ddots & \ddots \end{bmatrix}, \quad \mathbf{X}^{(\mathrm{L})} = \begin{bmatrix} \ddots & \ddots & \ddots & \\ \ddots & \mathbf{x}_0 & \mathbf{x}_{-1} & \ddots \\ \ddots & \mathbf{x}_1 & \mathbf{x}_0 & \ddots \\ & \ddots & \ddots & \ddots \end{bmatrix}.$$

Provided that suitable convergence conditions are met, the first (second) matrix defines a block-Toeplitz (block-Laurent) operator. The symbols of these two operators are defined naturally and coincide. From this point of view, $\{\mathbf{X}_N\}$ is a finite section of either $\mathbf{X}^{(\mathrm{T}),(\mathrm{L})}$. Both of these operators are uniquely determined by the bi-infinite series of $d \times d$ matrices $(\cdots, \mathbf{x}_{-1}, \mathbf{x}_0, \mathbf{x}_1, \cdots)$.

Generally, the symbol contains a plethora of information about the matrices and operators of interest. For example, we have the bound (see Ch. 6.1 of Ref. [3], for instance)

$$\|\mathbf{X}^{(\mathrm{T})}\| = \sup_{k\in[-\pi,\pi]} \|\mathbf{x}(e^{ik})\|. \tag{A.1}$$

In particular, we have $\|\mathbf{X}_N\| \leq \sup_{k\in[-\pi,\pi]}\|\mathbf{x}(e^{ik})\|$. This follows from the bound $\|\mathbf{X}_N\| \leq \|\mathbf{X}^{(\mathrm{T})}\|$ which, in turn, can be seen by noting that $\mathbf{X}_N$ may be computed as a projection of $\mathbf{X}^{(\mathrm{T})}$ onto the first N lattice sites. This bound has important consequences for matrix exponentials of block-Toeplitz matrices. Specifically, let $\mathbf{Y}_N$ denote an arbitrary block-Toeplitz matrix and consider the exponential $e^{t\mathbf{Y}_N}$. Then, we always have the bound

$$\left\| e^{t\mathbf{Y}_N} \right\| \leq e^{\Omega_N t}, \quad \forall t \in \mathbb{R}, \tag{A.2}$$

where $\Omega_N \equiv \alpha(\mathbf{X}_N)$ is the spectral abscissa of the the Hermitian matrix $\mathbf{X}_N \equiv (\mathbf{Y}_N + \mathbf{Y}_N^\dagger)/2$. The quantity Ω_N is known as the *numerical abscissa* of $\mathbf{Y}_N$. Since $\mathbf{X}_N$ is Hermitian, it follows that $|\Omega_N| \leq \|\mathbf{X}_N\|$ since the 2-norm of a Hermitian matrix is simply the largest (in absolute value) eigenvalue. Moreover, since $\mathbf{X}_N$ is also a block-Toeplitz matrix, Eq. (A.1) provides us with a system-size independent bound $|\Omega_N| \leq \Omega \equiv \sup_{k\in[-\pi,\pi]}\|\mathbf{x}(e^{ik})\|$, with $\mathbf{x}(z)$ the symbol associated to $\mathbf{X}_N$. The final result is that the matrix exponential is always bounded by a system size-independent quantity:

$$\left\| e^{t\mathbf{Y}_N} \right\| \leq e^{\Omega t}, \quad \forall t \in \mathbb{R}. \tag{A.3}$$

We remark that this bound is trivial in the case where $\Omega_N < 0$ so that $\left\| e^{t\mathbf{Y}_N} \right\| \leq 1$. Thus, it is only useful if $\Omega_N > 0$ (which allows room for $\| e^{t\mathbf{Y}_N} \|$ to grow beyond 1).

For more specific spectral characterizations, it is useful to consider separately the non-block and block cases.,

A.1 The Non-block Case ($d = 1$)

In the non-block case, the symbol is simply a complex-valued function $x(z)$. The following theorem characterizes the spectrum of the circulant matrix, Laurent operator, and Toeplitz operator associated to the symbol, in terms of the symbol itself.

Theorem A.1.1 (Thm. 7.1 in Ref. [4]) *Let* $\mathbf{X}$ *be either a circulant matrix, a Laurent operator, or a Toeplitz operator with continuous symbol* x*. Then*

(i) If $\mathbf{X} = \mathbf{X}_N$ *is circulant, then* $\sigma(\mathbf{X}) = \{x(e^{i2m\pi/N}) : m = 1, \ldots, N\}$.
(ii) If $\mathbf{X} = \mathbf{X}^{(L)}$ *is a Laurent operator, then* $\sigma(\mathbf{X}^{(L)}) = \bigcup_{k \in (-\pi,\pi]} x(e^{ik})$.
(iii) If $\mathbf{X} = \mathbf{X}^{(T)}$ *is a Toeplitz operator, then* $\sigma(\mathbf{X}^{(T)}) = \sigma(\mathbf{X}^{(L)}) \cup \{\lambda : \nu(\lambda, x) \neq 0\}$, *where*

$$\nu(\lambda, x) \equiv \frac{1}{2\pi i} \int_{-\pi}^{\pi} \frac{d}{dk} \ln(x(e^{ik}) - \lambda)\, dk \tag{A.4}$$

is the winding number *of the symbol about* λ.

The case when $\mathbf{X}_N$ is a Toeplitz matrix (which, physically, corresponds to OBCs in lattice models) is far more complicated. At a bare minimum, if $\mathbf{X}_N$ is banded (meaning that $\mathbf{x}_j = 0$ for all $|j| > R$ with R sufficiently large), then $\sigma(\mathbf{X}_N)$ clusters along curves in $\mathbb{C}$ as $N \to \infty$. In particular, the spectrum of $\mathbf{X}_N$ does not, in general, converge to that of $\mathbf{X}^{(L)}$. By contrast, there exists a complete characterization of the pseudospectrum:

Theorem A.1.2 (Thm. 7.2 of Ref. [4]) *Let* $\{\mathbf{X}_N\}$ *be a family of banded or semibanded Toeplitz matrices and let* λ *be any complex number with* $\nu(\lambda, x) \neq 0$*. Then for some* $M > 1$ *and all sufficiently large* N*, we have* $\left\| (\lambda \mathbb{1}_N - \mathbf{X}_N)^{-1} \right\| \geq M^N$*, and there exist normalized (pseudoeigen)vectors* $\vec{v}^{(N)}$ *satisfying* $\left\| (\mathbf{X}_N - \lambda \mathbb{1}_N)\vec{v}^{(N)} \right\| \leq M^{-N}$*, such that*

$$\frac{|v_j^{(N)}|}{\max_j |v_j^{(N)}|} \leq \begin{cases} M^{-j}, & \nu(\lambda, x) < 0, \\ M^{j-N}, & \nu(\lambda, x) > 0, \end{cases} \quad 1 \leq j \leq N.$$

The constant M can be taken to be any number for which $x(z) \neq \lambda$ in the annulus $1 \leq |z| \leq M$ (if $\nu(x, \lambda) < 0$) or $M^{-1} \leq |z| \leq 1$ (if $\nu(\lambda, x) > 0$).

Stated plainly, the winding number of the symbol defined both the ϵ-pseudospectrum (with ϵ exponentially small in system size) and the localization profiles of the associated pseudoeigenvectors. Finally, the large N limit of the pseudospectrum enjoys a full characterization via the following theorem:

Theorem A.1.3 (Thm. 7.3 in Ref. [4]) *Let $\mathbf{X}^{(T)}$ be a Toeplitz operator with continuous symbol x and let $\{\mathbf{X}_N\}$ be the associated family of Toeplitz matrices. Then for any $\epsilon > 0$,*

$$\lim_{N\to\infty} \sigma_\epsilon(\mathbf{X}_N) = \sigma_\epsilon(\mathbf{X}^{(T)}), \quad \textit{and thus} \quad \lim_{\epsilon\to 0}\lim_{N\to\infty} \sigma_\epsilon(\mathbf{X}_N) = \sigma(\mathbf{X}^{(T)}). \tag{A.5}$$

Accordingly, the two limits in the above equation do *not* commute in general. They do commute if the Toeplitz matrices and operator are normal (in particular, Hermitian).

A.2 Block Case ($d > 1$)

Things become considerably more complex in the block case. For the spectrum, points (i) and (ii) in Theorem A.1.1 still hold under the replacement $x(e^{ik}) \mapsto \sigma(\mathbf{x}(e^{ik}))$ since $\mathbf{x}(z)$ is now matrix-valued. One may predict that (iii) also still persists but with the replacement $x(e^{ik}) \mapsto \det \mathbf{x}(e^{ik})$ in Eq. (A.4), but that is incorrect. Instead, the spectrum of the block-Toeplitz operator is given by that of the corresponding Laurent operator plus all $\lambda \in \mathbb{C}$ where the matrix $\mathbf{x}(z) - \lambda \mathbb{1}_d$ has at least one non-vanishing *partial index*. Partial indices are defined in the context of the Wiener-Hopf factorization and may be thought of as a generalization of the winding number. As it turns out, the aforementioned generalization of the winding number involving $\det \mathbf{x}(z)$ is equal to the sum of the partial indices. Thus, there may be spectral points about which $\det \mathbf{x}(e^{ik})$ does not wind. Conversely, every point with nonzero winding must be in the spectrum.

Unfortunately, the characterization of the pseudospectra is similarly difficult. In particular, it necessitates defining the *associated symbol*, $\tilde{\mathbf{x}}(z) \equiv \mathbf{x}(z^{-1})$. The *associated block-Toeplitz operator* $\tilde{\mathbf{X}}^{(\mathrm{T})}$ is defined analogously. With this, we have the following.

Theorem A.2.1 (Adapted from Cor. 6.16 in Ref. [3]) *Let $\mathbf{X}^{(T)}$ and be a block-Toeplitz operator with continuous symbol $\mathbf{x}$ with corresponding family of block-Toeplitz matrices $\{\mathbf{X}_N\}$ and $\tilde{\mathbf{X}}^{(T)}$ be the associated block-Toeplitz operator. Then for any $\epsilon > 0$,*

$$\lim_{N\to\infty} \sigma_\epsilon(\mathbf{X}_N) = \sigma_\epsilon(\mathbf{X}^{(T)}) \cup \sigma_\epsilon(\tilde{\mathbf{X}}^{(T)}), \quad \textit{and thus}$$

$$\lim_{\epsilon\to 0}\lim_{N\to\infty} \sigma_\epsilon(\mathbf{X}_N) = \sigma(\mathbf{X}^{(T)}) \cup \sigma(\tilde{\mathbf{X}}^{(T)}). \tag{A.6}$$

This result Eq. (A.6) is the natural generalization of Eq. (A.5) to the block case.

A.3 Jordan Canonical Form of Corner-Modified Banded Block-Toeplitz Matrices

The above results have provided us a characterization of the pseudospectra of block-Toeplitz matrices. However, we have not yet described their spectra. To this end, let us now describe the methodology developed in Ref. [1] for casting corner-modified block-Toeplitz matrices (of which block-Toeplitz matrices are a special case) into Jordan canonical form. Since our sole application of this methodology was to QBHs, we will present the key concepts in the language of bosonic dynamical matrices. Specifically, we will focus on matrices of the form Eq. (2.83). To simplify the discussion, and to remain consistent with the notation of the previous sections, let us relabel the main matrices and operators of interest:

$$\mathbf{G}_N \equiv \mathbf{G}_N^{\text{OBC}}, \quad \mathbf{G}^{(\text{L})} \equiv \mathbf{G}^{\text{BIBC}}, \quad \mathbf{G}^{(\text{T})} \equiv \mathbf{G}^{\text{SIBC}} \tag{A.7}$$

As in Eq. (2.83), $\mathbf{G} = \mathbf{G}_N + \mathbf{B}$, with $\mathbf{B}$ the boundary modification. We also define the symbol

$$\mathbf{g}(z) \equiv \sum_{j=-R}^{R} \mathbf{g}_j z^j, \quad z \in \mathbb{C}, \tag{A.8}$$

whose restriction to the unit circle $|z| = 1$ we recognize as the Bloch dynamical matrix in Eq. (2.80). Notations for the various shift operators will remain as they were in the main text (e.g., as in Sect. 2.2). Finally, we will assume there is only one internal degree of freedom per lattice site.

The key to diagonalizing these matrices is to identify the so-called *bulk-boundary separation*. We define the *bulk* and *boundary projectors* as $\mathbf{P}_B \equiv \sum_{j=R+1}^{N-R} \vec{e}_j \vec{e}_j^{\,\dagger} \otimes \mathbb{1}_2$ and $\mathbf{P}_\partial = \mathbb{1}_{2N} - \mathbf{P}_B$, respectively. The goal is to solve the eigenvalue equation $\mathbf{G}_N \vec{\psi} = \omega \vec{\psi}$. Since $\mathbf{P}_B + \mathbf{P}_\partial = \mathbb{1}_{2N}$ and $\mathbf{P}_B \mathbf{B} = 0$, the eigenproblem is equivalent to the following "bulk-boundary system of equations" [5]:

$$\mathbf{P}_B \mathbf{G}_N \vec{\psi} = \omega \mathbf{P}_B \vec{\psi}, \tag{A.9}$$

$$\mathbf{P}_\partial (\mathbf{G}_N + \mathbf{B}) \vec{\psi} = \omega \mathbf{P}_\partial \vec{\psi}. \tag{A.10}$$

The diagonalization proceeds by first solving the *bulk equation*, Eq. (A.9), parametrically in ω, and then employing the resulting solutions as an Ansatz for the *boundary equation*, Eq. (A.10). One can show that, generically, such a strategy yields *all* of the eigenvectors of $\mathbf{G}$ and can also be applied for computing generalized eigenvectors [1].

For fixed $\omega \in \mathbb{C}$, the complete set of solutions to the bulk Eq. (A.9) breaks up into three different types of solutions (which we will derive momentarily). Solutions of the first type are obtained by restricting to the finite-lattice solutions of the translation-invariant equation $(\mathbf{G}^{(\mathrm{L})} - \omega)^n \vec{\psi} = 0$, for some suitable n, and thus arise from eigenvectors and generalized eigenvectors of $\mathbf{G}^{(\mathrm{L})}$. Specifically, these solutions take the form

$$\vec{\psi}_{\ell s} = \sum_{\nu=1}^{s_\ell} \vec{\zeta}_\nu(z_\ell) \otimes |u_{\ell s \nu}\rangle,$$

where the z_ℓ are the roots of the equation $\det(\mathbf{g}(z) - \omega) = 0$ with algebraic multiplicity s_ℓ, and the vectors $\vec{\zeta}_\nu(z_\ell)$ are as follows: for $\nu = 1$, $\vec{\zeta}_1(z_\ell) = \sum_{j=1}^{N} z_\ell^j \vec{e}_j$ represents a *generalized Bloch wave*, with possibly complex momentum $k_\ell = -i\log(z_\ell)$; for $\nu > 1$ the $\vec{\zeta}_\nu(z_\ell)$ are proportional to $\partial_z^{\nu-1}\vec{\zeta}_\nu(z_\ell)$, and hence contain amplitudes with a power-law pre-factor to the exponential weight z^j. The other two types of solutions that can arise are localized on the boundary of the system and are no longer controlled by $\mathbf{G}^{(\mathrm{L})}$ and the corresponding (non-unitary) translation symmetry. Rather, they emerge entirely due to the truncation from the bi-infinite lattice to a finite one. We will denote these left ($-$) and right ($+$) localized *emergent solutions* by $\vec{\psi}_\ell^{\pm}$, with $\ell = 1, \ldots, s_0 \equiv 2R - \frac{1}{2}\sum_{\ell=1}^{n} s_\ell$. Here, s_0 is the multiplicity of $z = 0$ as a root of $\det(\mathbf{g}(z) - \omega) = 0$. Finally, we remark that there may exist exceptional, isolated values of ω, which physically correspond to dispersion-less "flat bands" and whose associated eigenvectors are not included among the previous three types of solutions. While we refer to [1] for more discussion, flat bands will not be encountered in the models under consideration in this paper.

The complete set of solutions to the bulk equation may thus be parameterized as follows:

$$\vec{\psi}_{\vec{\alpha}} = \sum_{\ell=1}^{n}\sum_{s=1}^{s_\ell} \alpha_{\ell s}\vec{\psi}_{\ell s} + \sum_{\ell=1}^{s_0} \alpha_\ell^- \vec{\psi}_\ell^- + \sum_{\ell=1}^{s_0} \alpha_\ell^+ \vec{\psi}_\ell^+, \tag{A.11}$$

where $\vec{\alpha} \equiv [\alpha_{11}, \ldots, \alpha_{ns_n}, \alpha_1^-, \ldots, \alpha_{s_0}^-, \alpha_1^+, \ldots, \alpha_{s_0}^+]^T \in \mathbb{C}^{4R}$. Using $\vec{\psi}_{\vec{\alpha}}$ as an Ansatz for the boundary equation, Eq. (A.10), leads to the identity

$$\mathbf{P}_\partial(G - \omega\mathbb{1}_{2N})\vec{\psi}_{\vec{\alpha}} = \sum_b \vec{e}_b(B(\omega)\vec{\alpha})_b, \quad b \in \{1, \ldots, R, N-R+1, \ldots, N\}. \tag{A.12}$$

Here, the *boundary matrix*[2] $B(\omega)$ has elements $B_{bs}(\omega)$ that consist of 2×1 blocks and are given by $B_{bs}(\omega) = \vec{e}_b{}^\dagger \, (\mathbf{G} - \omega \mathbb{1}_{2N}) \, \vec{\Psi}$, with

$$\vec{\Psi} \equiv [\vec{\psi}_{11}, \ldots, \vec{\psi}_{ns_n}, \vec{\psi}_1^-, \ldots, \vec{\psi}_{s_0}^-, \vec{\psi}_1^+, \ldots, \vec{\psi}_{s_0}^+]^T. \tag{A.13}$$

Equation (A.12) tells us that if $B(\omega)\vec{\alpha} = 0$, then $\vec{\psi}_{\vec{\alpha}}$ solves *both* the bulk and boundary equations and hence is an eigenvector of $\mathbf{G}_N$ with eigenvalue ω, as desired.

For diagonalizable matrices, the above procedure yields a Bloch-like diagonal basis. However, G may fail to be diagonalizable, in which case the generalized eigenvectors of G are needed in addition to the eigenvectors to complete a basis. One can calculate some generalized eigenvectors in Bloch-like form by repeating the above procedure to determine $\ker(\mathbf{G} - \omega \mathbb{1}_{2N})^p$ for various powers p and each eigenvalue ω. However, there is a constraint $p < (N-1)/R \equiv p_{\max}$ on how large p can be because, for $p \geq p_{\max}$, $(\mathbf{G} - \omega \mathbb{1}_{2N})^p$ need not be a corner-modified block-Toeplitz matrix. If there are any, generalized eigenvectors of rank greater than $p_{\max} - 1$ may have to be determined by means other than the above bulk-boundary separation. As it turns out, the BKC, when considered in certain parameter regimes, offers examples of this peculiar phenomenon.

Let us now derive the three types of solutions that make up the Ansatz Eq˙ (A.11). Consider the bi-infinite matrix (Laurent operator) $\mathbf{G}^{(\mathrm{L})}$. Then if $\mathbf{G}^{(\mathrm{L})}\vec{\psi} = \omega\vec{\psi}$, it follows that $\vec{\psi}^N \equiv \mathbf{P}_{1,N}\vec{\psi}$, with $\mathbf{P}_{1,N} = \sum_{j=1}^N \vec{e}_j \vec{e}_j^\dagger \otimes \mathbb{1}_2$, is a solution of the bulk equation. In the generic case where $\det \mathbf{g}_{\pm R} \neq 0$, this method yields the complete set of solutions to the bulk equations, i.e., $\ker \mathbf{P}_B(\mathbf{G}_N - \omega \mathbb{1}_{2N}) = \mathbf{P}_{1,N} \ker(\mathbf{G}^{(\mathrm{L})} - \omega \mathbf{1})$. For the time being, we restrict ourselves to this case.

If we were interested in diagonalizing $\mathbf{G}^{(\mathrm{L})}$ on its own, we would restrict to only the eigenvectors that are normalizable on the corresponding Hilbert space. Crucially, this does *not* capture the full kernel of $\mathbf{G}^{(\mathrm{L})} - \omega\mathbf{1}$, however: since we only consider the finite-lattice projections, the non-normalizable elements of $\ker(\mathbf{G}^{(\mathrm{L})} - \omega\mathbf{1})$ also provide solutions to the bulk equation. Furthermore, in the space of all bi-infinite sequences, the left and right translation operators $\mathbf{V}$ and $\mathbf{V}^{-1}$ are no-longer unitary and so these operators need not have spectra restricted to the unit circle.

The translation invariance of $\mathbf{G}^{(\mathrm{L})}$ manifests as the vanishing commutators $[\mathbf{G}^{(\mathrm{L})}, \mathbf{V}] = [\mathbf{G}^{(\mathrm{L})}, \mathbf{V}^{-1}] = 0$. Hence, it is possible to construct simultaneous eigenvectors of $\mathbf{G}^{(\mathrm{L})}$, $\mathbf{V}$, and $\mathbf{V}^{-1}$. The simultaneous eigenvectors of $\mathbf{V}$ and $\mathbf{V}^{-1}$ are given by $\vec{f}_1(z) \equiv \sum_{j\in\mathbb{Z}} z^j \vec{e}_j$, where z is an arbitrary, non-zero complex number. Explicitly, $\mathbf{V}\vec{f}_1(z) = z\vec{f}_1(z)$ and $\mathbf{V}^{-1}\vec{f}_1(z) = z^{-1}\vec{f}_1(z)$, which immediately lead to the identity

$$\mathbf{G}^{(\mathrm{L})}\vec{f}_1(z) \otimes \vec{u} = \vec{f}_1(z) \otimes \mathbf{g}(z)\vec{u}, \tag{A.14}$$

[2] Note that, despite being a matrix, we forgoe the bold notation for the boundary matrix as to avoid conflation with the boundary modification **B**.

where $\vec{u} \in \mathbb{C}^2$ is arbitrary. We see that for any $z \neq 0$ such that $\mathbf{g}(z)\vec{u} = \omega\vec{u}$, $\vec{f}_1(z)\otimes\vec{u}$ is an eigenvector of $\mathbf{G}^{(\mathrm{L})}$ with eigenvalue ω.

To continue, we define the complex characteristic polynomial $P(\omega, z) \equiv z^{4R}\det(\mathbf{g}(z) - \omega\mathbb{1}_2)$. We call an eigenvalue ω *regular* if $P(\omega, z)$ is not the zero polynomial. Otherwise, we say ω is *singular*. For the applications in this paper, it suffices to restrict to the eigenvalues ω that are regular. For a fixed ω, let $\{z_\ell\}_{\ell=1}^n$ denote the n distinct roots of $P(\omega, z)$ and $\{s_\ell\}_{\ell=1}^n$ denote their corresponding multiplicities. Generically, $\mathbf{g}(z_\ell)$ will have s_ℓ eigenvectors $\{\vec{u}_{\ell s}\}_{s=1}^{s_\ell}$ satisfying $\mathbf{g}(z_\ell)\vec{u}_{\ell s} = \omega\vec{u}_{\ell s}$, in which case, the vectors

$$\vec{\zeta}_1(z_\ell) \otimes \vec{u}_{\ell s} \equiv \sum_{j=1}^{N} z_\ell^j \vec{e}_j \otimes \vec{u}_{\ell s} = \mathbf{P}_{1,N}\vec{f}_1(z_\ell) \otimes \vec{u}_{\ell s}, \tag{A.15}$$

are solutions to the bulk equation, and akin to Bloch waves with complex momentum.

When the symbol $\mathbf{g}(z_\ell)$ has less than s_ℓ eigenvectors, the remaining solutions are constructed from the generalized eigenvectors of the left and right translation operators. The sequences

$$\vec{f}_n(z) \equiv \frac{1}{(n-1)!}\partial_z^{n-1}\vec{f}_1(z)$$

span the kernel of $(\mathbf{V} - z)^s$ for $\nu = 1, \ldots, s$. Furthermore,

$$\mathbf{G}^{(\mathrm{L})}\vec{f}_n(z) \otimes \vec{u} = \frac{1}{(n-1)!}\partial_z^{n-1}\vec{f}_1(z) \otimes \mathbf{g}(z)\vec{u}.$$

One can then show that the sequence $\vec{\psi} \equiv \sum_{n=1}^{\nu}\vec{f}_n(z)\vec{u}_n$ satisfy

$$\mathbf{G}^{(\mathrm{L})}\vec{\psi} = \sum_{n=1}^{\nu}\sum_{m'=1}^{\nu}\vec{f}_m(z) \otimes [\mathbf{g}_\nu(z)]_{mm'}\vec{u}_{m'},$$

where $\mathbf{g}_\nu(z)$ is an upper-triangular block-Toeplitz matrix with non-zero blocks

$$[\mathbf{g}_\nu(z)]_{mm'} = \frac{1}{(m'-m)!}\partial_z^{m'-m}\mathbf{g}(z), \quad 1 \leq m \leq m' \leq \nu. \tag{A.16}$$

It can then be shown that the eigenspace of $\mathbf{G}^{(\mathrm{L})}$ corresponding to eigenvalue ω is a direct sum of n vector spaces spanned by generalized eigenvectors of $\mathbf{V}^{\pm 1}$ of the form

$$\Psi_{\ell s} = \sum_{\nu=1}^{s_\ell}\Phi_{z_\ell,\nu}\vec{u}_{\ell s\nu},$$

where the linearly independent vectors $\{\vec{u}_{\ell s\nu}\}$ are chosen in such a way that $\mathbf{g}_{s_\ell}(z_\ell)\vec{u}_{\ell s} = \omega \vec{u}_{\ell s}$, with $\vec{u}_{\ell s} = [\vec{u}_{\ell s 1}, \ldots, \vec{u}_{\ell s s_\ell}]^T$. With these, we obtain $\sum_{\ell=1}^{n} s_\ell$ solutions to the bulk equation given by

$$\vec{\psi}_{\ell s} = \sum_{\nu=1}^{s_\ell} \vec{\zeta}_\nu(z_\ell)\vec{u}_{\ell s\nu}, \quad \vec{\zeta}_\nu(z_\ell) = \mathbf{P}_{1,N}\vec{f}_\nu(z_\ell)$$

If $g_{\pm R}$ are not invertible, then there exists $2s_0 \equiv 4R - \sum_{\ell=1}^{n} s_\ell$ additional boundary localized solutions to the bulk equation, where s_0 is the multiplicity of $z = 0$ as a root of the characteristic polynomial $P(\omega, z)$ for a given regular eigenvalue ω. We will now demonstrate how to construct the left ($j = 1$) localized solutions. Since these solutions emerge due to the truncation of the bi-infinite lattice to a finite one, we consider the semi-infinite dynamical matrix $\mathbf{G}^{(\mathrm{T})}$ and the and unilateral shift operators $\mathbf{T}$ and $\mathbf{T}^\dagger$ (Eq. (2.31)). The corresponding half-infinite bulk projector is

$$\mathbf{P}_B^- \equiv \sum_{j=R+1}^{\infty} \vec{e}_j \vec{e}_j^{\,\dagger} \otimes \mathbb{1}_2 = \mathbf{T}^{\dagger R}\mathbf{T}^R \otimes \mathbb{1}_2.$$

Now, suppose there is a vector $\vec{\Upsilon}^-$, that solves the half-infinite bulk equation

$$\mathbf{P}_B^- \left(\mathbf{G}^{(\mathrm{T})} - \omega \mathbf{1}_-\right)\vec{\Upsilon}^- = 0.$$

Then one can verify that $\vec{\psi} = \mathbf{P}_{1,N}\vec{\Upsilon}^-$ is a solution to the bulk equation. The emergent solutions are precisely those derived from the half-infinite bulk equation and not the bi-infinite eigenvalue problem. Since $\mathbf{T}\mathbf{T}^\dagger = \mathbf{1}_-$, we may write $\mathbf{P}_B^- \left(\mathbf{G}^{(\mathrm{T})} - \omega \mathbf{1}_-\right) = \mathbf{T}^{\dagger R}\mathbf{K}^-(\omega, \mathbf{T})$, where $\mathbf{K}^-(\omega, z)$ is the matrix polynomial $\mathbf{K}^-(\omega, z) \equiv z^R\left(\mathbf{g}(z) - \omega \mathbb{1}_2\right)$. Thus, the s_0 left-localized emergent solutions to the bulk equation are determined by the kernel of the matrix $\mathbf{K}_{s_0}^-(\omega, z_0 = 0) \equiv \mathbf{K}^-(\omega)$, with $\mathbf{K}_\nu^-(\omega, z)$ constructed exactly as in Eq. (A.16). Given a basis $\{\vec{u}_s^-\}_{s=1}^{s_0}$ for $\ker \mathbf{K}^-(\omega)$, with $\vec{u}_s^- = [\vec{u}_{s1}^-, \vec{u}_{s2}^-, \ldots, \vec{u}_{ss_0}^-]^T$, we can construct s_0 left localized solutions to the bulk equation given by

$$\vec{\psi}_s^- = \sum_{j=1}^{s_0} \vec{e}_j \otimes \vec{u}_{sj}^-.$$

The remaining s_0 right-localized solutions, with support on $j = N$, can be found in an analogous way. Explicitly, they can be constructed using the kernel vectors $\{\vec{u}_s^+\}_{s=1}^{s_0}$ of the matrix $\mathbf{K}^+(\omega) = \boldsymbol{\tau}_3[\mathbf{K}^-(\omega)]^\dagger \boldsymbol{\tau}_3$. That is, if $\vec{u}_s^+ = [\vec{u}_{s1}^+, \vec{u}_{s2}^+, \ldots, \vec{u}_{ss_0}^+]^T$, then the vectors

$$\vec{\psi}_s^+ = \sum_{j=1}^{s_0} \vec{e}_{N-s_0+j} \otimes \vec{u}_{sj}^+, \quad s = 1, \ldots, s_0,$$

provide right-localized solutions to the bulk equation.

Appendix B
Miscellaneous Technical Calculations

B.1 Existence of a Bosonic Eigenbasis for Dynamically Stable QBHs

Let $\mathbf{G}$ be the dynamical matrix of a dynamically stable QBH. In this appendix, we will prove that there exists a *bosonic eigenbasis* for $\mathbf{G}$, i.e., an eigenbasis satisfying Eq. (2.63), in the case where there are no ZMs.

Suppose we have the set of eigenvalues $\omega_n \in \mathbb{R}$ of $\mathbf{G}$ with eigenvectors $\vec{v}_n$. If $\omega_n \neq \omega_m$, then pseudo-Hermiticity provides

$$\vec{v}_n^\dagger \boldsymbol{\tau}_3 \mathbf{G} \vec{v}_m = \vec{v}_n^\dagger \mathbf{G}^\dagger \boldsymbol{\tau}_3 \vec{v}_m \Rightarrow (\omega_n - \omega_m) \vec{v}_n^\dagger \boldsymbol{\tau}_3 \vec{v}_m^\dagger = 0, \tag{B.1}$$

so that $\vec{v}_n^\dagger \boldsymbol{\tau}_3 \vec{v}_m = 0$. Now, if ω_n is a (possibly degenerate) non-zero eigenvalue with d_n eigenvectors $\vec{v}_{n,j}$, $j = 1, \ldots, d_n$, then

$$(\mathbf{M}_n)_{jk} = \omega_n^{-1} \, \vec{v}_{n,j}^\dagger \boldsymbol{\tau}_3 \mathbf{G} \vec{v}_{n,k} = \omega_n^{-1} \, \vec{v}_{n,j}^\dagger \mathbf{H} \vec{v}_{m,k} \tag{B.2}$$

is a $d_n \times d_n$ matrix. From the final equality, we observe that $\mathbf{M}_n$ is (i) Hermitian, thanks to Hermiticity of $\mathbf{H}$, and (ii) invertible, since we assume $\mathbf{G}$, and hence $\mathbf{H}$, lacks zero frequency eigenvectors. Let $\vec{\alpha}_\ell$ and $r_\ell \neq 0$, $\ell = 1, \ldots, d_n$, orthonormal eigenvectors and eigenvalues of $\mathbf{M}$, respectively. We then define the d_n vectors

$$\vec{\psi}_{n,\ell}^{s_\ell} \equiv \frac{1}{\sqrt{|r_\ell|}} \sum_{j=1}^{d_n} (\vec{\alpha}_\ell)_j \vec{v}_{n,j}, \quad \ell = 1, \ldots, d_n, \quad s_\ell \equiv \mathrm{sgn}(r_\ell). \tag{B.3}$$

It is easy to see that these are eigenvectors of $\mathbf{G}$ corresponding to eigenvalue ω_n. Furthermore,

V. P. Flynn, *Effective Non-Hermiticity and Topology in Markovian Quadratic Bosonic Dynamics*, Springer Theses, https://doi.org/10.1007/978-3-031-52045-7

$$\vec{\psi}_{n,\ell}^{s_\ell\,\dagger}\boldsymbol{\tau}_3\vec{\psi}_{n,\ell'}^{s_{\ell'}} = \frac{1}{\sqrt{|r_\ell||r_{\ell'}|}}\sum_{j,k=1}^{d_n}(\vec{\alpha}_\ell)_j^*(\vec{\alpha}_{\ell'})_k\vec{v}_{n,j}^\dagger\boldsymbol{\tau}_3\vec{v}_{n,k} \tag{B.4}$$

$$= \frac{1}{\sqrt{|r_\ell||r_{\ell'}|}}\sum_{j,k=1}^{d_n}(\vec{\alpha}_\ell)_j^*(\vec{\alpha}_{\ell'})_k(\mathbf{M}_n)_{jk} \tag{B.5}$$

$$= \frac{1}{\sqrt{|r_\ell||r_{\ell'}|}}\vec{\alpha}_\ell^\dagger\mathbf{M}_n\vec{\alpha}_{\ell'} \tag{B.6}$$

$$= \frac{1}{\sqrt{|r_\ell||r_{\ell'}|}}r_\ell\delta_{\ell\ell'} = s_\ell\delta_{\ell\ell'}. \tag{B.7}$$

We then form the bosonic basis by taking $\vec{\psi}_n^+$ to be the (necessarily N) eigenvectors $\vec{\psi}_{n,\ell}^{s_\ell}$, with $s_\ell = 1$. Here, n is understood to now be a multi-index (n, ℓ). We complete the basis with the remaining N eigenvectors $\vec{\psi}_n^- = \boldsymbol{\tau}_1(\vec{\psi}_n^{s_\ell})^*$ which correspond to frequencies $-\omega_n$.

B.2 Eigendecomposition of the BKC Under Various BCs

The goal of this appendix will be to diagonalize (or cast in to Jordan normal form, in the non-diagonalizable cases) the dynamical matrix (Eq. (3.26)) of the BKC Hamiltonian (Eq. (3.25)) under various BCs. To accomplish this, we will utilize the techniques of Appendix A.3.

B.2.1 Open BCs

First, note that the the internal matrices $\mathbf{g}_{\pm 1}$ commute. A basis of simultaneous eigenvectors is thus given by $\vec{v}_\pm \equiv (1/\sqrt{2})[1, \pm 1]^T$. This allows us to write

$$-i\mathbf{G}^{\rm OBC} = \frac{1}{2}\left(\gamma_+\mathbf{T}^\dagger - \gamma_-\mathbf{T}\right)\otimes\vec{v}_+\vec{v}_+^\dagger + \frac{1}{2}\left(\gamma_-\mathbf{T}^\dagger - \gamma_+\mathbf{T}\right)\otimes\vec{v}_-\vec{v}_-^\dagger, \quad \gamma_\pm \equiv J \pm \Delta$$

When $J = \Delta$ ($\gamma_- = 0$), we see that the generalized eigenvectors are constructed from those of $\mathbf{T}$ and $\mathbf{T}^\dagger$. Specifically, $\vec{\chi}_{1k} = (-i\gamma_+)^{-k+1}\vec{e}_k \otimes \vec{v}_-$ and $\vec{\chi}_{2k} = (i\gamma_+)^{-k+1}\vec{e}_{N+k-1} \otimes \vec{v}_+$, with $k = 1, \ldots, N$ in both cases, provide two length-N Jordan chains at eigenvalue $\omega = 0$.

Henceforth, we restrict to the case $J \neq \Delta$. Thus, the problem reduces to diagonalizing an $N \times N$ Toeplitz matrix of the form $\mathbf{X} = \left(a\mathbf{T} + b\mathbf{T}^\dagger\right)/2$, with $a, b \in \mathbb{R} \setminus \{0\}$. The symbol $x(z) = (az + bz^{-1})/2$ and the corresponding characteristic polynomial $P(z, \omega) = z\,(x(z) - \omega) = (az^2 + b)/2 - \omega z$. The roots are $z_\pm = (1/a)\left(\omega \pm \sqrt{\omega^2 - ab}\right)$, which satisfy $z_- = c/z_+$ in terms of $c \equiv b/a$. These roots only coalesce when $\omega = \omega_\pm \equiv \pm\sqrt{ab}$.

For the case $\omega \neq \omega_\pm$, the two bulk eigenstates are $\vec{\zeta}_1(z_\pm)$ which yields the boundary matrix

$$B(\omega) = \frac{1}{2}\begin{bmatrix} -b & -b \\ z_+^{N-1}(b - 2\omega z_+) & (c/z_+)^{N-1}(b - 2\omega z_+^{-1}) \end{bmatrix}, \qquad c \equiv b/a.$$

It can be quickly checked that $B(-\omega)$ is similar to $B(\omega)$ and so the spectrum is necessarily symmetric about $\omega = 0$. The condition for a nontrivial kernel ($\det B(\omega) = 0$) reduces to the equation

$$z_+^{2N-2}(b - 2\omega z_+) = c^N \left(a - 2\omega z_+^{-1}\right).$$

The $2N$ roots (of which only N are distinct) are given by $z_+ = \pm\sqrt{c}e^{im\pi/(N+1)}$, with $m = 1, \ldots N$. The corresponding N distinct eigenvalues are $\omega_m = \text{sgn}(a)\sqrt{ab}\cos(m\pi/(N+1))$. Note that $\omega_m \neq \omega_\pm$ and so we need not address the case of two coalescing roots. Taking the roots $z_+ = \sqrt{c}e^{im\pi/(N+1)}$ yields the kernel vector $\vec{\alpha} = [1, -1]$. The (unnormalized) eigenvectors are then

$$\vec{\psi}_m = \vec{\zeta}_1(z_m) - \vec{\zeta}_1(c/z_m) = \sum_{j=1}^{N} c^{j/2} \sin\left(\frac{m\pi j}{N+1}\right)\vec{e}_j, \quad (\mathbf{X} - \omega_m \mathbb{1}_N)\,|\psi_m\rangle = 0.$$

With these solutions, we define

$$\vec{\phi}_m^{\pm} \equiv \sum_{j=1}^{N} (-\sigma)^{j/2} e^{\pm jr} \sin\left(\frac{m\pi j}{N+1}\right)\vec{e}_j \otimes \vec{v}_\pm, \quad \omega_m \equiv \sqrt{t^2 - \Delta^2}\cos\left(\frac{m\pi}{N+1}\right),$$

where $r = 1/2\ln(\gamma_+/|\gamma_-|)$. These satisfy

$$\mathbf{G}^{\text{OBC}}\vec{\phi}_m^{\pm} = \begin{cases} \omega_m \vec{\phi}_m^{\pm}, & \text{sgn}(J - \Delta) = 1, \\ \pm\omega_m \vec{\phi}_m^{\pm}, & \text{sgn}(J - \Delta) = -1. \end{cases}$$

These eigenvectors can then be combined to form the eigenvectors used to construct the normal modes described in Sect. 3.3.3.2.

B.2.2 Twisted BCs

B.2.2.1 The Parameter Regime $s = 1, \varphi = \pi/2, J \neq \Delta$

Instead of diagonalizing $\mathbf{G}_T \equiv \mathbf{G}(1, \pi/2)$ directly, we will first perform a unitary rotation $\mathbf{G}'_T \equiv \mathbf{\Sigma}\mathbf{G}_T\mathbf{\Sigma}^\dagger = \mathbf{G}^{\mathrm{OBC}'} + \mathbf{B}'$ in terms of the the matrix $\mathbf{\Sigma}$ defined in Eq. (2.2). The rotated dynamical matrix has a very simple structure in this basis;

$$\mathbf{G}^{\mathrm{OBC}'} = \mathbf{T} \otimes \mathbf{g}'_1 + \mathbf{T}^\dagger \otimes \mathbf{g}_{-1}, \qquad \mathbf{B}' = \vec{e}_N \vec{e}_1^{\,\dagger} \otimes \mathbf{b}'_1 + \vec{e}_1 \vec{e}_N^{\,\dagger} \otimes \mathbf{b}'_{-1}, \tag{B.8}$$

$$\mathbf{g}'_1 = -\frac{i}{2}\begin{bmatrix} \gamma_- & 0 \\ 0 & \gamma_+ \end{bmatrix} = \sigma_2 \mathbf{g}'^{\dagger}_{-1}\sigma_2, \quad \mathbf{b}'_1 = -\frac{i}{2}\begin{bmatrix} 0 & \gamma_- \\ -\gamma_+ & 0 \end{bmatrix} = \sigma_2 \mathbf{b}'^{\dagger}_{-1}\sigma_2 = \mathbf{b}'_{-1}, \tag{B.9}$$

where again $\gamma_+ = J + \Delta$ and $\gamma_- = J - \Delta$. The relevant symbol is

$$\mathbf{g}'(z, z^{-1}) = \mathbf{g}'_1 z + \mathbf{g}_{-1} z^{-1} = -\frac{i}{2}\begin{bmatrix} \gamma_- z - \gamma_+ z^{-1} & 0 \\ 0 & \gamma_+ z - \gamma_- z^{-1} \end{bmatrix}. \tag{B.10}$$

The characteristic polynomial $P(\omega, z) \equiv z^2 \det(\mathbf{g}'(z, z^{-1}) - \omega \mathbb{1}_2)$ has four roots

$$z_1 = \frac{1}{\gamma_-}\left(i\omega - \sqrt{\gamma_+\gamma_- - \omega^2}\right), \quad z_2 = \frac{1}{\gamma_+}\left(i\omega - \sqrt{\gamma_+\gamma_- - \omega^2}\right), \tag{B.11}$$

$$z_3 = \frac{1}{\gamma_-}\left(i\omega + \sqrt{\gamma_+\gamma_- - \omega^2}\right), \quad z_4 = \frac{1}{\gamma_+}\left(i\omega + \sqrt{\gamma_+\gamma_- - \omega^2}\right), \tag{B.12}$$

which are all distinct as long as $\omega \notin \mathcal{S} \equiv \{\pm\sqrt{\gamma_+\gamma_-}, \pm(\gamma_+ + \gamma_-)/2\}$. We will first assume that $\omega \notin \mathcal{S}$. With this, we can easily find the bulk solutions

$$\vec{\psi}_1 = \vec{\zeta}_1(z_1) \otimes \begin{bmatrix} 1 \\ 0 \end{bmatrix}, \quad \vec{\psi}_2 = \vec{\zeta}_1(z_2) \otimes \begin{bmatrix} 0 \\ 1 \end{bmatrix}, \quad \vec{\psi}_3 = \vec{\zeta}_1(z_3) \otimes \begin{bmatrix} 1 \\ 0 \end{bmatrix},$$
$$\vec{\psi}_4 = \vec{\zeta}_1(z_4) \otimes \begin{bmatrix} 0 \\ 1 \end{bmatrix}, \tag{B.13}$$

from which we can construct the boundary matrix $B(\omega)$ given by

$$B(\omega) = \frac{i}{2}\begin{bmatrix} -\gamma_+ & -\gamma_- z_1^N e^{-2Nr'} & -\gamma_+ & -\gamma_-(-z_1)^{-N} \\ \gamma_+ z_1^N & -\gamma_- & \gamma_+ e^{2Nr'}(-z_1)^{-N} & -\gamma_- \\ z_1^N R & -\gamma_- z_1 e^{-2r'} & e^{2Nr'}(-z_1)^{-N} S & \gamma_- z_1^{-1} \\ \gamma_+ z_1 & z_1^N e^{-2Nr'} R & -\gamma_+ e^{2r'} z_1^{-1} & (-z_1)^{-N} S \end{bmatrix},$$

$$S \equiv 2i\omega - \gamma_- z_1, \quad R \equiv \gamma_+ z_1^{-1} + 2i\omega,$$

where $r' = r$ for $J > \Delta$ and $r' = r - i\pi/2$ for $\Delta > J$. The condition for ω to be an eigenvalue is $\det B(\omega) = 0$. From the expression for z_1, we can see that $\omega = i(\gamma_- z_1 - \gamma_+ z_1^{-1})$. Inserting this into $B(\omega)$ and taking the determinant introduces 4 fictitious roots of $\det B(\omega) = 0$, which we will identify after finding all of the roots

$$\left(z_1^{2N} + e^{2Nr'}\right)^2 \left(\gamma_+ + \gamma_- z_1^2\right)^2 = 0.$$

If $z_1 = \pm i\sqrt{\gamma_+/\gamma_-}$, then $\omega = \pm\sigma\sqrt{\gamma_+\gamma_-} \in \mathcal{S}$, with $\sigma = \operatorname{sgn}(\gamma_-)$, which must be considered separately. The remaining roots are $z_1 = \pm z_m \equiv \pm e^{r'} e^{ik_m}$, where $k_m = (m + 1/2)\pi/N$ and $m = 0, \ldots, 2N - 1$. This gives the $2N$ potential eigenvalues $\omega_m \equiv \sigma\sqrt{\gamma_+\gamma_-}\sin(k_m) = \sqrt{t^2 - \Delta^2}\sin(k_m)$, with $m = 0, \ldots, 2N - 1$.

Now, we must split into separate cases: if N is even, $\omega_m \neq \pm\sqrt{\gamma_+\gamma_-}$ for all m and so we have all $2N$ eigenvalues of $\mathbf{G}_T{}'$, and hence for $\mathbf{G}_T$. If N is odd, then when $m = (N - 1)/2$, $\omega_m = \sigma\sqrt{\gamma_+\gamma_-}$ and when $m = (3N - 1)/2$, $\omega_m = -\sigma\sqrt{\gamma_+\gamma_-}$. Since these are in $\mathcal{S}$, we must handle these separately. We do this after finding the eigenvectors for the remaining eigenvalues. The kernel vectors of $B(\omega_m)$ are

$$\vec{\alpha}_m = [e^{-(N+2)r'}, i(-1)^m, 0, 0]^T, \qquad \vec{\beta}_m = [0, 0, e^{-(N+2)r'}, i(-1)^{N-1-m}]^T,$$

with degeneracy arising due to the fact that each $\omega_m \neq \pm\sqrt{\gamma_+\gamma_-}$ is doubly degenerate. The degenerate eigenvectors of $\mathbf{G}_T{}'$ corresponding to eigenvalue ω_m are $\vec{\alpha}_m^T\vec{\Psi}$ and $\vec{\beta}_m^T\vec{\Psi}$, with $\vec{\Psi} \equiv [\vec{\psi}_1, \vec{\psi}_2, \vec{\psi}_3, \vec{\psi}_4]^T$. Rotating back via the unitary transformation $\mathbf{\Sigma}$ gives the eigenvectors of $\mathbf{G}_T$ corresponding to eigenvalue $\sigma\omega_m$ as

$$\vec{\psi}_{m,\sigma} = \frac{1}{\sqrt{N}}\sum_{j=1}^{N} e^{ijk_m}\vec{e}_j \otimes \vec{\xi}_m(j), \qquad \vec{\xi}_m(j) = \boldsymbol{\sigma}_1^m \begin{bmatrix} \sinh\left[\left(j - \frac{N+2}{2}\right)r'\right] \\ \cosh\left[\left(j - \frac{N+2}{2}\right)r'\right] \end{bmatrix}. \tag{B.14}$$

For N even, the above procedure exhausts all possibilities. For N odd, we consider the case $\omega = \sigma\sqrt{\gamma_+\gamma_-}$ explicitly and note that the case $\omega = -\sigma\sqrt{\gamma_+\gamma_-}$ can be handled in an analogous way. In this case, the characteristic polynomial has two distinct roots $z_1 = ie^{r'}$ and $z_2 = -1/z_1$. The corresponding eigenvectors of $\mathbf{g}(z_j)$ are $\vec{u}_1 = [1, 0]^T$ and $\vec{u}_2 = [0, 1]^T$ giving two bulk solutions

$$\vec{\psi}_{1,1} = \vec{\zeta}_1(z_1) \otimes \begin{bmatrix} 1 \\ 0 \end{bmatrix}, \qquad \vec{\psi}_{2,1} = \vec{\zeta}_1(z_2) \otimes \begin{bmatrix} 0 \\ 1 \end{bmatrix}.$$

The remaining two bulk solutions arise from the eigenvectors of $\mathbf{g}_1'(z_j)$ where

$$\mathbf{g}_1'(z) = \begin{bmatrix} \mathbf{g}'(z) & \partial_z\mathbf{g}'(z) \\ 0 & \mathbf{g}'(z) \end{bmatrix}.$$

These yield two more bulk solutions

$$\vec{\psi}_{1,2} = \vec{\zeta}_2(z_1) \otimes \begin{bmatrix} 1 \\ 0 \end{bmatrix}, \qquad \vec{\psi}_{2,2} = \vec{\zeta}_2(z_2) \otimes \begin{bmatrix} 0 \\ 1 \end{bmatrix}.$$

The boundary matrix at $\omega = \sigma\sqrt{\gamma_+\gamma_-}$ is then

$$B(\sigma\sqrt{\gamma_+\gamma_-}) = \frac{i}{2}\begin{bmatrix} -\gamma_+ & -\gamma_-(-z_1)^{-N} & 0 & -N\gamma_-(-z)^{1-N} \\ \gamma_+ z_1^N & -\gamma_- & N\gamma_+ z_1^{N-1} & 0 \\ -\gamma_+ z_1^{N-1} & -\sigma\gamma_-/z_1 & (N+1)\gamma_- z_1^N & -\gamma_- \\ \gamma_+ z_1 & -\gamma_-(-z_1)^{1-N} & \gamma_+ & \sigma(N+1)\gamma_+(-z_1)^{-N} \end{bmatrix}.$$

Then $\det B(\sigma\sqrt{\gamma_+\gamma_-}) \propto 1 + (-1)^N = 0$ for N odd. The kernel is one dimensional and is spanned by

$$\vec{\alpha} = [e^{-(N+2)r'}, i(-1)^{(N-1)/2}, 0, 0]^T.$$

Hence, the eigenvector corresponding to $\sigma\sqrt{\gamma_+\gamma_-}$ is $\vec{\psi}_{(N-1)/2,\sigma}$ where $\vec{\psi}_{m,\sigma}$ is exactly as in Eq. (B.14). Similarly, the eigenvector corresponding to $-\sigma\sqrt{\gamma_+\gamma_-}$ is $\vec{\psi}_{(3N-1)/2,\sigma}$.

B.2.2.2 Dynamical Phase Boundaries

In this section, we determine analytically the dynamical phase boundaries in boundary parameter space. An important assumption of this derivation is that certain phase boundaries are characterized by the emergence of ZMs and that the phase diagram is symmetric about $\varphi = \pi/2$. Thus, we will uncover the conditions on s and φ for $\mathbf{G}(s, \varphi)$ to possesses zero as an eigenvalue.

As in the preceding Appendix, we will rotate via the unitary $\mathbf{\Sigma}$ and study the unitarily equivalent matrix $\mathbf{G}'(s, \varphi)$. In contrast to the preceding section, however, we keep φ arbitrary and restrict to the non-open case $s \in (0, 1]$. Since the bulk ($\mathbf{G}^{\text{OBC}'}$) is unchanged, and the roots of the characteristic polynomial $P(\omega = 0, z)$ are distinct, we have the same four bulk solutions $\vec{\psi}_j$, $j = 1, 2, 3, 4$, given in Eqs. (B.8)–(B.13). On the other hand, the boundary modification is now given by

$$\mathbf{B}'(s, \varphi) = \vec{e}_N \vec{e}_1^{\dagger} \otimes \mathbf{b}'_1(s, \varphi) + \vec{e}_1 \vec{e}_N^{\dagger} \otimes \mathbf{b}'_{-1}(s, \varphi),$$

$$\mathbf{b}'_1 = -\frac{is}{2}\begin{bmatrix} \gamma_- \cos(\varphi) & \gamma_- \sin(\varphi) \\ -\gamma_+ \sin(\varphi) & \gamma_+ \cos(\varphi) \end{bmatrix} = \boldsymbol{\sigma}_2 \mathbf{b}'^{\dagger}_{-1} \boldsymbol{\sigma}_2.$$

Since the BC is different, the boundary matrix becomes

$$B(\omega = 0) = \frac{i}{2}\left[\vec{c}_1(z_1)\ \vec{c}_2(z_1^{-1})\ \vec{c}_1(-z_1)\ \vec{c}_2(-z_1^{-1})\right],$$

where

$$\vec{c}_1(z) \equiv [-\gamma_+\left(1 - sz^N\cos(\varphi)\right), s\gamma_+ z^N\sin(\varphi), fz \\ \times\left(z^N - s\cos(\varphi)\right), \gamma_+ zs\sin(\varphi)]^T,$$

$$\vec{c}_2(z) \equiv [-sfz^N\sin(\varphi), -f\left(1 - sz^N\cos(\varphi)\right), \\ -sfz\sin(\varphi), \gamma_+ z\left(z^N - s\cos(\varphi)\right)]^T.$$

Demanding that the determinant vanishes, we obtain the conditions

$$\cos(\varphi^\pm) = \frac{1}{2}\begin{cases}(s + s^{-1})\mathrm{sech}(Nr), & N \text{ even},\\ \pm 2\,\mathrm{sech}(Nr), & N \text{ odd}.\end{cases} \tag{B.15}$$

For N even, this specifies one angle $\varphi^+ = \varphi^-$ in the interval $[0, \pi]$, in fact, smaller than $\pi/2$. On the other hand, for N odd, there are two distinct angles $\varphi^\pm$ symmetric about each side of $\pi/2$. Thus, both phase boundaries host ZMs for N odd and just the left boundary for N even.

When Eq. (B.15) is satisfied, the kernel of $B(0)$ can be determined analytically. The cases $s \neq 1$ and $s = 1$ must be handled separately. We begin by taking $s \neq 1$. For N even, ker $B(0)$ is two-dimensional and spanned by the vectors

$$\vec{\alpha} = \frac{1}{s - s^{-1}}\left[\left(4 - (s + s^{-1})^2\mathrm{sech}^2(Nr)\right)^{1/2} e^{-(N+2)r}, \right. \\ \left. (s + s^{-1})\tanh(Nr), 0, s - s^{-1}\right],$$

$$\vec{\beta} = \frac{1}{s - s^{-1}}\left[(s + s^{-1})\tanh(Nr), \left(4 - (s + s^{-1})^2\mathrm{sech}^2(Nr)\right)^{1/2} e^{(N+2)r}, s^{-1}\right. \\ \left. -s, 0\right]^T.$$

For N odd and $\varphi = \varphi^\pm$, ker $B(0)$ is one-dimensional and spanned by

$$\boldsymbol{\alpha}_\pm = \left[\left(\frac{s \mp 1}{s \pm 1}\right) e^{-(N+2)r}, \frac{s \mp 1}{s \pm 1}, e^{-(N+2)r}, 1\right]^T.$$

For $s = 1$, the analogous kernel vectors for N even are

$$\vec{\alpha} = \left[e^{-(N+2)r}, 1, 0, 0\right]^T, \quad \vec{\beta} = \left[0, 0, e^{-(N+2)r}, 1\right]^T,$$

whereas for N odd are

$$\vec{\alpha}_+ = \left[0, 0, e^{-(N+2)r}, 1\right]^T, \quad \vec{\alpha}_- = \left[e^{-(N+2)r}, -1, 0, 0\right]^T.$$

The important thing to note is that these calculations reveal that the dimension of the zero-mode subspace is one (two) for N odd (even). The four-fold symmetry of the spectra of bosonic dynamical matrices implies that the *algebraic multiplicity of the zero eigenvalue must always be even.* This confirms that for N odd, there must be a Jordan chain of length two at zero, along the phase boundaries ($s > 0$). An additional symmetry of the even chain implies that all non-zero eigenvalues of $\mathbf{G}(s, \varphi)$ are at least doubly degenerate, implying that the zero eigenvalue has algebraic multiplicity four. Thus, the even chain possesses two length-two Jordan chains at zero, along the left phase boundary. Alternatively, this can be concluded by checking that the dimension of kernel of the boundary matrix of $\mathbf{G}^2$ at zero frequency is four.

B.2.2.3 The Parameter Regime $s = 1, \varphi \in (0, \pi), J = \Delta$.

At $\varphi = \pi/2$, $\mathbf{G}_T$ is non-diagonalizable when $J = \Delta$. The Jordan chains can be constructed by inspection and are given by

$$\vec{\chi}_{1k} = \left(\frac{i}{t}\right)^k \vec{e}_k \otimes \vec{v}_-, \quad k = 1, \ldots, N, \tag{B.16}$$

$$\vec{\chi}_{2k} = \left(\frac{i}{t}\right)^k \begin{cases} i\vec{e}_{k+1} \otimes \vec{v}_- + (-1)^{k+1}\vec{e}_{N+1-k} \otimes \vec{v}_+, & 1 \le k < N, \\ -\vec{e}_1 \otimes \vec{v}_+, & k = N, \end{cases} \tag{B.17}$$

for N even, and

$$\vec{\chi}_{1k} = \left(\frac{i}{t}\right)^k \begin{cases} 2\vec{e}_1 \otimes \vec{v}_-, & k = 1, \\ \vec{e}_k \otimes \vec{v}_- + i(-1)^k |N+2-k\rangle \otimes \vec{v}_+, & 2 \le k \le N+1, \end{cases} \tag{B.18}$$

$$\vec{\chi}_{2k} = \left(\frac{i}{t}\right)^k \left(i\vec{e}_{k+1} \otimes \vec{v}_- + (-1)^{k+1}\vec{e}_{N-k+1} \otimes \vec{v}_+\right), \quad k = 1, \ldots, N-1$$

$$, \tag{B.19}$$

for N odd. Specifically, these satisfy $\mathbf{G}_T \vec{\chi}_{jk} = \mathbf{G}_T \vec{\chi}_{j(k-1)}$, with $k \neq 1$ and $\mathbf{G}_T \vec{\chi}_{j1} = 0$ for $j = 1, 2$. It is interesting to note that for N even there are two length-N Jordan chains, whereas for N odd there is a Jordan chain of length $N+1$ and one of length $N-1$.

For $\varphi \neq \pi/2$ we define $\mathbf{G}_T(\varphi) \equiv \mathbf{G}(1,\varphi)$. Again, we simplify the problem by first diagonalizing $\mathbf{G}_T{}'(\varphi) \equiv \mathbf{\Sigma}\mathbf{G}_T(\varphi)\mathbf{\Sigma}^\dagger$. In this case, $\gamma_- = 0$ and $\gamma_+ = 2J$, and the corner modification takes the form

$$\mathbf{B}'(\varphi) = \mathbf{\Sigma} V(1,\varphi)\mathbf{\Sigma}^\dagger = \vec{e}_N\vec{e}_1^{\,\dagger} \otimes \mathbf{b}'_1(\varphi) + \vec{e}_1\vec{e}_N^{\,\dagger} \otimes \mathbf{b}'_{-1}(\varphi),$$

$$\mathbf{b}'_1(\varphi) \equiv iJ\begin{bmatrix} 0 & 0 \\ \sin(\varphi) & -\cos(\varphi) \end{bmatrix}, \quad \mathbf{b}'_{-1}(\varphi) \equiv iJ\begin{bmatrix} \cos(\varphi) & 0 \\ \sin(\varphi) & 0 \end{bmatrix}.$$

In particular, we note that $\det \mathbf{g}'_1 = \det \mathbf{g}'_{-1} = 0$ and so we expect emergent solutions to the bulk equation. When $J = \Delta$, the symbol is

$$\mathbf{g}(z) = iJ\begin{bmatrix} z^{-1} & 0 \\ 0 & -z \end{bmatrix}.$$

The roots of the characteristic polynomial are $z_1 = it/\omega$ and $z_2 = -1/z_1$ wish coalesce only for $\omega = \pm t$. The eigenvectors are $\vec{u}_1 = [1,0]^T$ and $\vec{u}_2 = [0,1]^T$ which provide two bulk solutions

$$\vec{\psi}_1 = \vec{\zeta}_1(z_1) \otimes \begin{bmatrix} 1 \\ 0 \end{bmatrix}, \qquad \vec{\psi}_2 = \vec{\zeta}_1(z_2) \otimes \begin{bmatrix} 0 \\ 1 \end{bmatrix}.$$

The remaining two bulk solutions come from the kernels of the matrices

$$\mathbf{K}^-(\omega) = \begin{bmatrix} \mathbf{g}'_{-1} & -\omega\mathbb{1}_2 & \mathbf{g}'_1 & 0 \\ 0 & \mathbf{g}'_{-1} & -\omega\mathbb{1}_2 & \mathbf{g}'_1 \\ 0 & 0 & \mathbf{g}'_{-1} & -\omega\mathbb{1}_2 \\ 0 & 0 & 0 & \mathbf{g}'_{-1} \end{bmatrix}, \qquad \mathbf{K}^+(\omega) \equiv \begin{bmatrix} \mathbf{g}'_1 & 0 & 0 & 0 \\ -\omega\mathbb{1}_2 & \mathbf{g}'_1 & 0 & 0 \\ \mathbf{g}'_{-1} & -\omega\mathbb{1}_2 & \mathbf{g}'_1 & 0 \\ 0 & \mathbf{g}'_{-1} & -\omega\mathbb{1}_2 & \mathbf{g}'_1 \end{bmatrix},$$

which are spanned by $\vec{u}_- = [0,1,0,0,0,0,0,0]^T$ and $\vec{u}_+ = [0,0,0,0,0,0,1,0]^T$ respectively. With these, the two additional bulk solutions

$$\vec{\psi}_- = \vec{e}_1 \otimes \begin{bmatrix} 0 \\ 1 \end{bmatrix}, \qquad \vec{\psi}_+ = \vec{e}_N \otimes \begin{bmatrix} 1 \\ 0 \end{bmatrix}.$$

The corresponding boundary matrix is

$$B(\omega) = it\begin{bmatrix} z_1^N\cos(\varphi) - 1 & 0 & 0 & \cos(\varphi) \\ z_1^N\sin(\varphi) & 0 & z_2 & \sin(\varphi) \\ 0 & 0 & 0 & z_2 \\ z_1\sin(\varphi) & z_2(z_2^N - \cos(\varphi)) - \cos(\varphi) & 0 & \end{bmatrix},$$

where we have used $\omega = it/z_1$. The condition for a vanishing determinant is

$$(z_1^N \cos(\varphi) - 1)(z_2^N - \cos(\varphi)) = 0.$$

For N even, the roots are doubly degenerate and given by $z_1 = z_m\,(\cos(\varphi))^{-1/N} e^{2m\pi i/N}$, with $m = 1, \ldots, N$. For N odd, the roots are $z_m = (\cos(\varphi))^{-1/N} e^{im\pi/N}$, with $m = 1, \ldots 2N$. In both cases we let $(\cos(\varphi))^{-1/N} \equiv e^{-i\pi/N} |\cos(\varphi)|^{-1/N}$, for $\varphi \in (\pi/2, \pi)$. The eigenvalues are then given by $\omega_m = it/z_m$. Equivalently, the spectrum $\sigma(\mathbf{G}_T(\varphi))$ is related to the periodic and anti-periodic cases as

$$\sigma(\mathbf{G}_T(\varphi)) = |\cos(\varphi)|^{1/N} \begin{cases} \sigma(\mathbf{G}(1,0)), & \varphi \in (0, \pi/2], \\ \sigma(\mathbf{G}(1,\pi)), & \varphi \in (\pi/2, \pi), \end{cases}$$

with $\mathbf{G}(1,0)$ ($\mathbf{G}(1,\pi)$) the dynamical matrix of the chain under periodic (anti-periodic) BCs with $J = \Delta$. Note that $|\omega_m| < t$ for all m and $\varphi \in (0, \pi)$ and so we need not address the case $\omega_m = \pm t$.

Now, for N even the kernel of $B(\omega)$ is 2 dimensional and spanned by

$$\vec{\alpha}_m = [e^{2m\pi i/N}(\cos(\varphi))^{1/N}, 0, \tan(\varphi), 0]^T, \qquad \vec{\beta} = [0, 1, 0, 0]^T.$$

After rotating back to the $(a, a^\dagger)$ basis, the (doubly degenerate) eigenvectors of $\mathbf{G}_T(\varphi)$ for N even, corresponding the eigenvalue ω_m, are

$$|\psi_{m,1}\rangle = \mathcal{N}_{m,1}\left(z_m^{-1}\vec{\zeta}_1(z_m)\vec{v}_+ + i\tan(\varphi)\vec{e}_1 \otimes \vec{v}_-\right),$$

$$|\psi_{m,2}\rangle = \mathcal{N}_{m,2}\vec{\zeta}_1(-z_m^{-1}) \otimes \vec{v}_-,$$

with $\mathcal{N}_{m,\ell}$, $\ell = 1, 2$ normalization constants. For N even the kernel of B is one-dimensional and is spanned by $\vec{\alpha}_{m/2}$, for m even, and $\vec{\beta}$, for m odd. Hence, the eigenvector of $\mathbf{G}_T(\varphi)$ for N odd corresponding to the eigenvalue ω_m is given, up to a normalization constant, by

$$\vec{\psi}_m = \mathcal{N}_m \begin{cases} z_m^{-1}\vec{\zeta}_1(z_m) \otimes \vec{v}_+ + i\tan(\varphi)\vec{e}_1 \otimes \vec{v}_-, & m \text{ even}, \\ \vec{\zeta}_1(-z_m^{-1}) \otimes \vec{v}_-, & m \text{ odd}. \end{cases}$$

B.3 An Isospectral Mapping Between the PDC and the DBKC

In this appendix, we will establish an isospectral mapping between the PDC and the DBKC. Let $\mathbf{G}^{\text{DBKC}}(\kappa, J, \Delta)$ and $\mathbf{G}^{\text{PDC}}(\mu_F, J_F, \Delta_F)$ denote the dynamical matrices of the DBKC (with $\mu = \Gamma = 0$) and the PDC under OBCs, respectively.

Note that we have distinguished the FKC hopping, pairing, and onsite potential with a subscript F. Define the momentum-space translation operator

$$\mathbf{\Lambda}(\delta k) \equiv \mathrm{diag}(e^{-i\delta k}, \ldots, e^{-iN\delta k}) \otimes \mathbb{1}_2. \tag{B.20}$$

With this, one may verify

$$\mathbf{\Lambda}\left(\frac{\pi}{2}\right) \mathbf{G}^{\mathrm{PDC}}\left(0, -\frac{J}{2}, -\frac{\Delta}{2}\right) \mathbf{\Lambda}^{-1}\left(\frac{\pi}{2}\right) = i\mathbf{G}^{\mathrm{DBKC}}(0, J, \Delta) \tag{B.21}$$

That is, the dynamical matrices are unitarily equivalent up to a phase when $\mu_F = \kappa = 0$. In particular, the $\mu_F = 0$ OBC rapidity spectrum for the PDC is equal to i times that of the DBKC with the identifications $\kappa = 0$, $J_F \leftrightarrow -J/2$, and $\Delta_F \leftrightarrow -\Delta/2$. This allows us to establish Eq. 9.7 from the known spectral properties of the BKC.

B.4 Persistence of MBs in a Disordered DBKC

Owing to the robustness of ϵ-pseudospectra, MBs are robust to weak perturbations (recall property (v) in Sect. 8.3.1). As a concrete example, we consider the DBKC with weak disorder. Explicitly, we take $\Gamma = 0$ and allow the model parameters to have spatial dependence $(J, \Delta, \mu, \kappa) \mapsto (J_j, \Delta_j, \mu_j, \kappa_j)$. For closed-form solutions, we focus on the case $\Delta_j = J_j$ for all j. With this, two approximate ZMs are given by

$$\gamma_L^z = x_1 + \sum_{j=2}^{N}\left(\prod_{\ell=1}^{j-1} \delta_-^{(\ell)}\right) x_j, \quad \gamma_R^z = \sum_{j=1}^{N-1} \frac{J_N}{J_j}\left(\prod_{\ell=j+1}^{N} \delta_+^{(\ell)}\right) p_j + p_N,$$
$$\delta_\pm^{(\ell)} \equiv -\frac{\kappa_\ell \pm \mu_\ell}{J_\ell}, \tag{B.22}$$

which satisfy

$$\mathcal{L}^\star(\gamma_L^z) = -J_N\left(\prod_{\ell=1}^{N} \delta_-^{(\ell)}\right) x_N, \quad \mathcal{L}^\star(\gamma_R^z) = J_N\left(\prod_{\ell=1}^{N} \delta_+^{(\ell)}\right) p_1. \tag{B.23}$$

As in the disorder-free case (i.e., when Δ and J site-independent), the corresponding approximate SGs γ_L^s and γ_R^s are obtained by taking $\kappa_j \mapsto -\kappa_j$. The pair $(\gamma_{L(R)}^z, \gamma_{L(R)}^s)$ are proper MBs (i.e., they are exponentially localized and are either approximately conserved or generate an approximate symmetry) when the disorder is weak enough, say, $\max_\ell |\delta_\pm^{(\ell)}| < 1$.

B.5 Exact Time-Evolution of Bosonic Parity of a Dissipative Harmonic Oscillator Prepared in a Cat State

In this appendix, we present the explicit calculation of the cat-state parity dynamics in the pure steady-state DBKC. As we have seen, this model can be seen as a set of independent damped quantum harmonic oscillators in the normal-mode basis. This decoupling allows us to reduce the problem to that of computing the parity dynamics of a single-mode cat-state under damped harmonic motion. The multimode generalization then follows naturally.

Let $|\alpha\rangle$, with $\alpha \in \mathbb{C}$ denote a single-mode coherent state and define the single-mode cat state $|\mathcal{C}_\phi(\alpha)\rangle = \mathcal{N}_\phi(\alpha)\left(|\alpha\rangle + e^{i\phi}|-\alpha\rangle\right)$. Here, $\mathcal{N}_\phi(\alpha)$ is a normalization constant. The dynamical problem at hand is to compute the expectation value of parity $P = e^{i\pi a^\dagger a}$ in the state $\rho_{\alpha,\phi}(t)$, with $\rho(0) = |\mathcal{C}_\phi(\alpha)\rangle\langle\mathcal{C}_\phi(\alpha)|$ evolved in time via the Lindbladian

$$\mathcal{L}(\rho) = -i[\omega a^\dagger a, \rho] + 2\kappa\left(a\rho a^\dagger - \frac{1}{2}\{a^\dagger a, \rho\}\right) \tag{B.24}$$

Here ω is the oscillator frequency and κ is the damping rate. We may compute $\rho_{\alpha,\phi}(t)$ exactly utilizing the results of Ref. [6, 7]. Firstly,

$$\rho_{\alpha,\phi}(t) = e^{t\mathcal{L}}\left(\rho_{\alpha,\phi}(0)\right) = \mathcal{N}_\phi(\alpha)^2\left(\sigma_\alpha(t) + \sigma_{-\alpha}(t) + e^{i\phi}\chi_\alpha(t) + e^{-i\phi}\chi_{-\alpha}(t)\right)$$

where

$$\sigma_\alpha(t) \equiv e^{t\mathcal{L}}(|\alpha\rangle\langle\alpha|), \quad \chi_\alpha(t) \equiv e^{t\mathcal{L}}(|\alpha\rangle\langle-\alpha|)$$

The terms $\sigma_{\pm\alpha}(t)$ can be quoted directly as

$$\sigma_{\pm\alpha}(t) = |\pm\alpha(t)\rangle\langle\pm\alpha(t)| \quad \alpha(t) = e^{-(\kappa+i\omega)t}\alpha \tag{B.25}$$

On the other hand, $\chi_{\pm\alpha}(t)$ are slightly more complicated,

$$\chi_\alpha(t) = D(t)\left(\sum_{k=0}^{\infty}\frac{(2e^{-\kappa t}\sinh(\kappa t))^k}{k!}a^k\chi_\alpha(0)a^{\dagger k}\right)D(t)^\dagger$$

with $D(t) = e^{-(\kappa+i\omega)ta^\dagger a}$. Now,

$$a^k\chi_\alpha(0)a^{\dagger k} = \alpha^k\chi_\alpha(0)(-\alpha^*)^k = (-|\alpha|^2)^k\chi_\alpha(0)$$

which leads us to

$$\chi_\alpha(t) = \exp\left(-2\kappa|\alpha|^2 e^{-\kappa t}\sinh(\kappa t)\right)D(t)\chi_\alpha(0)D(t)^\dagger$$

The remaining time-dependence may be computed as

$$D(t)\chi_\alpha(0)D(t)^\dagger = \exp\left(-2|\alpha|^2 e^{-\kappa t}\sinh(\kappa t)\right)|\alpha(t)\rangle\langle-\alpha(t)|.$$

Finally,

$$\chi_\alpha(t) = \exp\left(-4|\alpha|^2 e^{-\kappa t}\sinh(\kappa t)\right)|\alpha(t)\rangle\langle-\alpha(t)|,$$

with $\chi_{-\alpha}(t)$ following accordingly.

With the exact time dependence of $\rho(t)$ computed, we can now evaluate the expectation value of parity. A particularly useful identity is $P|\alpha\rangle = |-\alpha\rangle$. Using this, we will compute $P_1^\alpha(t) = \mathrm{tr}[P\sigma_\alpha(t)]$ and $P_2^\alpha(t) = \mathrm{tr}[P\chi_\alpha(t)]$ to obtain

$$\langle P\rangle(t) = \mathcal{N}_\phi(\alpha)^2(P_1^\alpha(t) + P_1^{-\alpha}(t) + e^{i\phi}P_2^\alpha(t) + e^{-i\phi}P_2^{-\alpha}(t))$$

Proceeding,

$$P_1^\alpha(t) = \mathrm{tr}[P|\alpha(t)\rangle\langle\alpha(t)|] = \langle\alpha(t)|-\alpha(t)\rangle = \exp\left(-2|\alpha|^2 e^{-2\kappa t}\right).$$

Letting $f_\alpha(t) = \exp\left(-4|\alpha|^2 e^{-\kappa t}\sinh(\kappa t)\right)$, we also have

$$P_2^\alpha(t) = f(t)\mathrm{tr}[P|\alpha(t)\rangle\langle-\alpha(t)|] = f_\alpha(t)\langle-\alpha(t)|-\alpha(t)\rangle = f_\alpha(t).$$

Putting this all together,

$$\langle P\rangle(t) = \frac{e^{-2|\alpha|^2} + \cos(\phi)e^{-2|\alpha|^2(1-e^{-2\kappa t})}}{1+\cos(\phi)e^{-2|\alpha|^2}}. \tag{B.26}$$

References

1. E. Cobanera, A. Alase, G. Ortiz, L. Viola, J. Phys. A **50**, 195204 (2017)
2. A. Alase, E. Cobanera, G. Ortiz, L. Viola, Phys. Rev. B **96**, 195133 (2017)
3. A. Böttcher, B. Silbermann, in *Introduction to Large Truncated Toeplitz Matrices* (Springer Science & Business Media, Berlin, 2012)
4. L.N. Trefethen, M. Embree, in *Spectra and Pseudospectra: The Behavior of Nonormal Matrices and Operators* (Princeton University Press, Princeton, 2005)
5. A. Alase, E. Cobanera, G. Ortiz, L. Viola, Phys. Rev. Lett. **117**, 076804 (2016)
6. K. Fujii, in *Quantum Mechanics*, Chap. 7, ed. by P. Bracken (InTech, 2012). arXiv:1209.1437 (quant-ph)
7. H.J. Korsch, arXiv:1908.01187v2 (quant-ph) (2019)

Vincent P. Flynn, Ph.D.

Dartmouth College
Department of Physics and Astronomy
6127 Wilder Laboratory, Room 201
Hanover, New Hampshire 03755-3528
Email: vincent.p.flynn@dartmouth.edu
Phone: (475) 882-0674

Education

Sept 2017–May 2023 **Dartmouth College**
Ph.D. in Physics
Advisor: Prof. Lorenza Viola
Ph.D. Thesis: "*Effective Non-Hermiticity and Topology in Markovian Quadratic Bosonic Dynamics*."
To be published in the "Springer Theses" book series (2024).

Aug 2013–May 2017 **University of Connecticut**
B.S., *Cum Laude*, Double Major in Physics and Mathematics
Cum. GPA: 3.705/4.000

Research Experience

June 2023–Present Theoretical modeling of exotic non-Hermitian physics in experimental microwave photonic arrays.
Principal Investigator: Prof. Mattias Fitzpatrick

June 2018–Present Investigating the consequences and manifestations of non-trivial topology in closed and open quantum many-body systems.
Ph.D. and Postdoctoral Advisor: Prof. Lorenza Viola

July 2016–May 2017 Conducted and analyzed resonant inelastic X-ray scattering experiments on Ytterbium Hexaboride performed at Argonne National Lab in order to better understand the electronic structure of strongly correlated materials.
Undergraduate Advisor: Prof. Jason Hancock

Mar 2016–May 2017 Computational investigation of negative thermal expansion in Scandium Trifluoride using techniques from structural analysis and molecular dynamics.

Undergraduate Advisor: Prof. Jason Hancock

Publications

In preparation:

- **Vincent P. Flynn**, Emilio Cobanera, and Lorenza Viola, "*Topologically distinct classes of metastable Markovian dynamics*", In preparation for *Journal of Physics A* (2024)
- Joshuah Heath, **Vincent P. Flynn**, Andrew Cupo, Emilio Cobanera, and Lorenza Viola, "*Engineering a quantized integer thermal Hall response in non-interacting bosonic Hamiltonians*", In preparation for *Physical Review B* (2024).
- Mariam Ughrelidze, **Vincent P. Flynn**, Emilio Cobanera, and Lorenza Viola, "*The interplay of finite- and infinite-size stability in quadratic bosonic Lindbladians*", In preparation for *Physical Review A* (2024).

Journal publications:

- **Vincent P. Flynn**, Emilio Cobanera, and Lorenza Viola, "*Topological zero modes and edge symmetries of metastable Markovian bosonic systems*", Physical Review B **108**, 214312 (2023). [Citations as of Mar 5, 2024: **1**, per Google Scholar]
- **Vincent P. Flynn**, Emilio Cobanera, and Lorenza Viola, "*Topology by Dissipation: Majorana Bosons in Metastable Quadratic Markovian Dynamics*", Physical Review Letters **127**, 245701 (2021) [Citations as of Mar 5, 2024: **50**, per Google Scholar].
- Qiao-Ru Xu, **Vincent P. Flynn**, Abhijeet Alase, Emilio Cobanera, Lorenza Viola, and Gerardo Ortiz, "*Squaring the fermion: The threefold way and the fate of zero modes*", Physical Review B **102**, 125127 (2020). Editor's Suggestion. [Citations as of Mar 5, 2024: **27**, per Google Scholar]
- **Vincent P. Flynn**, Emilio Cobanera, and Lorenza Viola, "*Restoring particle conservation in quadratic bosonic Hamiltonians with dualities*", Europhysics Letters **131**, 40006 (2020) [Citations as of Mar 5, 2024: **13**, per Google Scholar]
- **Vincent P. Flynn**, Emilio Cobanera, and Lorenza Viola, "*Deconstructing effective non-Hermitian dynamics in quadratic bosonic Hamiltonians*", New Journal of Physics **22**, 083004 (2020) [Citations as of Mar 5, 2024: **40**, per Google Scholar]

- Donal Sheets, **Vincent Flynn**, Jungho Kim, Mary Upton, Diego Casa, Thomas Gog, Zach Fisk, Priscilla Rosa, Maxim Dzero, Jian-Xin Zhu, Ignace Jarrige, and Jason Hancock, "*Exploring itinerant states in divalent hexaborides using rare-earth L edge resonant inelastic x-ray scattering*", Journal of Physics **32**, 135601 (2019) [Citations as of Mar 5, 2024: **2**, per Google Scholar]

  Total citations as of Mar 5, 2024: **134**, per Google Scholar.

Presentations

November 2023	Poster: "*Majorana bosons and topologically protected metastable phases of quasi-free Markovian systems*" presented at the "Topological phases and strong correlations in many-body systems and light-matter hybrids" conference at ITAMP, Harvard University.
September 2023	Invited Seminar: "*Signature of genuine SPT physics in metastable Markovian bosonic systems*" presented for the Quantum Information and Foundations Group of the The Spanish National Research Council.
June 2023	Poster: "*Majorana bosons and topologically protected metastable phases of quasi-free Markovian systems*" presented at the "Many-Body Quantum Systems: From Quantum Computing and Simulation to Metrology and Coherent Light-Matter Hybrids" Gordon Research conference.
June 2021	Poster: "*Topological quasi-steady states and the role of Noether's theorem in metastable driven-dissipative bosonic systems*" presented at the "Topology meets Quantum Optics" conference held by the Centro de Ciencias de Benasque Pedro Pascual
March 2021	Talk: "*Restoring number conservation in quadratic bosonic Hamiltonians with dualities: Applications for quantum simulation and topological classification.*" at March Meeting 2021. Link to abstract.
March 2020	Talk: "*Deconstructing Effective Non-Hermitian Dynamics in Quadratic Bosonic Hamiltonians*" at March Meeting 2020. Link to slides.
June 2019	Poster: "*Exploiting mildly broken translation invariance to study effective non-Hermitian dynamics in a bosonic chain*" presented at the international workshop on "Engineering nonequilibrium Dynamics of open quantum Systems" at the Max Planck Institute for the Physics of Complex Systems in Dresden, Germany.

Course Research Projects

June 2018–Aug 2018	The Kibble-Zurek mechanism and the non-equilibrium dynamics of classical and quantum phase transitions (as part of Physics 109 - Statistical Mechanics II).
Mar 2018–May 2018	Classical simulations of quantum systems using tensor network techniques (as part of Physics 116 - Quantum Information Science).
Jan 2018–Mar 2018	Bosonization and its applications to the Tomonaga-Luttinger liquid (as part of Physics 113 - Microscopic Theory of Solids).

Teaching Experience

Apr 2022–May 2022	Guest lecturer for Physics 103 - (Graduate) Advanced Quantum Mechanics
Mar 2022–June 2022	T.A. for Physics 103 - (Graduate) Advanced Quantum Mechanics
Mar 2021–June 2021	T.A. for Physics 104 - (Graduate) Statistical Mechanics I
Mar 2021–Apr 2021	Mini-course on thermodynamics for Physics 104 - (Graduate) Statistical Mechanics I
Mar 2020–June 2020	T.A. for Physics 103 - (Graduate) Advanced Quantum Mechanics
Mar 2019–June 2019	T.A. for Physics 104 - (Graduate) Statistical Mechanics I
Feb 2019	Guest lecturer for Physics 82 - Mathematical Methods for Physicists
Jan 2019–Mar 2019	T.A. for Physics 82 - Introductory Mathematical Methods for Physicists
Nov 2018	Guest lecturer for Physics 100 - (Graduate) Mathematical Methods for Physicists
Oct 2018	Guest lecturer for Physics 19 - Relativistic and Quantum Physics.
Sept 2018–Dec 2018	T.A. for Physics 19 - Relativistic and Quantum Physics.

Mar 2018–June 2018	T.A. for Physics 19 - Relativistic and Quantum Physics.
Jan 2018–Mar 2018	T.A. for Physics 13 - Introductory Physics I.
Sept 2016–Dec 2016	Mentor for an incoming freshman physics student as a part of the first implementation of the UConn physics "Peer Mentor" program.
Sept 2015–May 2016	Physics and math tutor at the UConn quantitative learning center.

Awards, Honors, Fellowships, and Citations

May 2023	Nominee for the Dartmouth Guarini School for Graduate and Advanced Studies Hannah Croasdale Award, nominated by the Physics and Astronomy faculty.
Summer 2022	The Physics and Astronomy Graduate Research Award, awarded by the Dartmouth Physics and Astronomy faculty to a finishing or advanced Ph.D. student for outstanding research accomplishments as a graduate student and the commitment to continued work in physics, astronomy, or related sciences.
Summer 2021	The Gordon F. Hull Fellowship, awarded by the Dartmouth Physics and Astronomy faculty to a Ph.D. student based on exemplary teaching and research accomplishments as well as research promise.
Summer 2019	The Selamawit Tsehaye Teaching Award, awarded by Dartmouth Physics and Astronomy faculty to the graduate student best exemplifying dedication to the teaching of physics.
Spring 2018	*Dartmouth Center for the Advancement of Learning* outstanding graduate student teacher award, voted by undergraduate students.
Spring 2018	Voted best presentation by peers in Physics 116 - Quantum Information Science.
Winter 2018	Cited as highest performing student in Physics 105 - Electromagnetic Theory I by Prof. Barrett Rogers.

Summer 2016	*Mark E. Miller Award*: $2,000 stipend to conduct research with the UConn Physics Dept.
Spring 2016	Bernard Sippin '52 $7,500 scholarship recipient awarded by UConn Mathematics Dept. for excellence in mathematics.
2016 Academic Year	UConn *New England Scholar* for maintaining a GPA of at least 3.7/4.0 during the 2016 academic year.
2015 Academic Year	UConn *Babbidge Scholar* for maintaining a 4.0/4.0 cumulative GPA during the 2015 academic year.
Spring 2015–Spring 2017	UConn College of Liberal Arts and Sciences Dean's list (5 consecutive semesters).

Professional Memberships, Service, and Additional Experiences

Reviewer for *Physical Review Journals* and the *European Physical Journal.*

Member of the American Physical Society.

Technical Skills

Programming Languages/Software	Python, Mathematica, MATLAB, LaTeX, Qiskit/IBM Q Experience.
Operating Systems	Windows, macOS, Linux.

Languages

English (fluent), Italian (intermediate).

Other Skills and Interests

Active multi-instrumental musician, music performance and theory, philosophy of science and mathematics.

The manufacturer's authorised representative in the EU is Springer Nature Customer Service Centre GmbH, Europaplatz 3, 69115 Heidelberg, Germany. If you have any concerns regarding our products, please contact ProductSafety@springernature.com

Printed and bound by CPI Group (UK) Ltd, Croydon, CR0 4YY
01/10/2026
02225896-0003